AF404722

Electrostatics

Formalism of the electrostatic field in vacuum and matter

Online at: https://doi.org/10.1088/978-0-7503-5859-0

Electrostatics

Formalism of the electrostatic field in vacuum and matter

Ştefan Antohe and Vlad-Andrei Antohe
Faculty of Physics, University of Bucharest, Bucharest, Romania

IOP Publishing, Bristol, UK

© IOP Publishing Ltd 2023

All rights reserved. No part of this publication may be reproduced, stored in a retrieval system or transmitted in any form or by any means, electronic, mechanical, photocopying, recording or otherwise, without the prior permission of the publisher, or as expressly permitted by law or under terms agreed with the appropriate rights organization. Multiple copying is permitted in accordance with the terms of licences issued by the Copyright Licensing Agency, the Copyright Clearance Centre and other reproduction rights organizations.

Permission to make use of IOP Publishing content other than as set out above may be sought at permissions@ioppublishing.org.

Ştefan Antohe and Vlad-Andrei Antohe have asserted their right to be identified as the authors of this work in accordance with sections 77 and 78 of the Copyright, Designs and Patents Act 1988.

ISBN   978-0-7503-5859-0 (ebook)
ISBN   978-0-7503-5857-6 (print)
ISBN   978-0-7503-5860-6 (myPrint)
ISBN   978-0-7503-5858-3 (mobi)

DOI   10.1088/978-0-7503-5859-0

Version: 20230801

IOP ebooks

British Library Cataloguing-in-Publication Data: A catalogue record for this book is available from the British Library.

Published by IOP Publishing, wholly owned by The Institute of Physics, London

IOP Publishing, No.2 The Distillery, Glassfields, Avon Street, Bristol, BS2 0GR, UK

US Office: IOP Publishing, Inc., 190 North Independence Mall West, Suite 601, Philadelphia, PA 19106, USA

The authors dedicate this book to their families.

Contents

Preface

This textbook is the first one from a series of two books, covering the '*Electricity and Magnetism*' course unit held at Faculty of Physics within University of Bucharest. A complete training of a physicist cannot be fulfilled without deep knowledge on the complex mechanisms and phenomena related to classical electromagnetism. Therefore, the main scope of this course is a thorough understanding of the electric and magnetic phenomena, starting from the phenomenological analysis of the main laws governing electromagnetism. Essentially, this important part of physics can be concentrated into the complete system of Maxwell's equations, whose establishment is pursued as a leitmotif throughout a full academic year dedicated to 'Electricity and Magnetism' at the Faculty of Physics.

This work covers the first part of the 'Electricity and Magnetism' course, being focused on phenomena related to stationary electric charges (i.e., 'electrostatics') and representing a set of 14 lectures (of two hours each) held for students enrolled in different physics programs at University of Bucharest. Its content was developed in the last 30 years with consistent and regular updates required by various changes in curricula within Faculty of Physics and other science faculties. In particular, this textbook, entitled: '*Electrostatics: Formalism of the electrostatic field in vacuum and matter*', is dedicated to studying the electric field in both vacuum and condensed matter, being organized into a brief introduction and nine chapters. As the formalism of the electrostatic field requires adequate mathematical apparatus, the book is completed by two appendices detailing specific elements of algebra and vectorial analysis, where math operators (i.e., gradient, divergence, curl and Laplace) are introduced, starting from the physical premises that require their employment. The work ends with a third appendix displaying the numerical values of the most common universal constants. Each chapter and appendix includes a list of a few references, which all-together contributed to shaping the entire lecture as it is captured in this first textbook. The book is continued by a second part of the series, focused on phenomena related to moving electric charges, i.e., '*Electromagnetism and Special Methods for Electric Circuits Analysis*'.

The way the content is provided is considered by the authors a very efficient approach to gaining new knowledge. Moreover, the textbook contains numerous practical examples and exercises, written in *slanted* characters, that help towards better assimilation of the presented theoretical concepts. The authors strongly believe that this work would raise the attention of readers dealing with the theory and applications of classical electromagnetism, being particularly intended to serve as an academic course support for international students of various graduation levels (i.e., Bachelor, Engineer, Master and PhD), activating in the domains of physics and exact sciences in general.

Acknowledgments

During the elaboration of this work, the authors' thoughts headed respectfully towards their professors who placed the basis of this discipline at the Faculty of Physics within University of Bucharest from Romania. Also, the authors are deeply grateful to their colleagues from the Department of Electricity, Solid State Physics and Biophysics, as well as to the members of the Research and Development Center for Materials and Electronic & Optoelectronic Devices (MDEO), who sincerely engaged in various discussions and exchange of ideas on the theory of classical electromagnetism and its good teaching practices.

Bucharest—Măgurele, 2023

Professor PhD Ştefan Antohe
Associate Professor PhD Eng. Vlad-Andrei Antohe

Author biographies

Mr Ştefan Antohe, PhD
Professor Emeritus
University of Bucharest, Faculty of Physics,
Bucharest—Măgurele, Romania
Academy of Romanian Scientists (AOSR),
Ilfov Street 3, 050 045 Bucharest, Romania

Mr Ştefan Antohe graduated in 1977 within Faculty of Physics, University of Bucharest, Romania and he finished in 1994 a PhD program in physics at the same institution. He is currently **Professor Emeritus** of the University of Bucharest, **Head** of the Research and Development Center for Materials and Electronic & Optoelectronic Devices (MDEO) of the Faculty of Physics, as well as **Full Member** of the Academy of Romanian Scientists (AOSR).

His main research activities are dedicated to the physics of organic and inorganic semiconducting thin films and nanomaterials, with special emphasis on investigation of charge transport mechanisms within materials and interfaces. He also possesses great knowledge on the fabrication technologies and characterization of various electronic and optoelectronic devices, especially photovoltaic cells and transparent field-effect transistors. He has co-authored more than 250 scientific publications in ISI journals and 22 books or book chapters with national and international publishing houses, he has supervised several PhD and post-doctoral fellows, and also coordinated numerous research projects.

Mr Vlad-Andrei Antohe, PhD Eng.
Associate Professor
University of Bucharest, Faculty of Physics,
Bucharest—Măgurele, Romania

Mr Vlad-Andrei Antohe received in 2002 a Bachelor degree (BSc) in physics education and in 2005 a Master diploma (MSc) in physical electronics from Faculty of Physics, University of Bucharest, Romania. He also obtained in 2003 an Engineering degree (MEng) in electronics from Faculty of Electronics and Telecommunications, Polytechnic University of Bucharest, Romania. Then, he was awarded in 2012 with a PhD diploma in applied engineering sciences from Catholic University of Louvain, Belgium, where he continued his Post-Doctoral studies until 2016. Since then, he has been **Associate Professor** of University of Bucharest and **Scientific Collaborator** of Catholic University of Louvain.

Besides teaching activities with undergraduate and graduate students, his main research interests are in the areas of materials science and nanotechnology, with particular focus on the development and investigation of nanostructured materials and low-dimensional architectures, with the aim of generating novel structural arrangements tailored to specific desired properties. He also has expertize in the fabrication and characterization of electronic and optoelectronic devices based on inorganic, organic, or hybrid organic/inorganic nanostructured materials, such as photovoltaic cells, sensors and biosensors, magnetic media for various applications, as well as micro-batteries, micro-supercapacitors and other electrochemical systems. He has co-authored around 60 scientific publications, of which more than 15 ISI research papers are published in journals with high impact factor ranging approximately from 5 to 32.

Introduction

The first phenomenon leading to the emergence of *electricity* dates back in the 6th century BC (Before Christ), when the electrization through friction of macroscopic bodies was observed by *Thales of Miletus*[1]. He noticed that a yellow amber ('elektron' in Greek) stick attracts light bodies after its friction with the fur of a cat, saying in this way that by friction 'the amber becomes charged with electricity'. However, the first scientific publication documenting electric and magnetic phenomena was written much later, in 1600 AC (After Christ), by the English physicist *William Gilbert*. The work was entitled: *'De Magnete'*, and it is more of historical importance, as the correlation between electric and magnetic phenomena was not clearly explained at that time.

The laws of electricity and magnetism were practically discovered by experimentalists who had little or no knowledge on the modern theory of the atomic and molecular structure of matter. For instance, the law describing the electrostatic interactions between the differently charged bodies was discovered in 1785 by the French physicist *Ch A Coulomb*, who established for the first time that the force of interaction between two point charges placed in vacuum and separated by a distance d between their centers is proportional to the product of charges, reversely proportional with the square of d and always oriented along the line joining their centers. Moreover, the scientist knew that the electric charges could be positive or negative and that they could be separated one from another while the magnetic poles could not be separated. In 1799, the Italian physicist *A Volta* built the first voltaic pile (electric battery), opening thus the possibility to extend the electricity and magnetism experiments within many laboratories all over the world. Consequently, the forthcoming 19th century was represented by substantial progress in the fundamental understanding of electric and magnetic phenomena, as well as of their interaction. In this context, the first experimental investigation of the interaction between coils carrying electric currents was performed by the French physicist and mathematician *A M Ampère* during 1820–5, his work being pursued by the scientists: *H C Ørsted, J B Biot* and *F Savart*. Ampère found that two long parallel wires carrying electric currents in opposite directions repel each other, whereas when carrying electric currents in the same sense they attract each other. Furthermore, it has been also found that if a coil carrying an electric current is placed close to a compass needle, it experiences both a force and a couple. Under these circumstances, the group of scientists concluded that both, the magnet and coil carrying the current produce a magnetic field, taking into account the relation between electric and magnetic phenomena. Later on, in 1826 *G S Ohm* gave the law of electric conduction that carries his name. In 1831–3 *M Faraday* and *E H Lenz* gave the law of electromagnetic

---

[1] Thales (624–546 BC) was a Greek mathematician, astronomer and philosopher from Miletus (ancient Greek city on the western coast of Asia Minor). He was one of the Seven Sages of Greece, recognized as the first in western civilization engaged in scientific philosophy.

induction, but their names are being commonly used for other discoveries, such as: Faraday's law of electrolysis, or Joule–Lenz's law of electro-caloric effects.

The application of Faraday's ideas in electromagnetism, writing of the mathematical equations and development of the field concept, are due to the English physicist *James Clark Maxwell*. In his work, '*Treatise about Electricity and Magnetism*', written in 1873, he established the basis of the macroscopic theory of electromagnetic field. Maxwell foresaw theoretically the existence of electromagnetic waves and the 'displacement current' in 1862. With these concepts he created the electromagnetic theory of light in 1865. The experimental confirmation of Maxwell's theory was made by *H Hertz* in 1889, through his experiments on the propagation of electromagnetic waves. Essentially, the physicists of the 19th century (such as Ampère, Faraday and Maxwell) shaped the basis of classical electromagnetism, while the physicists and chemists of the 20th century contributed further to understanding the electromagnetism by introducing the atomic and molecular structure of matter.

Classical electromagnetism deals with the study of both stationary and moving charges. The study of *stationary charges* is the aim of **electrostatics**, while the study of *moving charges* is the aim of **electrodynamics**. Maxwell has shown that the laws of electromagnetism could be expressed in the form of four differential equations, referring to the components of electric and magnetic fields. The table of Maxwell's equations, as written in 1876 is appended below:

---

(1) $\quad \nabla \cdot \vec{D} = \rho$ $\qquad$ derived from the Gauss's law applied to the electric field;

(2) $\quad \nabla \cdot \vec{B} = 0$ $\qquad$ derived from the Gauss's law applied to the magnetic field;

(3) $\quad \nabla \times \vec{E} = -\dfrac{\partial \vec{B}}{\partial t}$ $\qquad$ derived from Faraday's and Lenz's laws of electromagnetic induction;

(4) $\quad \nabla \times \vec{H} = \vec{j} + \dfrac{\partial \vec{D}}{\partial t}$ $\qquad$ derived from the law of magnetic circuit in its particular case of Ampère's law.

Equation (1) represents the local form of Gauss's law (the theorem of electric flux) applied to the electric field and it signifies the direct relationship between the electric field and its own sources. Equation (2) comes from Gauss's law (the theorem of magnetic flux) applied to the magnetic field, showing that the magnetic field has no polar sources being a field with closed lines[2]. Equation (3) is derived from Faraday's and Lenz's laws of electromagnetic induction, considering just the transformational component of the induce electromotive force, showing that a time-dependent magnetic field is the source of a time-dependent electric field with the lines enclosing those of the magnetic field. Equation (4) is obtained from the law of magnetic circuit

---

[2] The magnetic charges (monopoles) cannot be separated.

in its particular case of Ampère's Law (taking into account only two sources for the magnetic field, namely, conduction current and displacement current). In the absence of the conduction current, it similarly shows that a time-dependent electric field is the source of a time-dependent magnetic field with the lines enclosing those of the electric field. Noteworthy, equations (3) and (4) reveal that the electromagnetic field is an ensemble of electric and magnetic fields able to generate reciprocally, that can leave the source giving rise to the propagating electromagnetic wave.

Maxwell's equations were written a long time before the development of quantum mechanics and relativity theory. Nevertheless, they have not undergone any changes to date. The importance of the classical theory of electromagnetism in modern physics can be explained by the fact that Maxwell's equations are perfectly compatible with relativity theory. Eventually, the changes arise from the quantum theory towards electromagnetic forces, which are important only at very small distances (10^{-10} cm), in other words at distances of a hundred times smaller than the atomic diameter. Therefore, the same laws of classical electromagnetism could also be used to study interactions at atomic and molecular level (or scale). However, for distances smaller than 10^{-10} cm, quantum electrodynamics has to be also taken into account for understanding the interactions at that level.

This book series comprises two textbooks (i.e., '***Electrostatics. Formalism of the electrostatic field in vacuum and matter***' and '***Electromagnetism and Special Methods for Electric Circuits Analysis***') entirely dedicated to studying the classical laws of electricity and magnetism, by progressively demonstrating and constructing the entire system of Maxwell's equations of the electromagnetic field presented above.

Further reading

[1] Von Laue M 2018 *History of Physics* (Franklin Classics Trade Press), ISBN: 978-0-35-320750-9

[2] Heilbron J L 2018 *The History of Physics: A Very Short Introduction* (Oxford: Oxford University Press), ISBN: 978-0-19-968412-0

Electrostatics
Formalism of the electrostatic field in vacuum and matter
Ştefan Antohe and Vlad-Andrei Antohe

Chapter 1

Electrostatic field in vacuum

The study of the electric and magnetic phenomena is more difficult than the study of other physical phenomena (such as the ones related to mechanics, heat or optics) because the human body does not have direct sensors for feeling electric and magnetic action forces. Only the electromagnetic waves with the wavelength in the range of 400–700 nm are directly sensed by the human eye. That is why the study of electromagnetic phenomena presumes knowledge about the ponderomotive actions (forces and torques) with which an electric field acts on its own sources.

1.1 Electrization state

The '*electrization state*' is the state of a body experienced by a force or/and a torque by an applied electric field. There are two types of electrization states, namely, electrization state *(i)* of a *charged body* and *(ii)* of a *polarized body*:

Electrization state of a charged body is the state of the body carrying an electric charge and characterized by the fact that it is experienced by an electric field with a ponderomotive action like the force.

Electrization state of a polarized body is the state of the body characterized by the fact that on one hand, an external electric field acts on it with supplementary ponderomotive actions like the torque, and on the other hand the body is able to generate an electric field even though it is neutral (it does not really carry an electric charge).

1.2 Electric charge. Definition. Properties

When a body caries a charge and the ponderomotive action experienced by an external electric field is just the force, it is said that the body is in the '*electrization state of a charged body*' and the parameter which characterizes this state is referred to as '*electric charge*'.

doi:10.1088/978-0-7503-5859-0ch1

© IOP Publishing Ltd 2023

1.2.1 Properties of the electric charge

The *electric charge* represents thus an essential physical quantity of matter, that is characterized by the following properties:

a. The electric charge is a *scalar quantity* which could be positive or negative. The objects charged with the same sign repel one another, whereas the objects charged with opposite sign attract one another.

b. The interaction between the electric charges is measured by the electric forces. The transmission of the electric interactions takes place by the electric field. Consequently, *the electric charge is the source of an electric field*. If the electric charge is stationary, it creates an electrostatic field which is the investigation subject of *electrostatics*. Hence, the electrostatics studies these electric states which are time- and space-independent and there are no energetic transformations involved.

c. The electric charge is a *primitive quantity*, namely, it can be determined experimentally. If in an uniform electric field $\overrightarrow{E_v}$ (created for example between the plates of a planar parallel capacitor), a charged body is introduced, measuring the force with which the body is experienced by the field $(\overrightarrow{F_e})$, its electric charge can be determined by:

$$q = \frac{\left| \overrightarrow{F_e} \right|}{\left| \overrightarrow{E_v} \right|} \tag{1.1}$$

If:

- $q > 0$, then $\overrightarrow{F_e} \uparrow \uparrow \overrightarrow{E_v}$ (the force is *omoparallel* with the electric field);
- $q < 0$, then $\overrightarrow{F_e} \uparrow \downarrow \overrightarrow{E_v}$ (the force is *antiparallel* with the electric field).

d. The electric charge characterizes a *localized property* of the bodies. A body is uncharged if neither of its sides has an electric charge.

e. The electric charge has a discrete set of definite values, namely, '*it is quantified*'. In the modern theory the atom consists of a central core or nucleus, featuring a diameter of about 10^{-12} cm, surrounded by a number of electrons. These electrons move round the nucleus in orbits whose diameter is about 10^{-8} cm and these orbits determine the size of the atom. The nucleus contains two types of particles: *(i) protons*, which are particles roughly 1.840 times heavier than the electrons, with a positive charge $(+e)$, and *(ii) neutrons*, of very nearly the same mass as protons, but with no electric charge. The number of electrons surrounding the nucleus is equal to the number of protons, and each electron has a negative charge $(-e)$ so that the atom as a whole is electrically neutral. The net residual charge on a neutral atom is less than $10^{-20}e$. This is remarkable, since apart from their electric behavior protons and electrons are totally dissimilar particles. Many elementary particles besides electrons and protons have been discovered by nuclear physicists, and all share the property of carrying charges $(\pm e)$ or zero

($|e| = 1.60219 \times 10^{-19}$ C). It follows that the total charge carried by any piece of matter *must be an integral multiple* of the electronic charge, i.e., e. A situation like this one, in which a physical quantity is not allowed to have a continuous range of values, but is restricted to a *set of definitive discrete values*, is generally referred to as a *quantum phenomenon*.

f. Conductors and insulators. For the purpose of electrostatic theory all substances can be divided in two fairly distinct classes: *(i) conductors* in which electric charge can flow easily from one place to another, and *(ii) insulators* in which the electric charge cannot move. In the case of solids, all metals and a few other substances such as carbon are conductors, and their electrical properties can be explained by assuming that a number of electrons are free to move around the whole volume of the solid instead of being rigidly attached to one atom.

g. Atoms which have lost one or more electrons in this way have a positive charge and are called ions. They remain fixed in position within a solid lattice. In solid substances of the second class, insulators, each electron is firmly attached to a particular atom or ion and cannot move from one to another. Typical solid insulators are sulfur, paraffin or mica.

1.2.2 Continuous charge distributions

At atomic scale the electric charge is quantified but at macroscopic scale the charge has a continuous distribution, namely, the electron has such a small charge value that the variations of the macroscopic charge from a point to another cannot be evaluated (measured). For the study of the macroscopic charge distributions, introducing the charge densities is needed.

Linear charge density
In order to characterize the electrization state of a very long charged body, the linear charge density is used. Considering the length element $\overrightarrow{\Delta l}$ from the charged body carrying the charge Δq (see figure 1.1), the linear charge density is defined by:

$$\lambda(\overrightarrow{r_0}) = \lim_{\Delta l \to 0} \frac{\Delta q}{\Delta l} = \frac{dq}{dl} \quad (\text{C m}^{-1}) \tag{1.2}$$

namely, the limit of ratio between the charge element Δq and length element Δl when it tends to zero.

Observation: Δl is centered onto point $P_0(\overrightarrow{r_0})$ and it must be very small to keep constant the total charge on the charged wire, but at the same time it must be long enough to keep the continuity of the charge onto it.

The linear charge density, $\lambda(\overrightarrow{r})$, is a scalar quantity which is a function of position (i.e., it is fully specified by its magnitude at each point). Scalar functions of position such as the charge density are called *scalar fields* (see appendix B.1). We shall often omit the argument of scalar fields, writing the linear charge density just as λ. Knowing the linear charge density given by equation (1.2), the total charge onto a charged body with the length L, can be calculated by the integral:

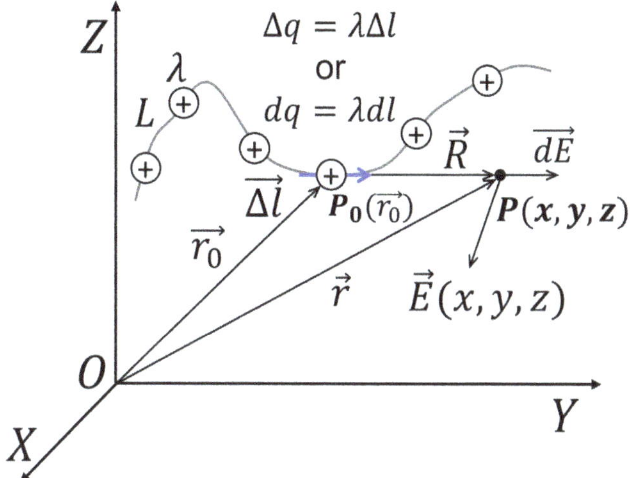

Figure 1.1. Linear charge density.

$$q = \int_L \lambda(\overrightarrow{r_0})\,dl = \int_L \lambda(x_0, y_0, z_0)\,dl \tag{1.3}$$

Relation (1.3) shows that the total charge q depends on the functional relation (1.2) for λ and on the shape of the charged body, respectively.

Surface charge density
A charged surface is characterized by the surface charge density defined as the limit of ratio between the elementary charge Δq and the surface element ΔS when it tends to zero (see figure 1.2):

$$\sigma(\overrightarrow{r_0}) = \lim_{\Delta S \to 0} \frac{\Delta q}{\Delta S} = \frac{dq}{dS} \ (\text{C m}^{-2}) \tag{1.4}$$

Observation: ΔS is centered onto point $P_0(\overrightarrow{r_0})$ and it must be very small to keep constant the total charge on the charged surface, but at the same time it must be large enough to keep the continuity of the charge onto it.

Similarly, the surface charge density $\sigma(\vec{r})$ is a scalar function dependent on the position. If the charge is uniformly distributed on the surface body, σ is constant (independent of the point). Knowing the surface charge density given by equation (1.4), the total charge onto a charged body can be calculated by surface integral:

$$q = \int_S \sigma(\overrightarrow{r_0})\,dS = \int_S \sigma(x_0, y_0, z_0)\,dS \tag{1.5}$$

when both, $\sigma(\overrightarrow{r_0})$ and S are known. Here we have introduced a shorthand notation for the surface integral, which represents effectively a double integral over the two components needed to specify the position vector $(\overrightarrow{r_0})$ within the surface S.

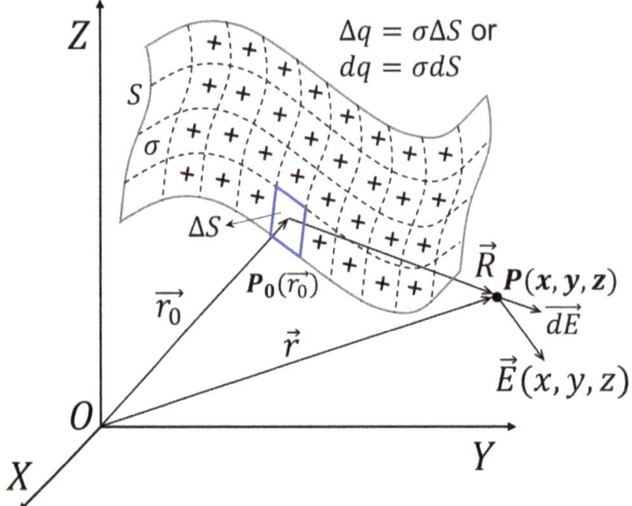

Figure 1.2. Surface charge density.

Volume charge density

A charge continuously distributed within the volume V is characterized by the volume charge density which represents the limit of ratio between the charge Δq contained within a small volume ΔV centered onto point $P_0(\vec{r_0})$ and the volume ΔV when it tends to zero (see figure 1.3):

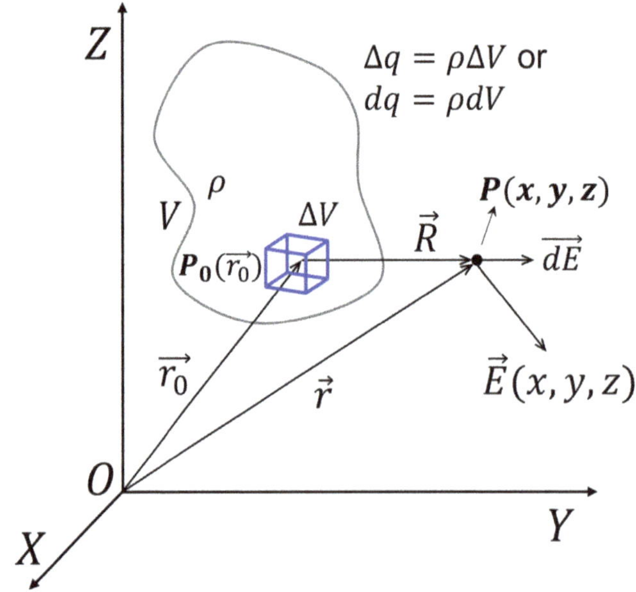

Figure 1.3. Volume charge density.

$$\rho(\vec{r_0}) = \lim_{\Delta V \to 0} \frac{\Delta q}{\Delta V} = \frac{dq}{dV} \quad (\text{C m}^{-3}) \tag{1.6}$$

Observation: ΔV is centered onto point $P_0(\vec{r_0})$ and it must be very small to keep constant the total charge on the charged volume, but at the same time it must be big enough to keep the continuity of the charge onto it.

Commonly, knowing the volume charge density, the total charge of the charged body can be calculated by the volume integral:

$$q = \int_V \rho(\vec{r_0})dV = \int_V \rho(x_0, y_0, z_0)dV \tag{1.7}$$

Here we have introduced a shorthand notation for the volume integral, which is effectively a triple integral over the three components (x_0, y_0, z_0) needed to specify the position vector $(\vec{r_0})$ within the volume V. In Cartesian coordinates dV becomes $dx_0dy_0dz_0$, i.e., the volume of the rectangular box enclosed by side lengths dx_0, dy_0 and dz_0, respectively.

Concluding, when we need to calculate the total charge carried by a body, we will have to resolve a line integral, a surface integral or a volume integral, depending on the type of the charge distribution. For this purpose we must know the charge density and the shape of the charged line, surface or volume. The line, surface or volume integral is easy to compute, as long as the line, surface or volume elements are written in the appropriate coordinate system. That is why knowledge on the *'three-orthogonal curvilinear coordinate systems'* is required at this point (see appendix A).

1.3 Electrostatic interactions. Coulomb's formula

Like in the case of any physical field, the introduction of the quantities describing the electrostatic field can be done if we know the ponderomotive actions with which the physical field acts on its own source. Then, in the case of the electrostatic field we will start with the force with which the electric field acts on a charge q placed at a point P in the field. But to do this, first of all we must study the interaction between the electric charges, namely *'Coulomb's formula'*, which gives the magnitude and direction of the force between two stationary particles, each of them carrying an electric charge.

1.3.1 Coulomb's formula

Inspired by the *'Cavendish experiment'*[1], a torque balance was employed for the first time in 1785 by *C-A Coulomb*[2] to measure the force experienced in vacuum between

---

[1] The *'Cavendish experiment'* was performed in 1797–8 by the British scientist Henry Cavendish, being the first experiment to measure the force of gravity between masses in the laboratory and the first to yield accurate values for the gravitational constant.
[2] *Charles-Augustin de Coulomb* (1736–806) was a French military engineer and physicist, famous for discovering *Coulomb's law*, although he did important work on friction, too.

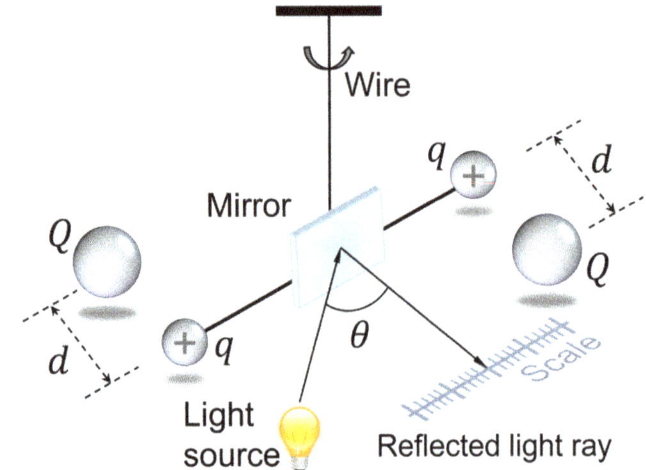

Figure 1.4. Coulomb's experiment.

two small charged bodies (see figure 1.4). The two small charged bodies were considered by Coulomb as two point charges. It is noteworthy that the two small charged bodies can be considered as mass points (point charges) when their sizes are smaller than the error which affects the measurement of the distances between them. In his experiment, the interaction between the charged particles produced a measurable twist in the torsion wire, which triggered rotation of the apparatus until equilibrium was reached. By accurately measuring the torsion angle, Coulomb confirmed the $1/r^2$ dependence of the electrostatic interaction force between the charged particles, with excellent precision.

The electrostatics essentially deals with: (i) point charges and (ii) stationary charges. In practical terms, while changing on one hand both, magnitudes and signs of the charges, and on the other hand the distance between them, Coulomb expressed the forces between the two stationary point charges placed in vacuum at the distance $\overrightarrow{r_{12}}$ between them (see figure 1.5) by the following relations:

$$\overrightarrow{F_{12}} = k_0 \frac{q_1 q_2}{r_{12}^3} \overrightarrow{r_{12}} \text{ the force acting from } q_1 \text{ to } q_2 \tag{1.8a}$$

$$\overrightarrow{F_{21}} = -k_0 \frac{q_1 q_2}{r_{12}^3} \overrightarrow{r_{12}} \text{ the force acting from } q_2 \text{ to } q_1 \tag{1.8b}$$

Relations (1.8) represent the mathematical statement of Coulomb's law, summarizing four facts:

a. Charges of the same sign repel each other, while charges of opposite signs attract each other;
b. The force acts along the line joining the centers of the two charged particles;
c. The force is proportional to the magnitude of each charge;
d. The force is reversely proportional to the square of the distance between the centers of the charged particles.

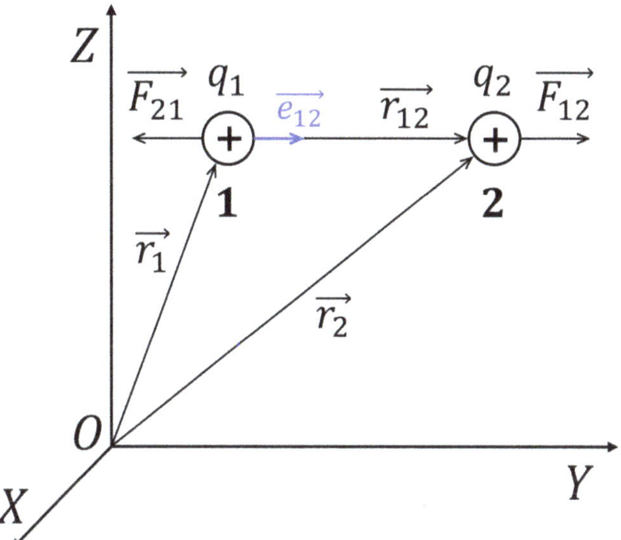

Figure 1.5. The forces measuring the interaction between two point charges.

The vector $\overrightarrow{F_{12}}$ in figure 1.5 represents the force on particle **2** (carrying the charge q_2) exerted by particle **1** (carrying the charge q_1). The line from q_1 to q_2 is represented by the vector $\overrightarrow{r_{12}}$ of length r_{12}, since the unit vector along the direction $\overrightarrow{r_{12}}$ can be written as $\overrightarrow{e_{12}} = \dfrac{\overrightarrow{r_{12}}}{r_{12}}$. Equations (1.8) are thus an inverse square law of force, although r_{12}^3 appears in the denominator.

Note, equation (1.8a) automatically accounts for the attractive or repulsive character of the force $\overrightarrow{F_{12}}$ if q_1 and q_2 contains the sign of the charge. When the charges q_1 and q_2 are both positive or negative, the force on q_2 is along the $\overrightarrow{r_{12}}$ (i.e., it is repulsive). In contrast, when a charge is positive and the other negative, the force is in the opposite direction to $\overrightarrow{r_{12}}$ (i.e., it is attractive). In the latter scenario, the scalar $k_0\dfrac{q_1q_2}{r_{12}^3}$ is negative and then the vector $\overrightarrow{F_{12}} = -k_0\dfrac{q_1q_2}{r_{12}^3}\overrightarrow{r_{12}}$ is in the opposite direction to $\overrightarrow{r_{12}}$, namely, q_1 attracts q_2.

To complete the statement of the force law, we must decide what units to use, and hence we have to determine the constant of proportionality in equation (1.8a). We shall use the International System (IS) of units which are preferred by most physicists and engineers applying electromagnetism to problems involving large scale objects[3]. In IS units, Coulomb's law can be written as:

$$\overrightarrow{F_{12}} = \frac{q_1q_2}{4\pi\varepsilon_0 r_{12}^3}\overrightarrow{r_{12}} \tag{1.9}$$

---

[3] A different system of units, called the *Gaussian System*, is almost universally used in atomic physics and solid state physics, but it is an unfortunate necessity for students to become reasonably familiar with both systems.

where: q_1 and q_2 are measured in *Coulombs* (C), r_{12} is measured in *meters* (m) and F_{12} is measured in *Newton* (N). In equation (1.9), k_0 was replaced by $\frac{1}{4\pi\varepsilon_0}$, representing a universal constant for vacuum, as the experiments showed that if other quantities have given units, k_0 has just only one value, being independent of any physical quantity or property (i.e., it is independent of the nature of the test charged materials). Thus, for vacuum k_0 is:

$$k_0 = \frac{1}{4\pi\varepsilon_0} \cong 9 \cdot 10^9 \text{ N m}^2 \text{ C}^{-2} \tag{1.10}$$

The constant ε_0 is called the *electrical permittivity of free space* (or vacuum) and it has the value:

$$\varepsilon_0 = 8.85 \times 10^{-12} \text{ C}^2 \text{ N}^{-1} \text{m}^{-2} \text{ (F m}^{-1} \text{ in IS units)} \tag{1.11}$$

If the two electric charges are placed in a certain medium, the force of interaction between them becomes:

$$\overrightarrow{F_{12}} = \frac{1}{4\pi\varepsilon} \frac{q_1 q_2}{r_{12}^3} \overrightarrow{r_{12}} \tag{1.12}$$

where: ε is the *absolute permittivity* of the medium, showing the influence of the medium over the interaction between the charges. Since the unit of absolute permittivity (ε) is dependent on the used units system, the non-dimensional constant ε_r (called *relative permittivity* or '*dielectric constant*') was introduced to characterize the electrical properties of the matter. ε_r is thus defined as the ratio between the absolute permittivity (ε) and the permittivity of vacuum (ε_0):

$$\varepsilon_r = \frac{\varepsilon}{\varepsilon_0} \tag{1.13}$$

Underlining, for a given medium, ε_r shows how many times the force of interaction between the two point charges placed in a vacuum is larger than the force between them, when they are placed at the same distance in that medium. For vacuum and air, $\varepsilon_r = 1$, while for other materials ε_r takes different values. Table 1.1 shows ε_r values of several most common media.

Table 1.1. Typical dielectric constants of several common materials.

Material	ε_r
Air/vacuum	1
Ebonite	2.5–3.2
Paper	2–2.6
Dried wood	2.5–5
Transformer oil	2.2
Water	80.1 (at 20 °C)
Toluene	2.4
Organic coating	4–8

1.3.2 Superposition principle for electrostatic forces

Electrostatic forces are '*two-body forces*', which means that the force between any pair of charges is unaltered by the presence of other charges in their neighborhoods. The forces between atoms are '*many-body forces*', since the force between two atoms depends on where other atoms are situated. The many-body nature of the forces arises because the distribution of the constituent charged particles within each atom is changed by the presence of other atoms. But if the distribution of all the charge within each atom would be specified, then the total force could be found by application of the superposition principle.

In a system containing many charges, the electrostatic force between any pair is given by Coulomb's law. To find the total force on any particle, one simply makes a vector sum of the forces it experiences due to all the others separately. This rule for the addition of electrostatic forces is known as the '*Superposition Principle*'[4]. Considering a system of $q_1, q_2, \ldots, q_n$ charges, having the position vectors $\vec{r_1}, \vec{r_2}, \ldots, \vec{r_n}$ with respect to the origin O of a coordinate system (see figure 1.6), the total force on q_j (charge having the position vector $\vec{r_j}$ with respect to O), is the vector sum $\vec{F_j}$ of the two forces $\vec{F_{ij}}$ which it experiences due to all charges from the system. Mathematically, the total force on q_j is:

$$\vec{F_j} = \frac{1}{4\pi\varepsilon_0} \sum_{i \neq j} \frac{q_i q_j}{r_{ij}^3} \vec{r_{ij}}$$

(1.14)

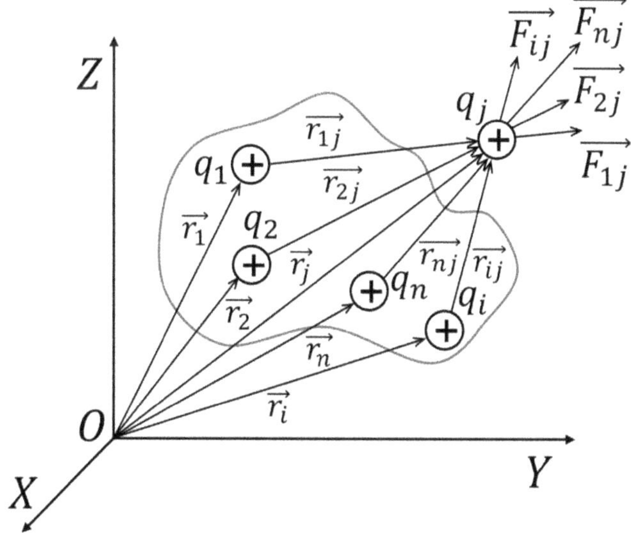

Figure text follows.

Figure 1.6. Superposition principle for electrostatic forces.

---

[4] A situation like this one exists in mechanics, when we need to calculate the total force on any mass point, due to all mass points from a system. Thus, we make the vector sum of forces it experiences due to all the others separately.

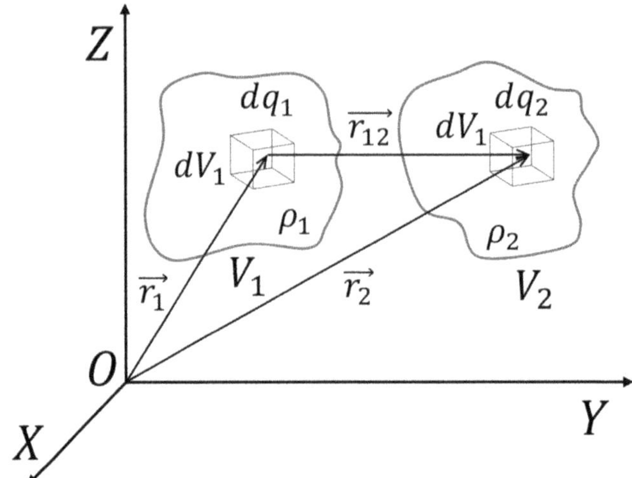

Figure 1.7. Force between two charged macroscopic bodies.

where $\overrightarrow{r_{ij}} = \overrightarrow{r_j} - \overrightarrow{r_i}$ is the vector joining the charges q_i and q_j, which can be written as a function of its components as:

$$\overrightarrow{r_{ij}} = (x_j - x_i)\overrightarrow{i} + (y_j - y_i)\overrightarrow{j} + (z_j - z_i)\overrightarrow{k} \tag{1.15}$$

A trivial example of the application of superposition principle is in working out the electrostatic forces exerted by atomic nuclei containing many protons, on the surrounding electrons. Nuclei are much smaller than the atoms, and for this purpose they can be regarded as point charges. The superposition principle then tells us that the attractive force between a nucleus containing z protons and an electron, is z times as great as that between a proton and an electron.

The superposition principle can be applied as well to calculate the electrostatic force between two charged macroscopic bodies having the volume charge densities $\rho_1(\overrightarrow{r_1})$ and $\rho_2(\overrightarrow{r_2})$, respectively (see figure 1.7). The electrostatic force between two volume elements dV_1 and dV_2, charged with the elementary charges $dq_1 = \rho_1(\overrightarrow{r_1})dV_1$ and $dq_2 = \rho_2(\overrightarrow{r_2})dV_2$, respectively, is:

$$\overrightarrow{dF_{12}} = \frac{dq_1 dq_2}{4\pi\varepsilon_0 r_{12}^3}\overrightarrow{r_{12}} \tag{1.16}$$

Taking into account the superposition principle described by equation (1.14), the force between the two charged bodies will be:

$$\overrightarrow{F_{12}} = \frac{1}{4\pi\varepsilon_0} \int_{V_1} \int_{V_2} \frac{\rho_1(\overrightarrow{r_1})\rho_2(\overrightarrow{r_2})}{r_{12}^3}\overrightarrow{r_{12}}dV_1 dV_2 \tag{1.17}$$

The double volume integral from equation (1.17) can be calculated when both the volume charge densities $\rho_1(\vec{r_1})$ and $\rho_2(\vec{r_2})$, as well as the shape of each charged body, are known.

1.4 The electrostatic field

According to the superposition principle, the total force experienced by a system of charged particles on a given charged particle is the vector sum of the forces exerted on the latter by all other charges from the system. Usually an enormous number of charged particles is present in real matter. When considering the forces acting on any of them, it is helpful to distract attention from the multitude of sources contributing to the net force by introducing the concept of the electric field. If a charge q_0 experiences a force $\vec{F}$ then the ratio $\dfrac{\vec{F}}{q_0}$ is called the *electric field* at the point where q_0 is located.

1.4.1 Definition of the electric field

The electric field of a discrete system of point charges acting on q_0 can be expressed in terms of the magnitudes of the other charges q_i in the neighborhood of q_0 and their relative position with respect to q_0. Let us assume that q_0 is a test charge which can be put anywhere, and that its magnitude is very small, so that it exerts negligible forces on the other charges. Also, the test charge must have an electrization state independent of time. Now we can evaluate the electric field at a point P, caused by a discrete system of charges q_i, by placing the test charge q_0 at P. In figure 1.8, the point P has a position vector $\vec{r_0}$ and the charges q_i have the position vectors $\vec{r_i}$ with respect to the origin O of the coordinate system. In agreement with the superposition principle, the force on the test charge q_0 is given by the vector sum of the forces $\vec{F_i}$ with which each charge q_i of the system acts on q_0:

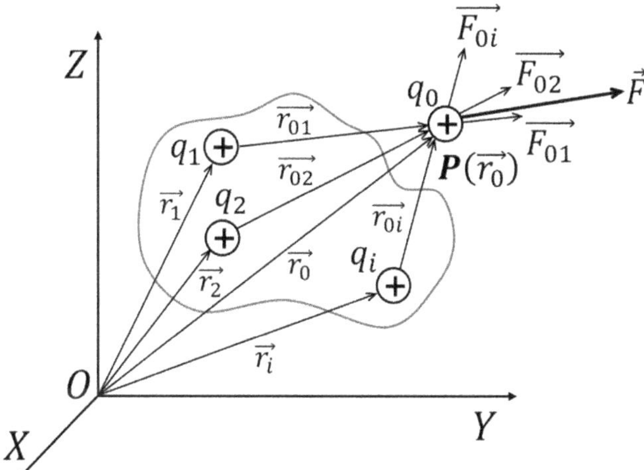

Figure 1.8. Vectors used in the definition of the electric field.

$$\vec{F} = \frac{1}{4\pi\varepsilon_0}\sum_{i=1}^{n}\frac{q_i q_0}{|\vec{r_0}-\vec{r_i}|^3}(\vec{r_0}-\vec{r_i}) = \frac{1}{4\pi\varepsilon_0}\sum_{i=1}^{n}\frac{q_i q_0}{r_{0i}^3}(\vec{r_{0i}}) \tag{1.18}$$

where: $(\vec{r_0}-\vec{r_i})$ is the vector $\vec{r_{0i}}$ joining q_i to the point P. We thus observe that q_0 is a common factor to all terms in the sum and it follows that:

$$\frac{\vec{F}}{q_0} = \frac{1}{4\pi\varepsilon_0}\sum_{i=1}^{n}\frac{q_i}{r_{0i}^3}(\vec{r_{0i}}) \tag{1.19}$$

The test charge q_0 does not appear on the right-hand side of equation (1.19). We can therefore allow q_0 to become vanishingly small, then we are quite sure that the presence of the test charge does not modify the position of the other charges. The electric field $\vec{E}(\vec{r_0})$ at the point P with position vector $\vec{r_0}$ is now:

$$\vec{E}(\vec{r_0}) = \lim_{q_0\to 0}\frac{\vec{F}}{q_0} \tag{1.20}$$

In the case of a discrete system of charges, the electric field $\vec{E}$ at the point P with the position vector $\vec{r_0}$ is:

$$\vec{E}(\vec{r_0}) = \frac{1}{4\pi\varepsilon_0}\sum_{i=1}^{n}\frac{q_i}{r_{0i}^3}\vec{r_{0i}} \tag{1.21}$$

Relation (1.21) shows that $\vec{E}(\vec{r_0})$ depends only on the magnitudes of the component charges q_i and on the relative position vector $\vec{r_{0i}}$ of each charge q_i from the system with respect to the P point, being independent of the test charge q_0. The test charges were introduced because they illustrate that the electric field is a force-per-unit charge, and because they help also to visualize the electric field lines. The function $\vec{E}(\vec{r_0})$ is itself a vector depending on the position vector $\vec{r_0}$ like the electric force $\vec{F}(\vec{r_0})$. Those functions of position which are themselves vectors are called *vector fields* (see appendix B.1). The dimensions of electric field are *[force]/[charge]* so in IS units, the electric field is measured in N C^{-1}. An equivalent unit, which will be explained when we come to deal with electrostatic potential is the V m^{-1}.

1.4.2 Electric field of a point charge

From equation (1.20), the magnitude of the electric field at a distance $\vec{r}$ from a positive point charge, $+q$, is $\vec{E} = \frac{q}{4\pi\varepsilon_0 r^3}\vec{r}$, pointing out away from the charge. In contrast, the field around a negative point charge, $-q$, is $\vec{E} = -\frac{q}{4\pi\varepsilon_0 r^3}\vec{r}$, pointing out towards the charge. In figure 1.9 the direction of the field around positive and negative charges is indicated by the arrows. These continuous lines, everywhere following the direction of the field, are called *lines of force* or *field lines*. The field lines may begin on positive charges and end on negative charges, but they may also go to infinity without being terminated, as shown in figure 1.9.

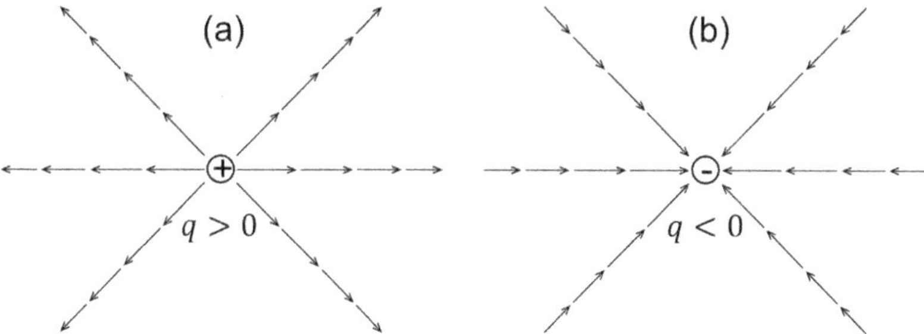

Figure 1.9. Field lines around positive (a) and negative (b) point charges.

Exercise: calculus of the electric field of a point charge

Supposing that the point charge Q is placed at the distance $\vec{r_0}$ with respect to the origin of the coordinate system, then the electric field generated at point P described by its relative position vector $\vec{r}$ with the same origin of the coordinate system (see figure 1.10) will be:

$$\vec{E} = \frac{Q\,\overrightarrow{\Delta r}}{4\pi\varepsilon_0 \left|\overrightarrow{\Delta r}\right|^3} \tag{1.22}$$

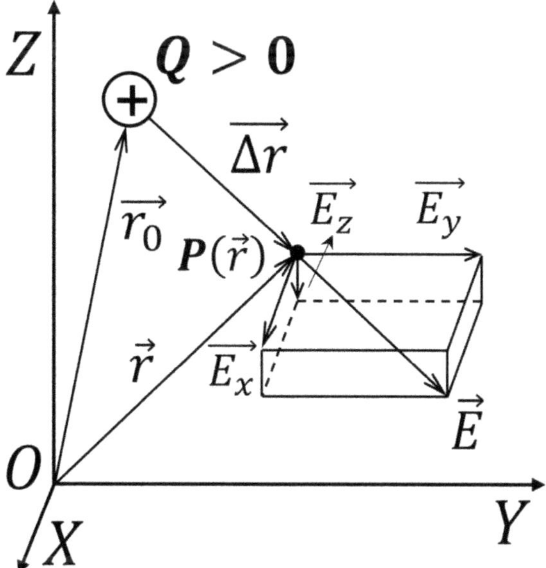

Figure 1.10. Electric field of a point charge Q placed in vacuum.

where:

$$\vec{r} = x\vec{i} + y\vec{j} + z\vec{k}\,;\; \vec{r_0} = x_0\vec{i} + y_0\vec{j} + z_0\vec{k}$$
$$\vec{\Delta r} = (x - x_0)\vec{i} + (y - y_0)\vec{j} + (z - z_0)\vec{k} \qquad (1.23)$$
$$\left|\vec{\Delta r}\right| = \left[(x - x_0)^2 + (y - y_0)^2 + (z - z_0)^2\right]^{1/2}$$

Notice that the field lines are close together near the point charges where the field is strong, and far apart at large distances where the field is weak.

Taking into account equation (1.22), the field components are:

$$E_x = \frac{Q(x - x_0)}{4\pi\varepsilon_0\left[(x - x_0)^2 + (y - y_0)^2 + (z - z_0)^2\right]^{3/2}}$$

$$E_y = \frac{Q(y - y_0)}{4\pi\varepsilon_0\left[(x - x_0)^2 + (y - y_0)^2 + (z - z_0)^2\right]^{3/2}}$$

$$E_z = \frac{Q(z - z_0)}{4\pi\varepsilon_0\left[(x - x_0)^2 + (y - y_0)^2 + (z - z_0)^2\right]^{3/2}}$$

Hence:

$$\vec{E} = E_x\vec{i} + E_y\vec{j} + E_z\vec{k}$$

and thus its module is:

$$E = \sqrt{E_x^2 + E_y^2 + E_z^2}$$

We can also draw the field lines to illustrate the electric field when there are many charges present. Field lines are continuous except where they terminate on positive or negative charges, and they never cross one another, since the direction of the field is unique at every point. One can often get a rough idea of the field around a distribution of charges simply by sketching the field lines, and without doing any mathematics. For example, figure 1.11 shows the field lines near a pair of point charges of equal magnitude, one positive and one negative. Such a pair of equal and opposite charges is called an *electric dipole*. Very close to each charge, the field is almost the same as for isolated point charges, but the field lines starting off at the positive charge curve round to finish at the negative charge. The space spectrum of the field lines is very complicated to plot, but in figure 1.11 only a cross-section through it is shown. This diagram gives an indication of the strength of the field as well as its direction, because field lines are *always densely packed* in regions where the field is strong. The total number of lines on the diagram is not significant, if twice as many lines were drawn terminating on each charge, regions of relatively high or low density of field lines would still correspond to regions of high or low field strength. The example of the electric dipole is an important one and later we will obtain a mathematical expression for the electric field at a large distance from a

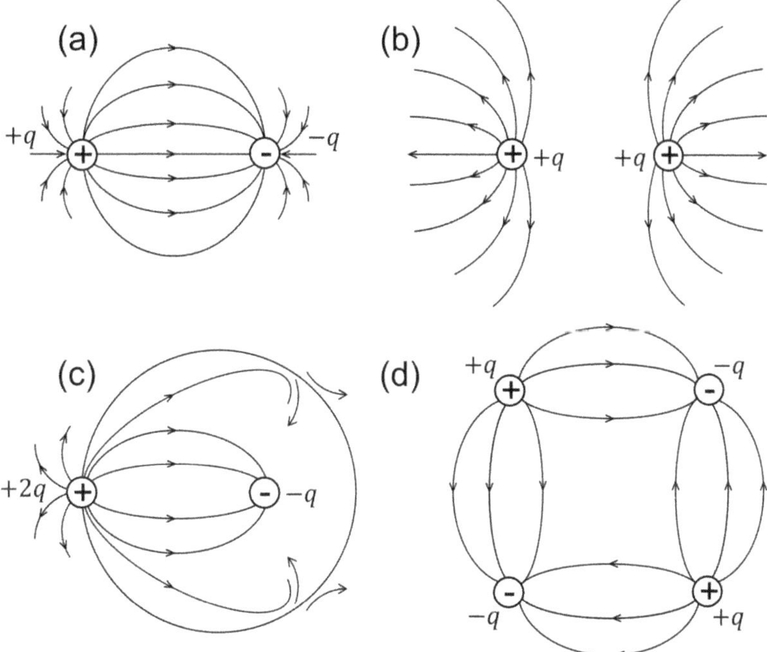

Figure 1.11. Field lines around different systems of charges.

dipole. The field lines near a pair of point charges of equal magnitude, having the same sign, are shown in figure 1.11(b).

1.4.3 Superposition principle for the electric field

The electric field generated by an assembly of point charges q_i was found from the superposition principle for forces, by making the vector sum of the forces acting on a test charge q_0 by each charge from system and dividing it by q_0. We have thus obtained the expression:

$$\vec{E}(\vec{r_0}) = \frac{1}{4\pi\varepsilon_0}\sum_{i=1}^{n}\frac{q_i}{r_{0i}^3}\vec{r_{0i}} = \sum_{i=1}^{n}\frac{1}{4\pi\varepsilon_0}\frac{q_i\,\vec{r_{0i}}}{r_{0i}^3} = \sum_{i=1}^{n}\vec{E_i} \tag{1.24}$$

which shows that the electric field created at point P, by a system of point charges is the vector sum of the fields due to each charge from the system, separately. This is the *superposition principle for the electric field*. In the case of *continuous charge distributions*, we can apply the same procedure to calculate the electric field. Considering the elementary charge dq onto an infinitesimal volume element, as a point charge (see figure 1.12), we shall write its elementary field at a point P:

$$\vec{dE} = \frac{dq\,\vec{R}}{4\pi\varepsilon_0 R^3} \tag{1.25}$$

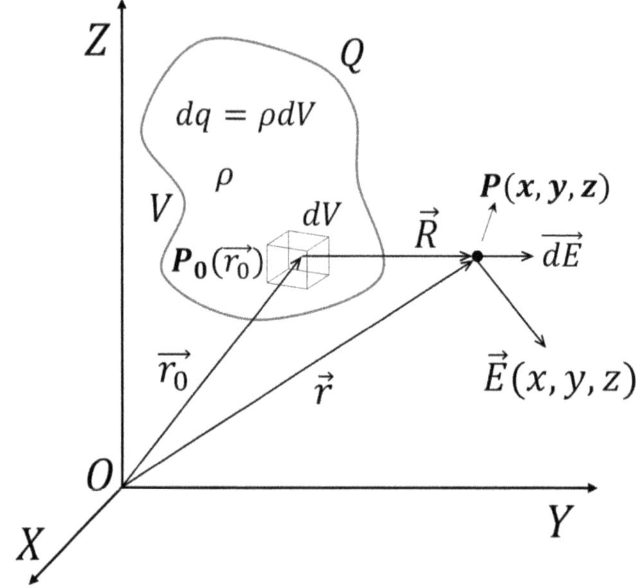

Figure 1.12. Electric field of a volume charge distribution.

and according to the superposition principle, the electric field generated at point P by the total charge of the charged body will be:

$$\vec{E} = \frac{1}{4\pi\varepsilon_0} \int_V \frac{\rho(x_0,\, y_0,\, z_0)dV\vec{R}}{R^3} \tag{1.26}$$

Similarly, in the case of a linear charge distribution (see figure 1.1), $dq = \lambda(x_0,\, y_0,\, z_0)dl$, and using the superposition principle the electric field strength will be given by the line integral:

$$\vec{E} = \frac{1}{4\pi\varepsilon_0} \int_L \frac{\lambda(x_0,\, y_0,\, z_0)dl\vec{R}}{R^3} \tag{1.27}$$

In the case of a surface charge distribution (see figure 1.2), $dq = \sigma(x_0,\, y_0,\, z_0)dS$, and the electric field strength will be given by the surface integral:

$$\vec{E} = \frac{1}{4\pi\varepsilon_0} \int_S \frac{\sigma(x_0,\, y_0,\, z_0)dS\vec{R}}{R^3} \tag{1.28}$$

Thus, for a complex system of charged bodies the electric field in vacuum will be:

$$\vec{E} = \frac{1}{4\pi\varepsilon_0}\left[\sum_{i=1}^{n} \frac{q_i\,\vec{r_{0i}}}{r_{0i}^3} + \int_L \frac{\lambda dl\vec{R}}{R^3} + \int_S \frac{\sigma dS\vec{R}}{R^3} + \int_V \frac{\rho dV\vec{R}}{R^3} \right] \tag{1.29}$$

Line, surface or volume integrals are easy to compute, if the line, surface or volume elements, respectively, are written in the appropriate three-orthogonal curvilinear coordinate system (see appendix A).

Exercise: calculus of the electric field of a continuous charge distribution

Let be the surface of a disk with the radius a and charged with an uniform surface charge distribution of density σ (see figure 1.13).

 a. *Find the electric field on the axis of the disk, at a distance h with respect to its plane (at point P), considering σ constant;*
 b. *What is the electric field at the same point P, when the surface density becomes σ = kr?*

 a. *Due to the particular symmetry of the surface distribution, the problem will be solved in a cylindrical coordinate system (see appendix A). In this context:*

$$dq = \sigma dS = \sigma r \, dr \, d\theta \tag{1.30}$$

and:

$$\overrightarrow{dE} = \frac{dq}{4\pi\varepsilon_0 R^3}\overrightarrow{R} \tag{1.31}$$

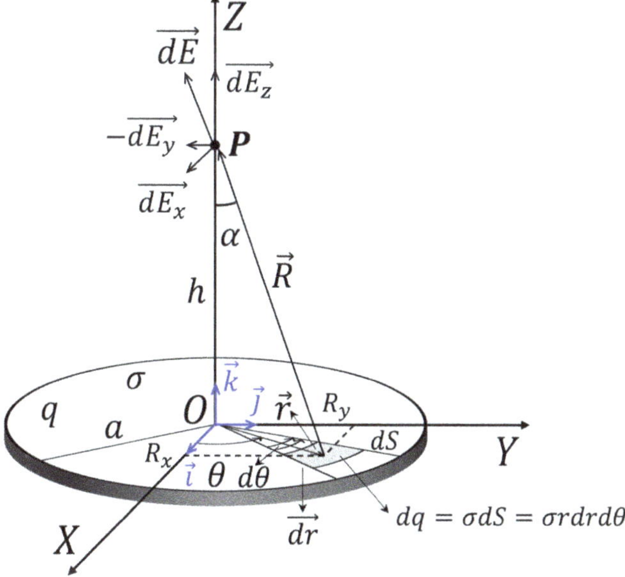

Figure 1.13. Electric field of a surface charge distribution, generated by a disk at point *P*, situated at a distance h with respect to the disk plane.

The electric field components become in cylindrical coordinates:

$$dE_x = +\frac{\sigma r dr d\theta r \cos\theta}{4\pi\varepsilon_0(r^2 + h^2)^{3/2}} \tag{1.32}$$

$$dE_y = -\frac{\sigma r dr d\theta r \sin\theta}{4\pi\varepsilon_0(r^2 + h^2)^{3/2}}$$

$$dE_z = +\frac{\sigma r dr d\theta h}{4\pi\varepsilon_0(r^2 + h^2)^{3/2}}$$

It can be observed that $E_x = E_y = 0$ after the θ integration of $\sin\theta$ and $\cos\theta$ on the $(0 - 2\pi)$ range, due to the fact that they represent their average value on a period. Then different to zero is only the normal component of the electric field to the plane of the disk which will be obtained by the integral:

$$E_z = \frac{\sigma h}{4\pi\varepsilon_0}\int_0^a\int_0^{2\pi}\frac{r dr d\theta}{(r^2 + h^2)^{3/2}} = \frac{\sigma h}{2\varepsilon_0}\int_0^a\frac{r dr}{(r^2 + h^2)^{3/2}}$$

$$= \frac{\sigma h}{2\varepsilon_0}\left(-\frac{1}{\sqrt{r^2 + h^2}}\Bigg|_0^a\right) \tag{1.33}$$

Consequently:

$$E_z = \frac{\sigma}{2\varepsilon_0}\left(1 - \frac{h}{\sqrt{a^2 + h^2}}\right) \tag{1.34}$$

Using equation (1.34), we may have three cases:

$$\begin{cases} h = 0 & \Rightarrow E = \dfrac{\sigma}{2\varepsilon_0} \\[2mm] h \to \infty & \Rightarrow E = 0 \\[2mm] a \to \infty & \Rightarrow E = \dfrac{\sigma}{2\varepsilon_0} \end{cases} \tag{1.35}$$

Observation: *according to the third case $(a \to \infty)$, one can see that the electric field of a large surface carrying the charge σ is perpendicular on the surface, being independent of the distance, $\vec{E} = \pm\frac{\sigma}{2\varepsilon_0}\vec{n}$ ($\vec{n}$ is normal to the surface).*

b. *Going further, we will replace $\sigma = kr$ in equation (1.34). Thus:*

$$E_z = \int_0^a\int_0^{2\pi}\frac{kr \cdot r dr d\theta h}{4\pi\varepsilon_0(r^2 + h^2)^{3/2}} = \frac{kh}{2\varepsilon_0}\int_0^a\frac{r^2 dr}{(r^2 + h^2)^{3/2}}$$

$$= \frac{kh}{2\varepsilon_0}\left[-\frac{r}{(r^2 + h^2)^{1/2}}\Bigg|_0^a + \int_0^a\frac{dr}{\sqrt{r^2 + h^2}}\right] \tag{1.36}$$

Hence:

$$E_z = \frac{kh}{2\varepsilon_0}\left[-\frac{a}{\sqrt{a^2 + h^2}} + \ln\frac{a + \sqrt{a^2 + h^2}}{h} \right]$$

$$= \frac{k}{2\varepsilon_0}\left[h\ln\frac{a + \sqrt{a^2 + h^2}}{h} - \frac{ah}{\sqrt{a^2 + h^2}} \right] \tag{1.37}$$

In this situation, two particular cases could be described:

$$\begin{cases} h \to 0 & \Rightarrow\ E_z = 0 \\ h \to \infty & \Rightarrow\ E_z = 0;\ \ \dfrac{dE_z}{dh} = 0 \Rightarrow h \text{ is maximum} \end{cases} \tag{1.38}$$

Finally, if all charge on the disk is q, then:

$$q = \int_0^a \int_0^{2\pi} kr \cdot r\,dr\,d\theta = 2\pi k \frac{r^3}{3}\bigg|_0^a = 2\pi k \frac{a^3}{3} \Rightarrow k = \frac{3q}{2\pi a^3} \tag{1.39}$$

Further reading

[1] Bleaney B I and Bleaney B 1965 *Electricity and Magnetism (Oxford Classic Texts in the Physical Sciences)* vol 1 3rd edn (Oxford: Oxford University Press)

[2] Blanpied W A 1971 *Modern Physics: An Introduction to Its Mathematical Language* (New York: Holt, Rinehart and Winston)

[3] Landsberg G 1987 *Cours élémentaire de physique—Tome 2, Electricité et magnétisme* (Moscow: Mir)

[4] Cross J 1987 *Electrostatics, Principles, Problems and Applications* (London: Taylor and Francis)

[5] Feynman R P, Leighton R B and Sands M 2011 *The Feynman Lectures on Physics, Vol II: The New Millennium Edition: Mainly Electromagnetism and Matter* (New York: Basic Books)

[6] Bécherrawy T 2012 Electrostatics in vacuum *Electromagnetism* ed T Bécherrawy (New York: Wiley) ch 2

[7] Purcell E M and Morin D J 2013 *Electricity and Magnetism* 3rd edn (Cambridge: Cambridge University Press)

[8] Bettini A 2016 Electrostatic field in a vacuum *A Course in Classical Physics 3 —Electromagnetism. Undergraduate Lecture Notes in Physics* (Berlin: Springer) pp 1–60

[9] Agrawal N 2020 *Electrostatics: Current and Capacitors (for IIT-JEE)* (Independently Published)

IOP Publishing

Electrostatics
Formalism of the electrostatic field in vacuum and matter
Ştefan Antohe and Vlad-Andrei Antohe

Chapter 2

The electrostatic potential

Now that we introduced the electrostatic field (or electric field) being a **measure of force** per unit of charge, we will introduce in this chapter the electrostatic potential which is a **measure of energy** per unit of charge. Essentially, the electrostatic potential represents the amount of work needed to move a unit of charge from a reference point to a specific point inside the field, without producing an acceleration. Typically, the reference point might be the Earth or a point at infinity, although any point could be used. The electrostatic potential is a very useful physical quantity preferred by physicists and engineers, because it simplifies solving the electrostatic problems, giving the possibility to have better knowledge of the electric field. In other words, as learned in chapter 1, the electric field generated by a stationary charge distribution is characterized by the vector field $\overrightarrow{E}$ showing how strong the electric field at one of its own points is. However, the electric field could be also characterized by a scalar function referred to as the *electrostatic potential*. This scalar quantity is a function of the position vector $\overrightarrow{r}$ and it effectively characterizes the '*potentiality*' of an electric field to make an available work.

Although the electrostatic potential has a precisely-described physical meaning (see section 2.3), it can be simply introduced in a mathematical sense, by first verifying that the electric field accepts a potential according with the fields theory, more precisely by verifying that the electrostatic field is a *potential vector field. The potential vector field represents a vector field resulting from a scalar field by the action of the gradient operator.* Thus, to introduce the electrostatic potential in the above-mentioned approach, particular knowledge on vector analysis is needed (see appendix B.3).

2.1 Electrostatic potential of a point charge

It was established in chapter 1 that the electric field generated by a stationary charge distribution is characterized by the electrostatic field $\overrightarrow{E}$. Taking into account the

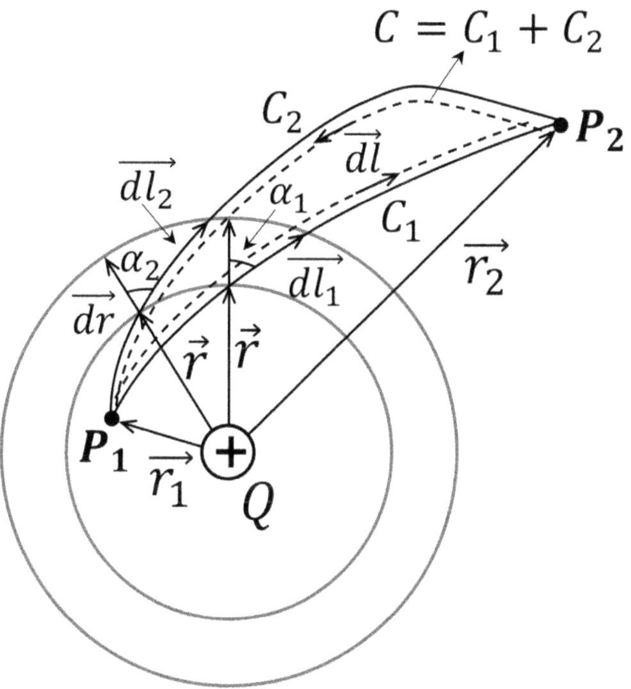

Figure 2.1. Schematic showing that the line integral of the electrostatic field between two points P_1 and P_2 is independent of the path of integration.

notions of a potential vector field and the potential of the field, discussed in appendix B, to be able to introduce a potential for the electrostatic field, first of all we must verify that the electrostatic field is a potential vector field. To do this, two points will be drawn in the electrostatic field of the generatrix charge Q and the line integral of the electrostatic field will be calculated along two paths C_1 and C_2 joining the points P_1 and P_2 (see figure 2.1).

In the case of the electrostatic field of point charge Q (see figure 2.1), the vector field is radially oriented and is given by:

$$\vec{E} = \frac{Q\vec{r}}{4\pi\varepsilon r^3} \tag{2.1}$$

The line integrals of the electric field $\vec{E}$ between points P_1 and P_2 along the two paths of integration C_1 and C_2, respectively, are equal:

$$\int_{P_1(C_1)}^{P_2} \vec{E}\ \overrightarrow{dl_1} = \int_{P_1(C_2)}^{P_2} \vec{E}\ \overrightarrow{dl_2} \tag{2.2}$$

Any spherical shell of the radius $\vec{r}$ and $\vec{r} + \overrightarrow{dr}$ will cut the paths of integration C_1 and C_2 giving rise to the length elements $\overrightarrow{dl_1}$ and $\overrightarrow{dl_2}$, then the corresponding elementary circulations will be:

$$(C_1) \quad \vec{E}(\vec{r})\,\overrightarrow{dl_1} = \left|\vec{E}\right|\left|\overrightarrow{dl_1}\right|\cos\alpha_1 = E\,dr = \frac{Q}{4\pi\varepsilon_0 r^2}dr$$
$$(C_2) \quad \vec{E}(\vec{r})\,\overrightarrow{dl_2} = \left|\vec{E}\right|\left|\overrightarrow{dl_2}\right|\cos\alpha_2 = E\,dr = \frac{Q}{4\pi\varepsilon_0 r^2}dr \tag{2.3}$$

Thus, the circulations on the two paths of integration C_1 and C_2 are equal:

$$\int_{P_1(C_1)}^{P_2} \vec{E}\,\overrightarrow{dl_1} = \int_{P_1(C_2)}^{P_2} \vec{E}\,\overrightarrow{dl_2}$$
$$\int_{P_1(C_1)}^{P_2} \vec{E}\,\overrightarrow{dl_1} = -\int_{P_2(C_2)}^{P_1} \vec{E}\,\overrightarrow{dl_2} \tag{2.4}$$

Considering now a closed path $C = C_1 + C_2$, represented by the dashed curve in figure 2.1, we will have:

$$\int_{P_1(C_1)}^{P_2} \vec{E}\,\overrightarrow{dl} + \int_{P_2(C_2)}^{P_1} \vec{E}\,\overrightarrow{dl} = \oint_C \vec{E}\,\overrightarrow{dl} = 0 \tag{2.5}$$

According to equation (2.5), the circulation of the electrostatic field on a closed path is zero. This property shows that *the electrostatic field $\vec{E}$ is a **potential vector field**, then it admits a potential, in other words, there is a scalar field V, called **electrostatic potential**, having the gradient equal with $\vec{E}$.*

The electrostatic field is a conservative field and then it may be derived from a scalar function such as the potential, applying the gradient operator. Indeed, the potential difference ΔV between two points at small distance apart, labeled by position vectors $\vec{r}$ and $\vec{r} + \overrightarrow{dr}$ is:

$$\Delta V = V(\vec{r} + \overrightarrow{dr}) - V(\vec{r}) = -\int_r^{r+dr} \vec{E}\,\overrightarrow{dl} \tag{2.6}$$

But, in the left side of equation (2.6) we have:

$$\Delta V = \int_r^{r+dr} dV = \int_r^{r+dr} \operatorname{grad} V \overrightarrow{dl} = \int_r^{r+dr} \nabla V \overrightarrow{dl} \tag{2.7}$$

Thus, from equations (2.6) and (2.7), it results:

$$\vec{E} = -\operatorname{grad} V = -\nabla V \tag{2.8}$$

According to the relationship for the electrostatic field of point charge Q:

$$\vec{E} = \frac{Q\vec{r}}{4\pi\varepsilon_0 r^3} \tag{2.9}$$

Observing that $\frac{\vec{r}}{r^3} = -\nabla\frac{1}{r}$ (see appendix B.2), then:

$$\vec{E} = -\frac{Q}{4\pi\varepsilon_0}\nabla\frac{1}{r} = -\nabla\frac{Q}{4\pi\varepsilon_0 r} = -\nabla V \tag{2.10}$$

So, the differential form of the relationship between field and potential is given by equation (2.8), existing at any point of an electrostatic field. Taking into account the equation (2.7), by integration along on an arbitrary path joining the reference point $P_1(\overrightarrow{r_1})$ and the current point $P(\overrightarrow{r})$, the relationship between the potential at a point and the electric field can be obtained:

$$V(P) = V(P_1) - \int_{P_1}^{P} \overrightarrow{E}\,\overrightarrow{dl} \tag{2.11}$$

Using the relation (2.11) and the condition to have a zero potential when the reference point is to infinity one obtains, for the electric potential of a point charge Q, the value:

$$V = -\int_{\infty}^{r} \frac{Q}{4\pi\varepsilon_0 r^2}dr = \frac{Q}{4\pi\varepsilon_0 r}; \quad V(\infty) = 0 \tag{2.12}$$

Then the potential of a point charge Q at a point P placed at the distance $\overrightarrow{r}$ with respect to the source, is proportional with the generatrix charge and reversely proportional with the distance from the charge to that point.

Around a point charge $+Q$ located at the origin, the potential is $V = \dfrac{Q}{4\pi\varepsilon_0 r}$. The spheres centered on the origin are surfaces of constant potential that are referred as *equipotential surfaces.*

2.2 Superposition principle for the potential

The superposition principle for the electric field generated by a discrete system of charges (see figure 2.2) states that:

$$\overrightarrow{E} = \sum_{i=1}^{n} \frac{q_i(\overrightarrow{r} - \overrightarrow{r_{0i}})}{4\pi\varepsilon_0 \mid \overrightarrow{r} - \overrightarrow{r_{0i}} \mid^3} = \sum_{i=1}^{n} \frac{q_i \overrightarrow{r_i}}{4\pi\varepsilon_0 \mid \overrightarrow{r_i} \mid^3} \tag{2.13}$$

But according to appendix B.2:

$$\frac{\overrightarrow{r_i}}{\mid \overrightarrow{r_i} \mid^3} = -\nabla_{\overrightarrow{r}}\left(\frac{1}{r_i}\right) \tag{2.14}$$

and:

$$\overrightarrow{r_i} = (x - x_{0i})\overrightarrow{i} + (y - y_{0i})\overrightarrow{j} + (z - z_{0i})\overrightarrow{k} \tag{2.15}$$

Thus:

$$\overrightarrow{E} = -\sum_{i=1}^{n}\nabla_r\left(\frac{q_i}{4\pi\varepsilon_0 r_i}\right) = -\nabla\left(\sum_{i=1}^{n}\frac{q_i}{4\pi\varepsilon_0 r_i}\right)$$
$$= -\nabla\left(\sum_{i=1}^{n} V_i\right) = -\nabla V \tag{2.16}$$

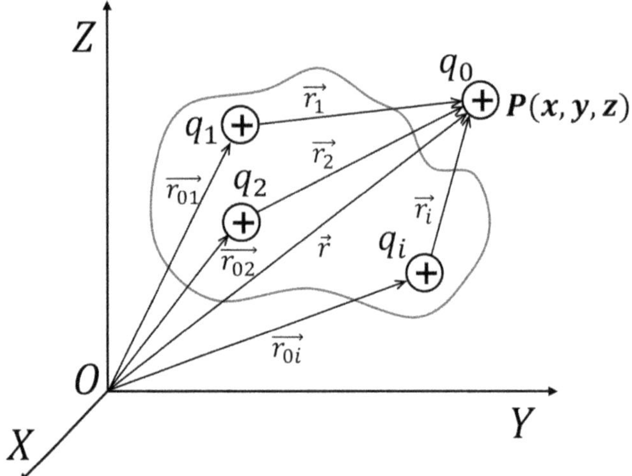

Figure 2.2. Discrete system of point charges used to demonstrate the superposition principle for the electrostatic potential.

Then:

$$V = \sum_{i=1}^{n} V_i = \sum_{i=1}^{n} \frac{q_i}{4\pi\varepsilon_0 r_i} \tag{2.17}$$

Relation (2.17) expresses the *superposition principle for the electrostatic potential,* stating that the electrostatic potential at a point $P(\vec{r})$ in the electric field of an assembly of charges q_i at positions $\vec{r_{0i}}$, is the algebraic sum of the potentials at that point, due to each charge separately. To find the potential due to a continuous charge distribution, the discrete summation in equation (2.17) is replaced by an integral, as:

(a) Linear charge distribution:

$$V = \frac{1}{4\pi\varepsilon_0} \int_L \frac{\lambda(x', y', z')dl}{r} \tag{2.18}$$

(b) Surface charge distribution:

$$V = \frac{1}{4\pi\varepsilon_0} \int_S \frac{\sigma(x', y', z')dS}{r} \tag{2.19}$$

(c) Volume charge distribution:

$$V = \frac{1}{4\pi\varepsilon_0} \int_V \frac{\rho(x', y', z')dV}{r} \tag{2.20}$$

Thus, for a complex system of charged bodies the potential will be:

$$V = \frac{1}{4\pi\varepsilon_0}$$

$$\cdot \left[\sum_{i=1}^{n} \frac{q_i}{r_i} + \int_L \frac{\lambda(x', y', z')dl}{r} + \int_S \frac{\sigma(x', y', z')dS}{r} + \int_V \frac{\rho(x', y', z')dV}{r} \right] \qquad (2.21)$$

Relation (2.21) remains valid with the convention requiring to have the potential zero to infinity [$V(\infty) = 0$]. Within the integrals of equation (2.21), the length, surface and volume elements, respectively, are positioned by the vector $\vec{r'}$.

Exercise: electrostatic potential for a continuous charge distribution

Let there be a sphere of radius a, uniformly charged with the volume charge density ρ. What is the potential at point P placed at a distance h with respect to its center (see figure 2.3), inside and outside of the spherical charge distribution?

Solution: *let's consider the point P along the OZ axis, at h distance from the center of the sphere. The volume element $dV = r^2 \sin\theta\, dr\, d\theta\, d\varphi$ carries the charge $dq = \rho r^2 \sin\theta\, dr\, d\theta\, d\varphi$ which can be considered as a point charge. Then, the potential of the electrostatic field generated by this elementary charge at point P is:*

$$dV = \frac{\rho r^2 \sin\theta\, dr\, d\theta\, d\varphi}{4\pi\varepsilon_0(r^2 + h^2 - 2rh\cos\theta)^{1/2}} \qquad (2.22)$$

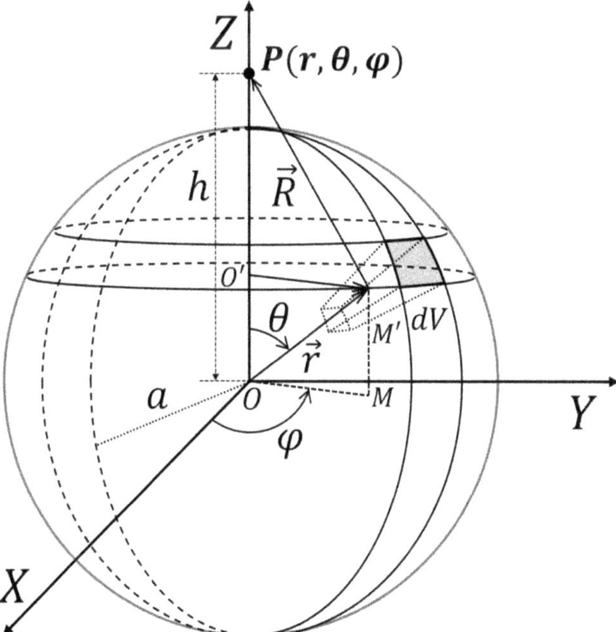

Figure 2.3. *A spherical charge distribution with volume charge density, ρ =constant.*

According to the superposition principle, the potential at point P will be:

$$V = \int_0^a \int_0^\pi \int_0^{2\pi} \frac{\rho r^2 \sin\theta \, dr \, d\theta \, d\varphi}{4\pi\varepsilon_0 (r^2 + h^2 - 2rh\cos\theta)^{1/2}}$$

$$= \frac{\rho}{4\pi\varepsilon_0} \int_0^a r^2 dr \int_0^\pi \frac{\sin\theta \, d\theta}{(r^2 + h^2 - 2rh\cos\theta)^{1/2}} \int_0^{2\pi} d\varphi \qquad (2.23)$$

$$= \frac{\rho}{2\varepsilon_0} \int_0^a r^2 I(r) \, dr$$

where:

$$I(r) = \int_0^\pi \frac{\sin\theta \, d\theta}{(r^2 + h^2 - 2rh\cos\theta)^{1/2}}$$

$$= \frac{1}{rh}(r^2 + h^2 - 2rh\cos\theta)^{1/2} \Big|_0^\pi = \frac{1}{rh}[(r + h) - |\, r - h\,|] \qquad (2.24)$$

At this point, we may discriminate between two situations:

(a) *Outside of the sphere $(h > a)$, when:*

$$I(r) = \frac{1}{rh}(r + h - h + r) = \frac{2}{h} \qquad (2.25)$$

(b) *Inside of the sphere $(h < a)$, when:*

$$\begin{cases} h > r \;\Rightarrow\; I(r) = \dfrac{2}{h} \\[2mm] h < r \;\Rightarrow\; I(r) = \dfrac{2}{r} \end{cases} \qquad (2.26)$$

Consequently the electrostatic potential will be:

(a) *Outside the sphere:*

$$V_e = \frac{\rho}{2\varepsilon_0} \int_0^a r^2 \frac{2}{h} dr = \frac{\rho a^3}{3\varepsilon_0 h} = \frac{Q}{4\pi\varepsilon_0 h} \qquad (2.27)$$

Observation: *far away from the spherical charge distribution, the potential behaves like the potential of a point charge equal with the total charge of the distribution, located at the origin of the sphere.*

(b) *Inside the sphere:*

$$\begin{aligned} V_i &= \frac{\rho}{2\varepsilon_0}\left[\int_0^h r^2 \frac{2}{h} dr + \int_h^a r^2 \frac{2}{r} dr\right] = \frac{\rho}{\varepsilon_0}\left[\int_0^h \frac{r^2}{h} dr + \int_h^a r \, dr\right] \\[2mm] &= \frac{\rho}{\varepsilon_0}\left[\frac{h^2}{3} + \frac{a^2}{2} - \frac{h^2}{2}\right] = \frac{\rho}{2\varepsilon_0}\left[a^2 - \frac{h^2}{3}\right] = \frac{Q}{8\pi\varepsilon_0 a^3}[3a^2 - h^2] \end{aligned} \qquad (2.28)$$

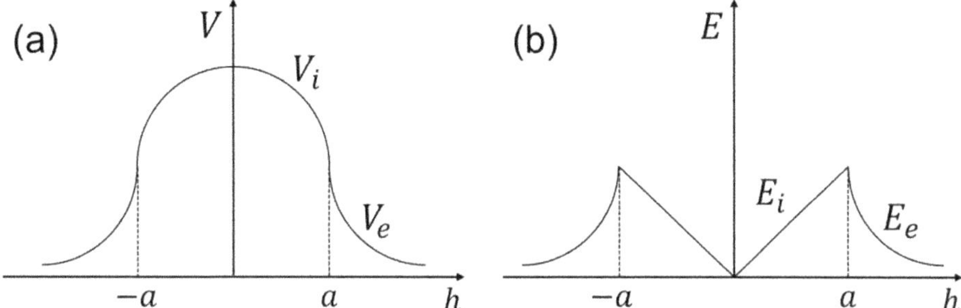

Figure 2.4. *Plots of V (a) and $\vec{E}$ (b) as a function of distance h between the center of the sphere and the point where they are calculated.*

While analyzing the latter two situations, it can be thus observed that inside the sphere the potential has a parabolic behavior, while outside the sphere it has a hyperbolic nature (see figure 2.4). Having the potential and knowing the relationship between the electric field and potential, the electric field inside and outside the sphere will be:

(a) *Inside the sphere:*

$$\vec{E_i} = -\nabla V_i = -\nabla_h \left[\frac{\rho}{2\varepsilon_0} \left(a^2 - \frac{h^2}{3} \right) \right]$$

$$= \frac{\rho}{2\varepsilon_0} \frac{2h\,\vec{e_h}}{3} = \frac{\rho h\,\vec{e_h}}{3\varepsilon_0} = \frac{\rho\vec{h}}{3\varepsilon_0}$$

(2.29)

(b) *Outside the sphere:*

$$\vec{E_e} = -\nabla V_e = -\nabla_h \left(\frac{\rho a^3}{3\varepsilon_0 h} \right) = \frac{\rho a^3\,\vec{e_h}}{3\varepsilon_0 h^2} = \frac{Q\,\vec{e_h}}{4\pi\varepsilon_0 h^2} = \frac{Q\vec{h}}{4\pi\varepsilon_0 h^3}$$

(2.30)

where $\vec{e_h} = \frac{\vec{h}}{h}$ is the versor of the vector $\vec{h}$.

Plots of V and $\vec{E}$, respectively, as a function of distance h between the center of the sphere and the point where they are calculated are shown in figure 2.4. It can be thus concluded that on the surface of the sphere the potential and the electric field continue if the dielectric constant is the same inside and outside the sphere:

$$\left. V_e \right|_{r=a} = \frac{\rho a^3}{3\varepsilon_0 a} = \left. V_i \right|_{r=a} = \frac{\rho}{2\varepsilon_0} \left(a^2 - \frac{a^2}{3} \right) = \frac{\rho a^2}{3\varepsilon_0}$$

(2.31)

$$\left. E_e \right|_{r=a} = \frac{Q}{4\pi\varepsilon_0 a^2} = \left. E_i \right|_{r=a} = \frac{Q}{4\pi\varepsilon_0 a^2} = \frac{\rho a}{3\varepsilon_0}$$

(2.32)

2.3 Physical meaning of the electrostatic potential

Considering a point charge q_0, moving without acceleration (uniformly) between points P_1 and P_2 from the electrostatic field $\vec{E}$ (see figure 2.5), the work done by the electric field to move the point charge q_0 from point P_1 to P_2 is equal to the difference of the potential energy associated with each point. If the work is done by the electrostatic field, the potential energy at end (final) point is smaller than its value to the start (initial) point. If the work is done by an external force, the potential energy at end (final) point is bigger than its value to the start (initial) point. Let's suppose that the point charge q_0 moves uniformly with a very low speed between points P_1 and P_2 due to the action of the external force against the electrostatic field (it was considered $\vec{F_{ext}} = -q\vec{E} = -\vec{F}$, such that the velocity is constant and kinetic energy also remains constant). The work done by the external force to move the point charge q_0 from P_1 to P_2 is:

$$L_{P_1 P_2} = \int_{P_1}^{P_2} \vec{F_{ext}}\,\vec{dl} = -q_0 \int_{P_1}^{P_2} \vec{E}\,\vec{dl} \tag{2.33}$$

Then:

$$\frac{L_{P_1 P_2}}{q_0} = -\int_{P_1}^{P_2} \vec{E}\,\vec{dl} = \int_{P_1}^{P_2} \nabla V \vec{dl} = \int_{P_1}^{P_2} dV = V_{P_2} - V_{P_1} \tag{2.34}$$

Thus, the ratio between the work done by the external force to move the point charge q_0 from P_1 to P_2 and the value of q_0 is equal to variation of the potential between the two points:

$$\frac{L_{P_1 P_2}}{q_0} = \Delta V = V(P_2) - V(P_1) \tag{2.35}$$

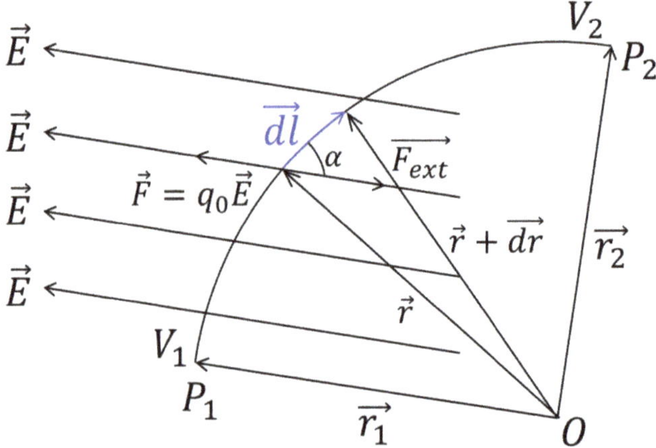

Figure 2.5. Physical meaning of the electrostatic potential.

Here, one can see that just the potential difference can be exactly measured and defined. The potential at point P_2 is:

$$V(P_2) = V(P_1) + \frac{L_{P_1P_2}}{q_0} = V(P_1) + \frac{1}{q_0} \int_{P_1}^{P_2} \overrightarrow{F_{\text{ext}}} \, \overrightarrow{dl} = V(P_1) - \int_{P_1}^{P_2} \overrightarrow{E} \, \overrightarrow{dl} \qquad (2.36)$$

equal with the potential at a reference point (P_1 here), adding the work done by external forces to move the unit charge from the reference point to target point. By convention, if the reference point is taken to infinite where the potential is zero, the potential at point P becomes:

$$V(P) = -\int_{\infty}^{P} \overrightarrow{E} \, \overrightarrow{dl} \qquad (2.37)$$

According to the equation (2.37), *the potential of the electrostatic field at a point P is the physical scalar quantity equal to the work done by the external forces to move the unit charge from infinite to that point.* If choosing the reference point to infinite lead to an infinite value for the potential, then the reference point cannot be taken to infinite.

2.4 The electric voltage

Let there be the electric field $\overrightarrow{E}$ in a vacuum and a path C drawn in the field (see figure 2.6). Then, the electric voltage between two points A and B along the path C is the physical scalar quantity characterizing the properties of the electric field along

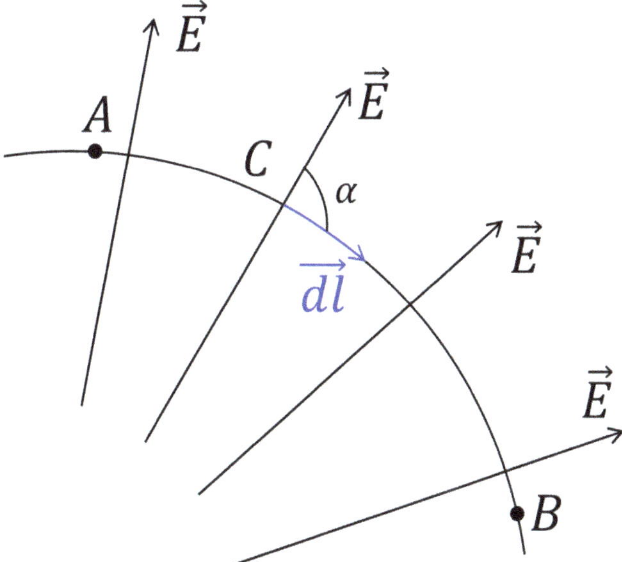

Figure 2.6. Definition of the electric voltage.

the path C and it is defined as the line integral of the electric field between point A and B along the path of integration C:

$$U_{AB} = \int_{A(C)}^{B} \vec{E}\,\vec{dl} = \int_{A(C)}^{B} Edl\cos\alpha \tag{2.38}$$

From this definition it results that the electric voltage depends on the sense of integration, $A \to B$ or $B \to A$, and then, $U_{AB} = -U_{BA}$:

$$\int_{A}^{B} \vec{E}\,\vec{dl} = -\int_{B}^{A} \vec{E}\,\vec{dl} \tag{2.39}$$

The sense of integration is called the reference sense for the voltage and is shown by a mark at the poles of a circuit:

$$U_{AB} = \int_{A}^{B} (E_x dx + E_y dy + E_z dz) \tag{2.40}$$

In the case of a potential vector field, as the field $\vec{E}$ considered here:

$$\vec{E} = -\nabla V \text{ and } dV = -\vec{E}\,\vec{dl} \tag{2.41}$$

Then:

$$U_{AB} = \int_{A}^{B} \vec{E}\,\vec{dl} = -\int_{A}^{B} dV = -(V_B - V_A) = V_A - V_B \tag{2.42}$$

$$U_{AB} = V_A - V_B = -\Delta V$$

In the electrostatic field, the electrical voltage between two points is equal to the potential difference in the sense of integration, in other words, it is equal to the minus of variation of the potential with respect to those points, $U_{AB} = -\Delta V$.

2.4.1 Physical meaning of the electric voltage

According to the definition of the electric voltage:

$$U_{AB} = \int_{A}^{B} \vec{E}\,\vec{dl} = \frac{1}{q}\int_{A}^{B} q\vec{E}\,\vec{dl} = \frac{1}{q}\int_{A}^{B} \vec{F}\,\vec{dl} = \frac{L_{AB(C)}}{q}; \quad \vec{F} = q\vec{E} \tag{2.43}$$

We can see that *the voltage between two points A and B in the electrostatic field is the scalar quantity equal with the work done by the *electric force*, $\vec{F} = q\vec{E}$, to move the unit of charge between the two points.

2.5 Equipotential surfaces

The surfaces defined by the equation $V(x, y, z) = $ constant are called equipotential surfaces. Considering two points on the equipotential surface at a very small distance between them (let be the length element $\vec{dl}$) the variation of the potential along this length element is:

$$dV = \nabla V \vec{dl} = -\vec{E}\ \vec{dl} = 0 \quad (V = \text{constant}) \tag{2.44}$$

Starting from this definition and from equation (2.44), several conclusions can be withdrawn:

(a) From relation (2.44) it results that the lines of the electric field are always normal to the equipotential surfaces.

(b) Considering the small length element along the electric field, $\vec{dl} \uparrow\uparrow \vec{E}$, results:

$$dV = -\vec{E}\ \vec{dl} = -Edl < 0 \tag{2.45}$$

showing that along the lines of the electric field the potential decreases, in other words, the electric field is oriented perpendicular to the equipotential surfaces from a higher potential to a lower potential. But $\nabla V = -\vec{E}$, then the gradient of the electric potential is also a vector perpendicular to the equipotential surface but has the sense in the sense of increasing of the potential from that point, in other words, from a lower potential to a higher potential.

(c) The electric field $\vec{E}$ is higher in the regions with dense equipotential surfaces. Nevertheless, considering two equipotential surfaces with the constant potential difference between them ($dV = V_2 - V_1 = \text{constant}$), see figure 2.7, on the base of relation:

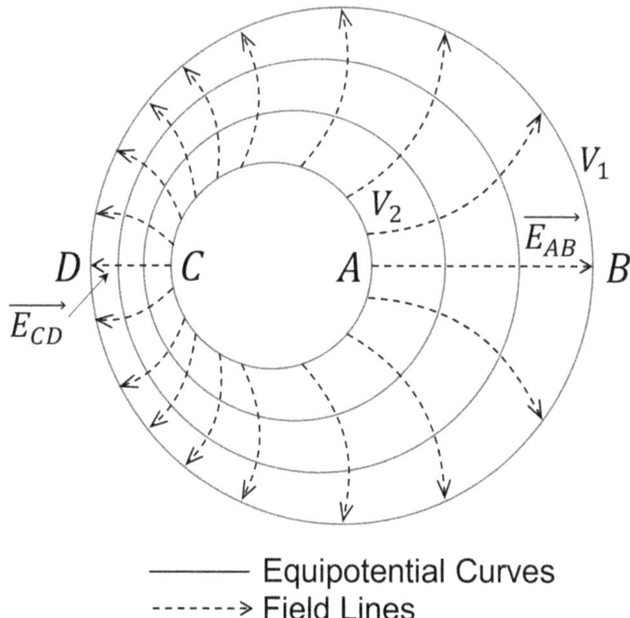

Figure 2.7. Equipotential surfaces and lines of the electric field.

$$dV = \left| \vec{E} \right| \cdot \left| \vec{dl} \right| = \text{constant} \qquad (2.46)$$

results:

$$E_{AB} \cdot d_{AB} = E_{CD} \cdot d_{CD} = \text{constant}$$
$$\frac{E_{CD}}{E_{AB}} = \frac{d_{AB}}{d_{CD}} > 1 \qquad (2.47)$$

showing that effectively the electric field strength is higher where the equipotential surfaces are closer.

(d) Inside of the conductors to electrostatic equilibrium, the electric field, $\vec{E}$, is zero, then the volume of the conductor and its surface form an equipotential region.

2.6 The theorem of the electrostatic potential

All the properties of the electrostatic field, described above, are contained in the *theorem of the electrostatic potential*, in its integral and differential form, as presented further.

2.6.1 Integral form of the electrostatic potential theorem

As already mentioned, the electrostatic field being conservative, no work at all is done when a test charge moves along a path which returns it to its original position. The work done by the electric field to move the test charge q from A to B is exactly canceled by the work done to return the charge from B to A (see figure 2.8):

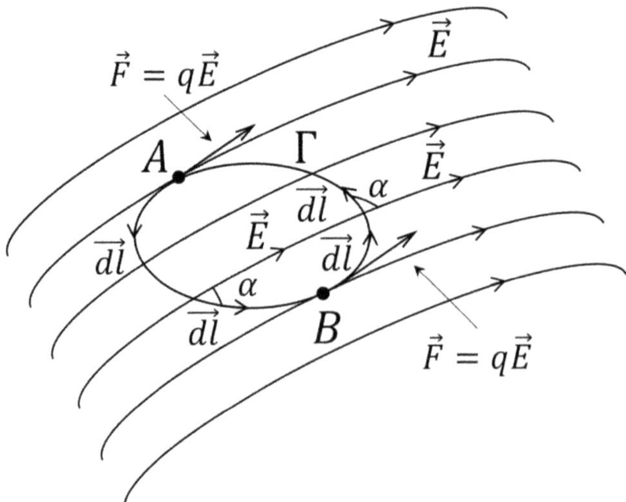

Figure 2.8. The closed path Γ drawn in the electrostatic field to demonstrate the theorem of the electrostatic potential.

$$\int_A^B q\vec{E}\ \vec{dl} + \int_B^A q\vec{E}\ \vec{dl} = W_i(A) - W_f(A) \qquad (2.48)$$

But:

$$W_i(A) = W_f(A) \Rightarrow \oint_\Gamma q\vec{E}\ \vec{dl} = 0 \Rightarrow \oint_\Gamma \vec{E}\ \vec{dl} = 0 \qquad (2.49)$$

Equation (2.49) represents the integral form of the electrostatic potential theorem, stating that: *the line integral along of any closed path drawn in an electrostatic field is zero*. This theorem is valid for any potential vector field, and for a stationary electric field. The electrostatic potential theorem has some important consequences, which should be mentioned here:

(a) In the electrostatic field there are no closed field lines. Indeed, if we suppose that in an electrostatic field there are closed field lines, as in figure 2.9, then the line integral along one of them must be different from 0 ($\oint_\Gamma \vec{E}\ \vec{dl} \neq 0$),

because it represents the sum of the scalar products $\vec{E}\ \vec{dl}$ always positive. So, supposing that the electrostatic field has closed field lines we find that the electrostatic potential theorem is not satisfied (is not true). *The electrostatic field is a potential vector field having only open lines of field.*

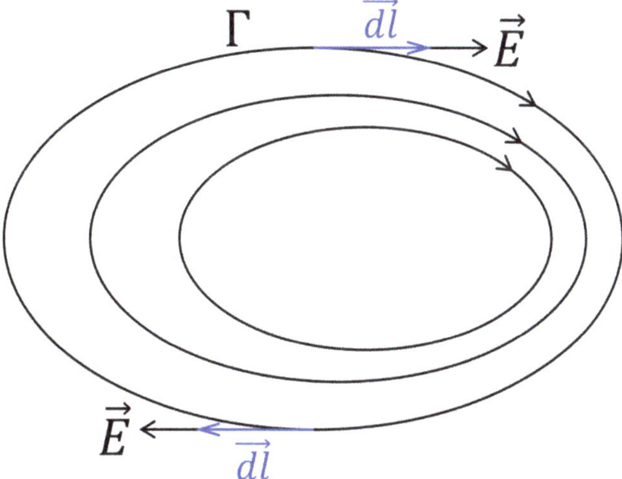

Figure 2.9. In the electrostatic field there are no closed field lines.

(b) Another scalar quantity which characterizes the properties of an electric field of a curve joining two points from the field, is the voltage U_{AB} defined within section 2.4 (see figure 2.10). It is defined as the line integral of $\vec{E}\ \vec{dl}$ along the path of integration joining the two points, equation (2.38):

$$U_{AB} = \int_A^B \vec{E}\ \vec{dl} = \int_A^B -\nabla V dl = -\int_A^B dV = -[V_B - V_A] = V_A - V_B \qquad (2.50)$$

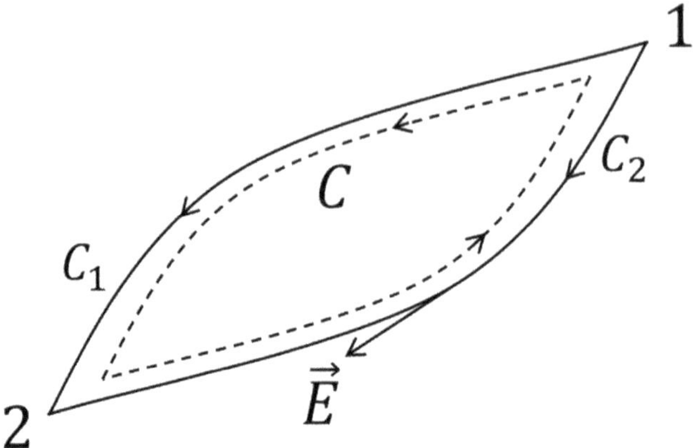

Figure 2.10. The paths drawn in the electric field $\vec{E}$.

As a consequence of the electrostatic potential theorem, it results that the voltage between two points from the electrostatic field is independent of the path of integration. Indeed, taking the closed curve $C = C_1 + C_2$ (dashed line) in agreement with the electrostatic potential theorem:

$$\oint_C \vec{E}\ \vec{dl} = \int_{1(C_1)}^{2} \vec{E}\ \vec{dl} + \int_{2(C_2)}^{1} \vec{E}\ \vec{dl} = 0 \tag{2.51}$$

But taking into account the definition of the electric voltage, equations (2.50) and (2.51), results:

$$U_{12} = \int_{1(C_1)}^{2} \vec{E}\ \vec{dl} = -\int_{2(C_2)}^{1} \vec{E}\ \vec{dl} = \int_{1(C_2)}^{2} \vec{E}\ \vec{dl} \tag{2.52}$$

Note, the voltage is independent of the path of integration.

2.6.2 Differential form of the potential theorem

We already know the integral form of the electrostatic potential theorem, and the potential of the electric field, defined by:

$$V_P = V_{P_0} - \int_{P_0}^{P} \vec{E}\ \vec{dl} \Rightarrow V_P - V_{P_0} = -\int_{P_0}^{P} \vec{E}\ \vec{dl} \Rightarrow dV = -\vec{E}\ \vec{dl}$$

$$\Rightarrow dV = -\left(E_x dx + E_y dy + E_z dz\right)$$

$$\Rightarrow E_x = -\frac{\partial V}{\partial x}; \quad E_y = -\frac{\partial V}{\partial y}; \quad E_z = -\frac{\partial V}{\partial z}$$

Concluding, at any point in the electrostatic field:

$$\vec{E} = -\nabla V = -\left(\frac{\partial V}{\partial x}\vec{i} + \frac{\partial V}{\partial y}\vec{j} + \frac{\partial V}{\partial z}\vec{k}\right) \tag{2.53}$$

Equation (2.53) represents *one expression of the local (differential) form of the electrostatic potential theorem.*

The second local form of the theorem expresses the content of the electrostatic potential theorem by a set of differential equations satisfied by the components of the electric field at a point. This will be possible just after the introduction of a new first-order differential operator referred to as the *curl of a vector field* (see appendix B.4). Recalling thus the electrostatic potential theorem, now we shall use *Stokes' theorem* (see appendix B.4.1) to transform the line integral (circulation) of the electric field along the closed path Γ, appearing in the integral form of the electrostatic potential theorem, into a surface integral over any surface S_Γ bounded by the path Γ, resulting:

$$\oint_\Gamma \vec{E}\,\vec{dl} = \int_{S_\Gamma} (\nabla \times \vec{E})\vec{ds} = 0 \tag{2.54}$$

The surface integral is zero when its integrant is zero:

$$\mathrm{curl}\,\vec{E} = \nabla \times \vec{E} = 0 \tag{2.55}$$

equation (2.55) represents *the second expression of the local (differential) form of the electrostatic potential theorem, i.e.,* **the curl of the electrostatic field is always and everywhere zero.**

The differential form of the electrostatic potential theorem should be obtained using the differential form of the relationship between electric field and electrostatic potential. Indeed, we know the fact that the electrostatic field is a potential vector field, namely, it results from a scalar field through the gradient operator, $\vec{E} = -\nabla V$. Then:

$$\nabla \times \vec{E} = -\nabla \times (\nabla V) = - \begin{vmatrix} \vec{i} & \vec{j} & \vec{k} \\ \dfrac{\partial}{\partial x} & \dfrac{\partial}{\partial y} & \dfrac{\partial}{\partial z} \\ \dfrac{\partial V}{\partial x} & \dfrac{\partial V}{\partial y} & \dfrac{\partial V}{\partial z} \end{vmatrix}$$

$$= -\vec{i}\left[\frac{\partial^2 V}{\partial y \partial z} - \frac{\partial^2 V}{\partial z \partial y}\right] - \vec{j}\left[\frac{\partial^2 V}{\partial z \partial x} - \frac{\partial^2 V}{\partial x \partial z}\right] - \vec{k}\left[\frac{\partial^2 V}{\partial x \partial y} - \frac{\partial^2 V}{\partial y \partial x}\right] = 0$$

Here, a new vector identity there is, stating that **always the curl of a gradient of a scalar field is zero**:

$$\nabla \times (\nabla \varphi) = 0 \tag{2.56}$$

We can see that both the differential form of relation between the electrostatic field and potential ($\vec{E} = -\nabla V$) and the differential form ($\nabla \times \vec{E} = 0$) are equivalent:

$$\vec{E} = -\nabla V \Leftrightarrow \nabla \times \vec{E} = 0 \tag{2.57}$$

Both, equations (2.57) show that the electrostatic field is a ***potential vector field***, conservative, characterized by the fact that its ***field lines are always open lines***. In other words, ***the electrostatic field is an irrotational field***. *Any vector field coming from the gradient of a scalar field is an irrotational field, namely, its curl is zero.*

Further reading

[1] Clayton P R, Keith W W and Syed N A 1997 *Introduction to Electromagnetic Fields* (New York: McGraw-Hill)

[2] Pandey S C 2010 *Course in Physics 4: Electrostatics and Current Electricity* (Chennai: Pearson Education)

[3] Feynman R P, Leighton R B and Sands M 2011 *The Feynman Lectures on Physics, Vol II: The New Millennium Edition: Mainly Electromagnetism and Matter* (New York: Basic Books)

[4] Murray J S and Politzer P 2011 The electrostatic potential: an overview *WIREs Comput. Mol. Sci.* **1** 153–63

[5] Purcell E M and Morin D J 2013 *Electricity and Magnetism* 3rd edn (Cambridge: Cambridge University Press)

[6] Bertrand C L 2013 *Electrostatics: Theory and Applications (Physics Research and Technology)* (Hauppauge, NY: NOVA Science Publishers)

[7] Sree Harsha N R, Prakash A and Kothari D P 2016 The electric potential *The Foundations of Electric Circuit Theory* (Bristol: IOP Publishing), ch 4 pp 4-1–14

[8] Griffiths D J 2017 *Introduction to Electrodynamics* 4th edn (Cambridge: Cambridge University Press)

[9] Agrawal N 2020 *Electrostatics: Current and Capacitors (for IIT-JEE)* (Independently Published)

Electrostatics
Formalism of the electrostatic field in vacuum and matter
Ştefan Antohe and Vlad-Andrei Antohe

Chapter 3

Flux of the electrostatic field

When a charged body has a high symmetry, the electric field can be easily evaluated by the application of Gauss' law, a very important theorem which relates the field on any closed surface to the net amount of charge enclosed within the surface. Using Gauss' law we shall be able to prove, for example, that the field outside an isolated and spherically symmetrical charged ion is exactly the same as if all its charge were concentrated at the center. An application to a macroscopic system is at the surface of a conductor, where Gauss' law leads to an expression for the surface charge density in terms of the external electric field. Gauss' law can be illustrated by considering a point charge q at the center of a sphere of radius r. The area of sphere is $4\pi r^2$ and the electric field on its surface is everywhere of magnitude $q/4\pi\varepsilon_0 r^2$. The product of the electric field and area is q/ε_0, i.e., it is independent of r. This is rather like what happens when a light bulb is placed at the center of a spherical lamp shade which is completely transparent. All the light from the bulb escapes, or in other words, the total amount of light passing through the shade is independent of its radius.

Gauss' law is the analogous result in electrostatics applied to a quantity called the *flux of the electric field*. The total flux out of a surface enclosing a charge q is q/ε_0 whatever the shape of the surface. We shall explain further what is meant by flux and then we shall demonstrate the Gauss' law rigorously.

3.1 The flux of a vector field

The *flux* may be defined for any vector field function of position but it is most easily visualized in the flow of fluids. Imagine that a fluid is flowing at a speed $\vec{v}$ through a small flat surface $\overrightarrow{dS}$ bounded by the contour of a tube, as shown in figure 3.1. The volume of fluid passing is larger when $\vec{v}$ is perpendicular to $\overrightarrow{dS}$ (see figure 3.1(a)), than when it is at an angle α with $\overrightarrow{dS}$ (see figure 3.1(b)). The flux of fluid through the physical area dS which is perpendicular to the direction of flow of the fluid (in other

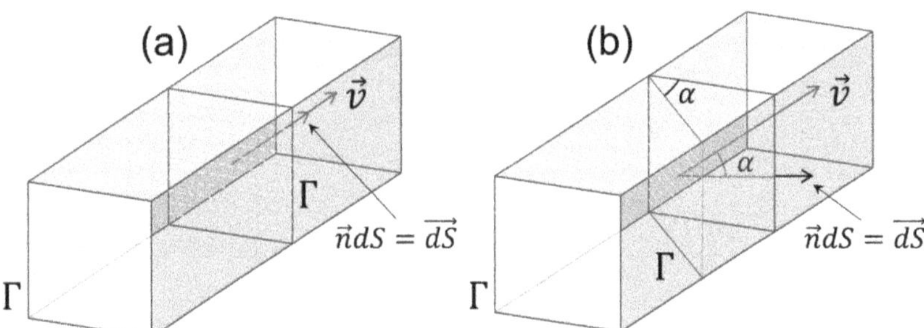

Figure 3.1. Schematic representing a column of fluid of length $\vec{v}$ passing through the area dS every second, where $\vec{v}$ and $\overrightarrow{dS}$ are perpendicular (a) and they form an angle α between them (b).

words perpendicular to the speed vector $\vec{v}$) is the product vdS of speed and area. If the normal to the surface dS is at an angle α to the direction of flow, the projected area on the perpendicular plane to direction of flow will be $dS_n = dS \cos \alpha$ and then the rate of flow of fluid through surface is:

$$vdS_n = vdS \cos \alpha = \vec{v}\cdot\overrightarrow{dS} \tag{3.1}$$

Thus, the surface vector dS is oriented (directed) along the normal to the surface. Similarly, for any vector field the flux may be defined as the scalar product of the vector field $\vec{G}$ and vector surface $\overrightarrow{dS}$.

Particularizing, in an electrostatic field there is of course nothing actually flowing, but the flux of an electric field $\vec{E}$ through $\overrightarrow{dS}$ can be defined in the same way, as $\vec{E}\cdot\overrightarrow{dS}$, the value of $\vec{E}$ being taken at the position of the surface $\overrightarrow{dS}$, representing practically the number of lines of the electric field cutting out the surface $\overrightarrow{dS}$. In terms of components, the flux of $\vec{E}$ through $\overrightarrow{dS}$ can be thus expressed as:

$$d\Phi_E = \vec{E}\cdot\overrightarrow{dS} = EdS \cos \alpha \tag{3.2}$$

where α is the angle between the vector electric field $\vec{E}$ and normal to the surface, the vector unit $\vec{n}$ perpendicular to the surface element dS, giving the orientation of surface vector $\overrightarrow{dS} = \vec{n}dS$.

So far, we have only defined the flux through a small flat surface, when $\vec{E}$ is constant at every point of the surface. To find the flux of the electric field through a surface S of arbitrary shape, we first approximate the shape of S by dividing it up into a lot of small flat surfaces dS_i. The flat surfaces are represented by vectors such as $\overrightarrow{dS}$ in figure 3.2, all pointing outwards from the surface side of the whole surface. Summing up the elementary flux $\vec{E}\,\overrightarrow{dS}$ over all surface S, one obtains the total flux of electric field through the surface S:

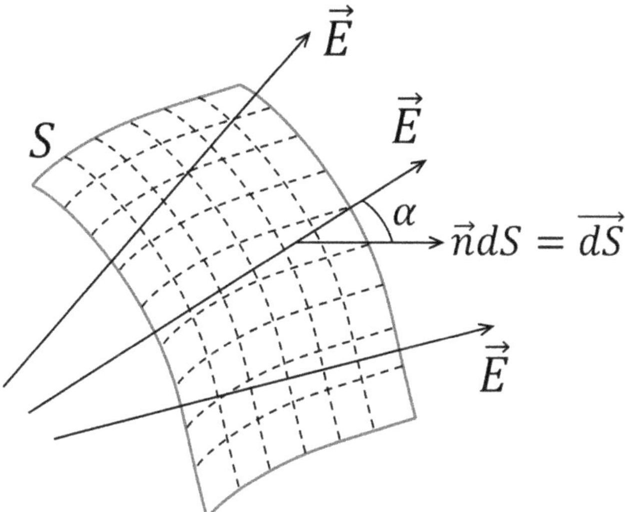

Figure 3.2. Splitting up a surface S into a lot of little flat surfaces $\vec{dS}$ in order to work out the flux of $\vec{E}$ through S.

$$\Phi_{\vec{E}} = \int_S \vec{E}\,\vec{dS} = \int_S EdS \cos \alpha \qquad (3.3)$$

3.2 Integral form of Gauss' law for electric field

Having now the notion of the flux of a vector field though a surface, we compute further the flux of the electric field to a closed surface, whatever will be its shape and the position of source of the electric field (the electric charge).

3.2.1 Flux of the electric field out of a closed surface

Let us now evaluate the flux through a closed surface S for the electric field $\vec{E}$ generated by a point charge q enclosed by S. We shall use the spherical coordinate system (see appendix A.3) referred to an origin at q, and concentrate attention on a surface element of area dS_n at the position with the spherical coordinates (r, θ, φ).

The surface element is represented by the vector $\vec{dS} = \vec{n}dS$ directed along the outward normal to S (see figure 3.3). The surface element is of such size and shape that its projection along the radius vector from P lies between the polar angles θ and $\theta + d\theta$, and the azimuthal angles φ and $\varphi + d\varphi$. As illustrated in figure 3.3, the projected area is:

$$dS_{\text{sphere}} = r^2 \sin \theta d\theta d\varphi = R^2 \sin \theta d\theta d\varphi \qquad (3.4)$$

The electric field of a point charge q located at P, at distance $\vec{r}$ is directed outwards along the radius vector and has the magnitude:

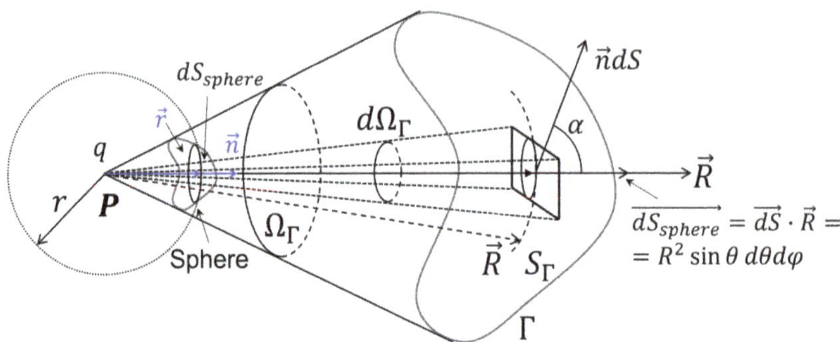

Figure 3.3. A cone starting from the origin and cutting out the sphere of the radius r, giving rise to a cross-sectional area of $r^2 \sin\theta\, d\theta d\varphi$ at a distance r from its apex.

$$\overrightarrow{E} = \frac{q}{4\pi\varepsilon_0 r^3}\overrightarrow{r} \tag{3.5}$$

Hence, the outward flux through $\overrightarrow{dS}$ is:

$$d\Phi_E = \overrightarrow{E}\,\overrightarrow{dS} = \frac{q\overrightarrow{r}\,\overrightarrow{dS}}{4\pi\varepsilon_0 r^3} = \frac{q\,dS_n}{4\pi\varepsilon_0 r^2} = \frac{q\,dS_{\text{sphere}}}{4\pi\varepsilon_0 r^2} = \frac{q}{4\pi\varepsilon_0}d\Omega \tag{3.6}$$

where the element $d\Omega$ is termed the *solid angle* of the cone with the apex at the origin and the base dS_{sphere}. Using equation (3.6), we can now transform the surface integral expression for the flux through the complete surface S into an integral over angular variables only. The range of variation required to cover the whole of the closed surface is $\theta = 0$ to π and $\varphi = 0$ to 2π. So, taking the limits of the infinitesimal area dS, the total flux through S is:

$$\Phi_E = \int_S \overrightarrow{E}\,\overrightarrow{dS} = \int_0^\pi \int_0^{2\pi} \frac{q}{4\pi\varepsilon_0 r^2}r^2 \sin\theta d\theta d\varphi = 4\pi\frac{q}{4\pi\varepsilon_0} = \frac{q}{\varepsilon_0} \tag{3.7}$$

In this derivation it has been assumed that the flux through $\overrightarrow{dS}$ is always positive, because the surface element $\overrightarrow{dS}$ was closed to be along the **outward** normal to S, and then the scalar product $\overrightarrow{E}\,\overrightarrow{dS}$ always gives the correct sign for the flux **out** of $\overrightarrow{dS}$.

    a. The charge q is **inside of a closed surface S**. Sometimes, however, when S has a complicated shape, the radius vector from q may pass from **outside to inside** the enclosing surface, as in figure 3.4. Where the field is directed inwards it makes a negative contribution to the net outward flux. So, for those parts of the surface where the field is directed inwards, the elementary flux through them will be:

$$d\Phi_E = -\frac{q}{4\pi\varepsilon_0}d\Omega \tag{3.8}$$

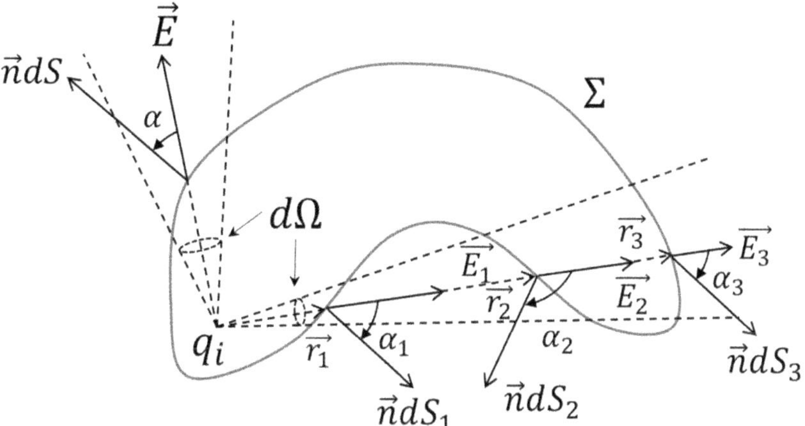

Figure 3.4. The radius vector from a <u>charge inside</u> a surface always makes one more outward crossing than the number of inward crossings *(an odd number of crossings)*.

Notice that however many times a radius vector may cross the surfaces, if the origin lies within S the radius vector must always make one more outward crossing than the number of inward crossing (an **odd** number of crossings). It follows for the net outward flux from figure 3.4, the value:

$$d\Phi_E = \frac{q}{4\pi\varepsilon_0}\left(\frac{\vec{r_1}\cdot\vec{dS_1}}{r_1^3} + \frac{\vec{r_2}\cdot\vec{dS_2}}{r_2^3} + \frac{\vec{r_3}\cdot\vec{dS_3}}{r_3^3}\right)$$

$$= \frac{q}{4\pi\varepsilon_0}\left(\frac{dS_{\text{sphere1}}}{r_1^2} - \frac{dS_{\text{sphere2}}}{r_2^2} + \frac{dS_{\text{sphere3}}}{r_3^2}\right) \tag{3.9}$$

$$= \frac{q}{4\pi\varepsilon_0}(d\Omega - d\Omega + d\Omega) = \frac{q}{4\pi\varepsilon_0}d\Omega$$

Then, the total flux through the surface S enclosing the charge q will be:

$$\Phi_{E_{q_i}} = \int_{\text{all surface}} \frac{q}{4\pi\varepsilon_0}d\Omega = \frac{q}{\varepsilon_0} \tag{3.10}$$

b. For a charge q which is **outside of a closed surface** S, on the other hand, the radius vector must cross S an **even** number of times, with equal contributions to the inward and outward fluxes, as illustrated in figure 3.5. The net outward **flux through a closed surface due to any charge not enclosed by the surface is therefore exactly zero:**

$$\Phi_{E_{q_e}} = \frac{q}{4\pi\varepsilon_0}(-d\Omega + d\Omega - d\Omega + d\Omega) = 0 \tag{3.11}$$

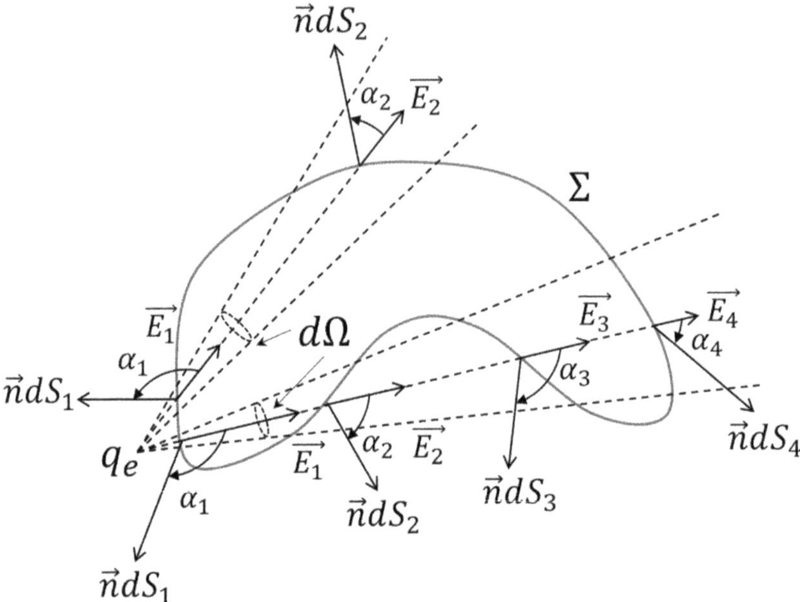

Figure 3.5. For a <u>charge outside</u> of the closed surface, the number of outward crossings is the same as the number of inward crossings (*an even number of crossings*).

c. For a charge q which is **on a closed surface** the solid angle opening to all the surface is a half of the whole solid angle (see figure 3.6). For this reason, **the net outward flux** through a closed surface due to any charge placed onto the surface will be:

$$\Phi_{E_{q_s}} = \int_{\text{all surface}} \vec{E}\,\vec{dS} = \frac{q_s}{2\varepsilon_0} \tag{3.12}$$

Now we can use the superposition principle to compute the outward flux through a closed surface S of the electric field $\vec{E}$ due to an arbitrary distribution of charges q_i enclosed by the surface (*inside the closed surface*), an arbitrary distribution of charges q_e not enclosed by surface (*outside the closed surface*) and an arbitrary distribution of charges q_s placed onto the surface (*on the closed surface*). Only those charges enclosed by S and placed onto S make a contribution to the net outward flux, then:

$$\Phi_E = \int_S \vec{E}\,\vec{dS} = \frac{1}{\varepsilon_0}\left(\sum_{i=1}^{n} q_i + \frac{1}{2}\sum_{s=1}^{n} q_s\right) \tag{3.13}$$

where the summations are restricted to the charges within S and the charges onto S, respectively.

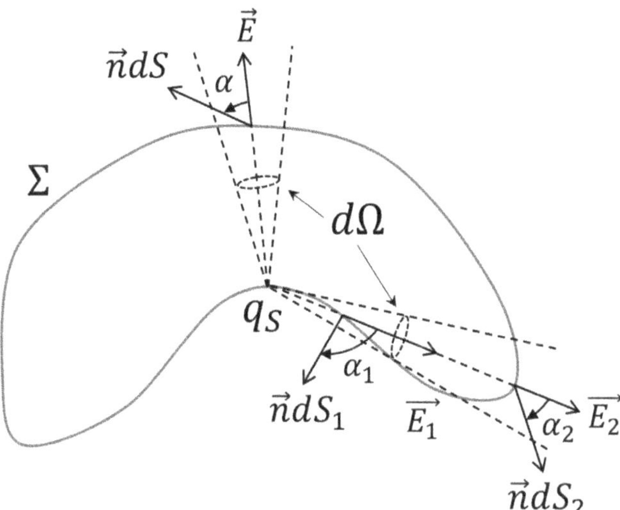

Figure 3.6. For a <u>charge onto the surface</u>, the solid angle opening to the surface is a half of the whole solid angle.

If the charge has a continuous distribution with a volume charge density $\rho(\vec{r})$, then the summation is replaced by a volume integral and the Gauss' law becomes:

$$\int_S \vec{E}\,\vec{dS} = \frac{1}{\varepsilon_0} \int_V \rho(\vec{r})dV \tag{3.14}$$

where the integral on the right side extends over the whole volume V enclosed by S. Here we have related a two-dimensional surface integral to a three-dimensional volume integral. To illustrate how to apply the relation to a particular coordinate system, let us write it out in full for a particular example as given below.

Exercise: Gauss' law application for a spherical charge distribution

Let there be a sphere of radius a, uniformly charged with the volume charge density ρ. What is the electric field and the potential at P point placed at a distance r with respect to its center (see figure 3.7) inside and outside of the spherical charge distribution?
Solution: *applying: Gauss' law to calculate the electric field inside and outside the charge distribution, we must use a spherical coordinate system (see appendix A.3).*

    a. ***Inside:*** *Let us consider a spherical surface Σ_i of radius $r < a$. Gauss' law relates the outward flux of electric field $\vec{E_i}$ through Σ_i to the total charge enclosed by Σ_i.*

$$\int_{\Sigma_i} \vec{E_i}\,\vec{dS} = \frac{1}{\varepsilon_0} \int_{V_{\Sigma_i}} \rho dV_i \tag{3.15}$$

Because $\vec{E_i}$ is independent of θ and φ, we have:

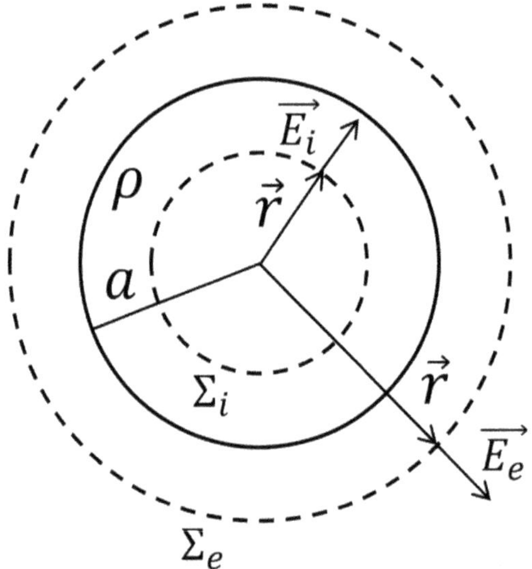

Figure 3.7. *A spherical charge distribution with the volume charge density,* $\rho = constant.$

$$E_i \int_0^\pi \int_0^{2\pi} r^2 \sin\theta d\theta d\varphi = \frac{1}{\varepsilon_0} \int_0^r \int_0^\pi \int_0^{2\pi} \rho r^2 \sin\theta dr d\theta d\varphi \tag{3.16}$$

Thus:

$$4\pi r^2 E_i = \rho \frac{4\pi r^3}{3\varepsilon_0} \tag{3.17}$$

Consequently:

$$\vec{E_i} = \frac{\rho}{3\varepsilon_0} \vec{r} \tag{3.18}$$

According to equation (3.18), inside the spherical charge distribution the electric field is proportional with the position vector $\vec{r}$.

b. ***Outside:*** *considering another spherical surface* Σ_e *having the radius* $r > a$, *Gauss' law leads to:*

$$\int_{\Sigma_e} \vec{E_e}\overrightarrow{dS} = \frac{1}{\varepsilon_0} \int_{\substack{all\ volume \\ of\ sphere}} \rho dV \tag{3.19}$$

$$E_e 4\pi r^2 = \frac{1}{\varepsilon_0} \int_0^a \int_0^\pi \int_0^{2\pi} \rho r^2 \sin\theta dr d\theta d\varphi$$

$$E_e 4\pi r^2 = \frac{4\pi a^3}{3\varepsilon_0}\rho \Rightarrow E_e = \frac{a^3\rho}{3r^2\varepsilon_0} \Rightarrow$$

$$\vec{E_e} = \frac{a^3\rho}{3\varepsilon_0 r^3}\vec{r} = \frac{q}{4\pi\varepsilon_0 r^3}\vec{r} \tag{3.20}$$

Knowing the electrostatic field outside and inside the charge distribution, the potential outside will be:

$$V_{\text{ext}} = -\int_{\infty}^{r} \overrightarrow{E_{\text{ext}}}\overrightarrow{dr} = -\int_{\infty}^{r} \frac{\rho a^3}{3\varepsilon_0 r^2}dr = \frac{\rho a^3}{3\varepsilon_0 r} = \frac{Q}{4\pi\varepsilon_0 r} \tag{3.21}$$

and the potential inside will be:

$$V_{\text{int}} = -\int_{\infty}^{r} \vec{E}\,\overrightarrow{dr} = -\int_{\infty}^{a} \overrightarrow{E_{\text{ext}}}\,\overrightarrow{dr} - \int_{a}^{r} \overrightarrow{E_{\text{int}}}\,\overrightarrow{dr}$$

$$= \frac{\rho a^3}{3\varepsilon_0 a} - \frac{\rho}{3\varepsilon_0}\left(\frac{r^2}{2} - \frac{a^2}{2}\right) = \frac{\rho}{2\varepsilon_0}\left(a^2 - \frac{r^2}{3}\right) \tag{3.22}$$

The relations for the potential were obtained also by using the superposition principle for the potential (see chapter 2). Nevertheless, the advantage of applying Gauss' theorem is evident. The continuity of the electric field and potential on the surface of the sphere can be easily verified. Concluding, outside the charge distribution, far from the center of sphere, the electric field is reversely proportional with the square of the distance between the center of the sphere and the point where the field is calculated. Also, we can note that far from the center of the sphere the electric field and the potential of the spherical distribution are the same as for a point charge equal to all charge of distribution q located at the origin of the sphere. A plot of E(r) dependence and potential V(r) is shown in figure 3.8, just replacing h from figure 2.4 with r distance

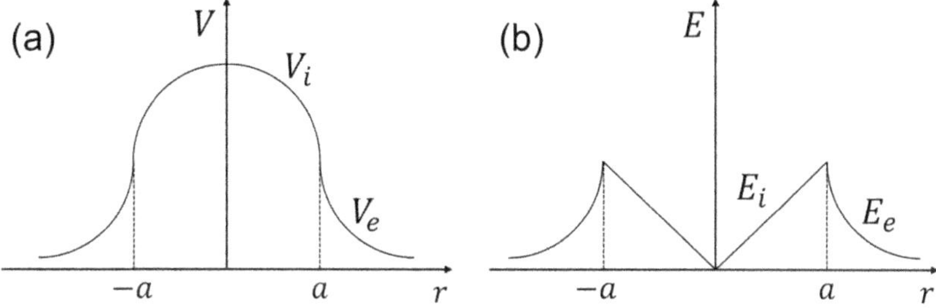

Figure 3.8. *Plots of V (a) and $\vec{E}$ (b) as a function of distance r between the center of the sphere and the point where they are calculated.*

from the center of the sphere to the point where the electric field and potential are expressed.

As has been already mentioned, Gauss' law can be used to calculate the field when the charge distribution is known and has a simple symmetry. But, Gauss' law is very important, because it also applies to calculate those charges or charge distributions which have created a well-known electric field, $\vec{E}$, by:

$$q = \varepsilon_0 \int_{\Sigma} \vec{E} \, \overrightarrow{dS} \tag{3.23}$$

Observation: Gauss' law is a direct result of the inverse square dependence of the electric field. Only with an inverse square law does a point charge generate a flux per unit solid angle which is independent of distance from the source. The analogy between Gauss' law and the flux of light through a closed surface surrounding a lamp also acts on the inverse square law. Light intensity diminishes as the inverse square of the distance from the source of light, and in this case the constancy of flux confined within a fixed solid angle, as for example in the beam of a flashlight, is a consequence of the conservation of energy.

After we study the properties of conductors at electrostatic equilibrium, we shall be able to show that the Coulomb's law is only an inverse square law. Gauss' law can be applied to any physical field having **an inverse square law of interaction**. If the field decreases, for example with r^5 ($\vec{r}$ to the fifth power), the flux through a closed surface will be:

$$\Phi = \int_0^{\pi} \int_0^{2\pi} \frac{q}{4\pi\varepsilon_0 r^5} r^2 \sin\theta \, d\theta \, d\varphi = \frac{q}{\varepsilon_0 r^3} \tag{3.24}$$

and it is namely dependent of the distance from the charge. Then Gauss' law is not valid.

3.3 Differential form of Gauss' law

The integral form of Gauss' law given by equation (3.14) represents the relation between an electric field on a closed surface and the sources of electric field (the total enclosed charge). But we should know what will be the direct relationship between the electric field and its own source at any point in the field. The answer to this question will be given by the local or differential forms of Gauss' law, which relate only the components of the electric field at a point with the source of the electric field present at that point, by a first order differential operator called *divergence* (see appendix B.5). The introduction of differential form of Gauss' law supposes again the knowledge of a new differential operator such as divergence. The divergence operator acts on a vector field, and the resultant is a scalar field. The introduction of the divergence of an arbitrary vector field $\vec{G}(\vec{r})$ written div $\vec{G}$ or $\nabla\vec{G}$ can be seen in appendix B.5.

The integral form of Gauss' law for the electric field of a continuous charge distribution is given by the relation (3.14), $\int_S \vec{E}\,\vec{dS} = \dfrac{1}{\varepsilon_0}\int_V \rho\,dV$. Taking into account that the electric field is a vector field, we can use the divergence theorem (Gauss–Ostrogradsky Theorem), see appendix B.5.1, equation (B.42), to transform the surface integral in a volume integral of divergence $\vec{E}$, resulting:

$$\int_V \nabla\vec{E}\,dV = \frac{1}{\varepsilon_0}\int_V \rho\,dV \tag{3.25}$$

equation (3.25) takes place if the integrants are equal, resulting in:

$$\nabla\vec{E} = \frac{\rho}{\varepsilon_0} \tag{3.26}$$

which represents the local or differential form of Gauss' law.

The differential form is the more useful one except when the electric field has a very simple symmetry, and we shall use it to solve some realistic problems. This equation relates the components of the electric field at a P point with its own source, the charge density existing at that point. From equation (3.26) it follows that the divergence of the electric field $\vec{E}$ is zero in all regions where there is no charge.

a) In the <u>neighborhood of a point charge</u>, for example, the field is non-uniform, but although the individual terms $\dfrac{\partial E_x}{\partial x}$, $\dfrac{\partial E_y}{\partial y}$, $\dfrac{\partial E_z}{\partial z}$ on the left side of equation (3.26) may be different from zero, their sum is always zero. To show this, let us place a point charge q at the origin. The position vector $\vec{r}$ has Cartesian coordinates x, y and z, and it may be written as:

$$\vec{r} = x\vec{i} + y\vec{j} + z\vec{k}$$

Then the electric field is:

$$\vec{E} = \frac{q\vec{r}}{4\pi\varepsilon_0 r^3} = \frac{q(x\vec{i} + y\vec{j} + z\vec{k})}{4\pi\varepsilon_0(x^2 + y^2 + z^2)^{3/2}}$$

Differentiating with respect to x (which is possible everywhere except right at the origin, where the field is singular) the x component of the electric field at $\vec{r}$, we have:

$$
\begin{aligned}
\frac{\partial E_x}{\partial x} &= \frac{q}{4\pi\varepsilon_0}\frac{\partial}{\partial x}\left(\frac{x}{(x^2 + y^2 + z^2)^{3/2}}\right)\\[2mm]
&= \frac{q}{4\pi\varepsilon_0}\left[\frac{1}{(x^2 + y^2 + z^2)^{3/2}} - \frac{\frac{3}{2}x \cdot 2x}{(x^2 + y^2 + z^2)^{5/2}}\right]\\[2mm]
&= \frac{q}{4\pi\varepsilon_0}\left[\frac{x^2 + y^2 + z^2 - 3x^2}{(x^2 + y^2 + z^2)^{5/2}}\right] = \frac{q(r^2 - 3x^2)}{(x^2 + y^2 + z^2)^{5/2}}\\[2mm]
&= \frac{q(r^2 - 3x^2)}{r^5}
\end{aligned}
\tag{3.27}
$$

The terms $\frac{\partial E_y}{\partial y}$ and $\frac{\partial E_z}{\partial z}$ have the same form, with x replaced by y and z, respectively. Hence:

$$\nabla\vec{E} = \frac{\partial E_x}{\partial x} + \frac{\partial E_y}{\partial y} + \frac{\partial E_z}{\partial z} = \frac{q}{4\pi\varepsilon_0}\left[\frac{3r^2 - 3(x^2 + y^2 + z^2)}{r^5}\right] = 0 \qquad (3.28)$$

b) For a better understanding of the differential form of Gauss' law, let us now consider a spherical charge distribution with the volume charge density ρ. The electric field inside of spherical distribution is in agreement with the integral form of Gauss' law:

$$\vec{E_i} = \frac{\rho}{3\varepsilon_0}\vec{r} \qquad (3.29)$$

The electric field outside the spherical distribution is:

$$\vec{E_e} = \frac{\rho a^3}{3\varepsilon_0 r^3}\vec{r} \qquad (3.30)$$

Inside:

$$\nabla\vec{E_i} = \frac{\rho}{3\varepsilon_0}\nabla\vec{r} = \frac{\rho}{\varepsilon_0} \qquad (3.31)$$

Indeed, at point P within the sphere, there is the charge ρ and then the field lines point away from the charge, if it is positive, or point towards the charge if it is negative, resulting in a net outward flux through volume unit from an infinitesimal volume element centered at point P.

Outside:

$$\nabla\vec{E_e} = \frac{\rho a^3}{3\varepsilon_0}\nabla\frac{\vec{r}}{r^3} = \frac{\rho a^3}{3\varepsilon_0}\left[\frac{\nabla\vec{r}}{r^3} + \vec{r}\nabla\frac{1}{r^3}\right] = \frac{\rho a^3}{3\varepsilon_0}\left[\frac{3}{r^3} - \frac{3r^2}{r^5}\right] = 0 \qquad (3.32)$$

In equation (3.32) we have used the properties of the divergence operator. Namely, when the divergence operator acts on the product of a vector field with a scalar field, it results in: $\nabla(U\vec{V}) = \vec{V}\nabla U + U\nabla\vec{V}$. We can see that outside, because there is no electric charge, $\nabla\vec{E} = 0$ showing that the inward flux equals the outward flux, and the net flux through any volume element is zero.

Observation: Because Coulomb's law has a straightforward physical incoming, we have chosen to adopt it as the fundamental law of electrostatics. Then, we have derived the integral and differential forms of Gauss' law as a consequence of Coulomb's law. But it is also possible to start at the other end of the logical chain, postulating at the very beginning that the electric field is a vector field which is the solution of equation (3.26). The integral from of Gauss' law can then be deduced very simply by integrating the equation (3.26) and taking into account the divergence theorem:

$$\int\limits_{V} \nabla\vec{E}\,dV = \int\limits_{V} \frac{\rho}{\varepsilon_0}dV \Rightarrow \int\limits_{S} \overrightarrow{E\,dS} = \frac{1}{\varepsilon_0}\int\limits_{V} \rho dV = \frac{q_i}{\varepsilon_0} \qquad (3.33)$$

Now the integral form of Gauss' law implies that Coulomb's law is valid since if a point charge is placed at the center of a sphere, only for an inverse square law of force is the flux of the electric field out of the sphere independent of its radius. Coulomb's law, the integral and differential forms of Gauss's law are equivalent statements, and any one of them may be taken as the fundamental law of electrostatics.

3.4 Poisson and Laplace equations

In section 3.3 we discussed the differential form of Gauss' law which relates the electric field with its own sources at any point, writing:

$$\nabla\vec{E} = \frac{\rho}{\varepsilon_0} \qquad (3.34)$$

where ρ is the total density, including also any polarization charges (all sources of an electrostatic field). Also, we saw that the electrostatic field is a potential vector field, therefore it may be characterized by a scalar field such as the electrostatic potential. The relationship between the electrostatic field and the electrostatic potential has also been studied both in integral and differential form (see section 2.6). Its differential form is:

$$\vec{E} = -\nabla V \qquad (3.35)$$

Now, we are going to find the relationship between the electrostatic potential of the electric field, and its own sources at any point of the field. In other words, we want to express Gauss' law in terms of the electrostatic potential. This aim can be attained simply by removing the electric field between the two differential equations (3.34) and (3.35). By substituting $\vec{E}$ from (3.35) into (3.34), results:

$$\nabla(\nabla V) = -\frac{\rho}{\varepsilon_0} \Rightarrow \nabla^2 V = \Delta V = -\frac{\rho}{\varepsilon_0} \qquad (3.36)$$

Equation (3.36) is known as *Poisson's equation*. The *Laplace operator* [div grad $= \nabla(\nabla) = \nabla^2 = \Delta$] is a second order differential operator. Its mathematical expression can be now very easy to find, even in a three-orthogonal curvilinear coordinate system knowing already both the differential operators, divergence and gradient (see appendix B.6).

The problem of finding the electrostatic field has now been reduced to the solution of Poisson's equation, a second order differential equation. A found solution to Poisson's equation has been already written down in a previous chapter, where the principle of superposition was invoked to show that:

$$V(r) = \frac{1}{4\pi\varepsilon_0} \int_\Gamma \frac{\lambda\,dl}{\left|\vec{r}-\vec{r'}\right|} + \frac{1}{4\pi\varepsilon_0} \int_S \frac{\sigma\,dS}{\left|\vec{r}-\vec{r'}\right|} + \frac{1}{4\pi\varepsilon_0} \int_V \frac{\rho\,dV}{\left|\vec{r}-\vec{r'}\right|} \qquad (3.37)$$

In equation (3.37), the potential at the point with position vector $\vec{r}$ is given in terms of the line charge density λ, surface charge density σ and volume charge density ρ at the point with position vector $\vec{r'}$. This expression for the potential is quite general, and therefore it is a solution to Poisson's equation. But it is not of much use when dielectrics and conducting materials are present, because induced surface charges and polarization charges then appear, of a magnitude which can only be evaluated when the field or potential is already known. The polarization charges must be accounted for, by introducing the electric displacement $\vec{D}$, and writing Gauss' law in terms of the free charge density, such as we shall see in the study of dielectrics. However, anticipating:

$$\nabla\vec{D} = \nabla(\varepsilon_0\varepsilon_r\vec{E}) = \rho_f \qquad (3.38)$$

when the space between two conductors is filled with a uniform dielectric of constant relative permittivity as it is in the case of a coaxial cable, this equation simplifies to:

$$\nabla\vec{E} = \frac{\rho_f}{\varepsilon_0\varepsilon_r} \qquad (3.39)$$

Again, $\vec{E}$ can be replaced by $(-\nabla V)$, leading to:

$$\nabla^2 V = \Delta V = -\frac{\rho_f}{\varepsilon_0\varepsilon_r} \qquad (3.40)$$

an equation still retaining the form of Poisson's equation for any medium different from vacuum. Equation (3.40) is the preferred one to be used for solving electrostatic problems, since it makes no reference to polarization charges or induced charges on conductors. The presence of conductors only imposes the condition that the potential V must be constant on any conducting surface.

Exercise 1:

In a spherical shell of radii a and b (a < b) the electric charge with the volume charge density ρ is uniformly distributed (see figure 3.9). Determine the intensity and potential of the electric field generated by this charge distribution at a point P placed at distance r from the center of the shell: (1) inside the shell (r < a), (2) in the shell (a < r < b), and (3) outside the shell (r > b), considering that in the three regions the relative electrical permittivity is equal to 1.

Solution: *taking into account that the charge is uniformly distributed in the spherical shell, the potential in all the regions will be independent of the θ and φ, depending just on the r coordinate. Then, the Laplace operator in spherical coordinates system will reduce to:*

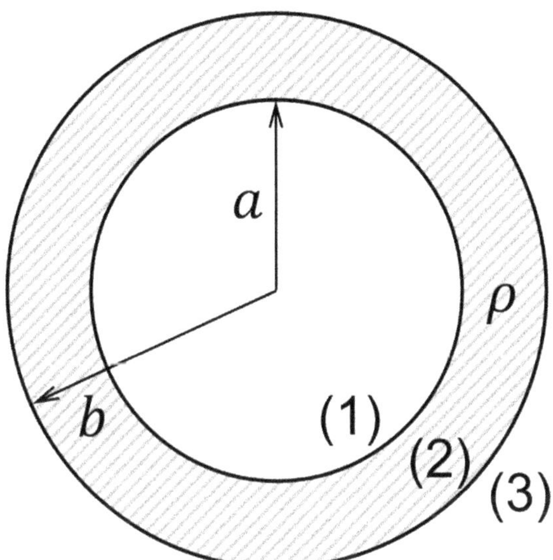

Figure 3.9. Spherical crown used to apply Poisson's equation.

$$\frac{1}{r^2}\frac{d}{dr}\left(r^2\frac{dV}{dr}\right)$$
(3.41)

(1) **Inside the shell** *(r < a) is valid the Laplace equation:*

$$\frac{1}{r^2}\frac{d}{dr}\left(r^2\frac{dV_1}{dr}\right) = 0$$
(3.42)

(2) **In the shell** *(a < r < b) is valid the Poisson equation:*

$$\frac{1}{r^2}\frac{d}{dr}\left(r^2\frac{dV_2}{dr}\right) = -\frac{\rho}{\varepsilon_0}$$
(3.43)

(3) **Outside the shell** *(r > b) is valid the Laplace equation:*

$$\frac{1}{r^2}\frac{d}{dr}\left(r^2\frac{dV_3}{dr}\right) = 0$$
(3.44)

After the first integration one obtains:

$$\frac{dV_1}{dr} = \frac{c_1}{r^2} \Rightarrow \overrightarrow{E_1} = -\frac{c_1}{r^2}\overrightarrow{e_r}$$

$$\frac{dV_2}{dr} = -\frac{\rho r}{3\varepsilon_0} + \frac{c_2}{r^2} \Rightarrow \overrightarrow{E_2} = \left(\frac{\rho r}{3\varepsilon_0} - \frac{c_2}{r^2}\right)\overrightarrow{e_r}$$
(3.45)

$$\frac{dV_3}{dr} = \frac{c_3}{r^2} \Rightarrow \overrightarrow{E_3} = -\frac{c_3}{r^2}\overrightarrow{e_r}$$

After the second integration one obtains:

$$V_1 = -\frac{c_1}{r} + c_1'$$

$$V_2 = -\frac{\rho r^2}{6\varepsilon_0} - \frac{c_2}{r} + c_2' \tag{3.46}$$

$$V_3 = -\frac{c_3}{r} + c_3'$$

Constants of integration, c_1, c_1', c_2, c_2', c_3, c_3', are obtained solving the system:

$$
\begin{aligned}
\vec{E_1}(r=0) &= 0 \\
V_3(r \to \infty) &= 0 \\
\vec{E_1}(r=a) &= \vec{E_2}(r=a) \\
V_1(r=a) &= V_2(r=a) \\
\vec{E_2}(r=b) &= \vec{E_3}(r=b) \\
V_2(r=b) &= V_2(r=b)
\end{aligned}
\Rightarrow
\begin{cases}
c_1 = 0 \\
c_1' = \dfrac{\rho(b^2 - a^2)}{2\varepsilon_0} \\
c_2 = \dfrac{\rho a^3}{3\varepsilon_0} \\
c_2' = \dfrac{\rho b^2}{2\varepsilon_0} \\
c_3 = \dfrac{\rho(a^3 - b^3)}{3\varepsilon_0} \\
c_3' = 0
\end{cases}
\tag{3.47}
$$

And then replacing the constants of integration given by (3.47) within relations (3.45) and (3.46), one obtains the electric field and the potential required for each region, as:

$$
\begin{cases}
\vec{E_1} = 0 \\
\vec{E_2} = \left(\dfrac{\rho r}{3\varepsilon_0} - \dfrac{\rho a^3}{3\varepsilon_0 r^2} \right) \vec{e_r} \\
\vec{E_3} = \dfrac{\rho(b^3 - a^3)}{3\varepsilon_0 r^2} \vec{e_r}
\end{cases}
\tag{3.48}
$$

and, respectively:

$$
\begin{cases}
V_1 = \dfrac{\rho(b^2 - a^2)}{2\varepsilon_0} \\
V_2 = -\dfrac{\rho r^2}{6\varepsilon_0} - \dfrac{\rho a^3}{3\varepsilon_0 r} + \dfrac{\rho b^2}{2\varepsilon_0} \\
V_3 = \dfrac{\rho(b^3 - a^3)}{3\varepsilon_0 r}
\end{cases}
\tag{3.49}
$$

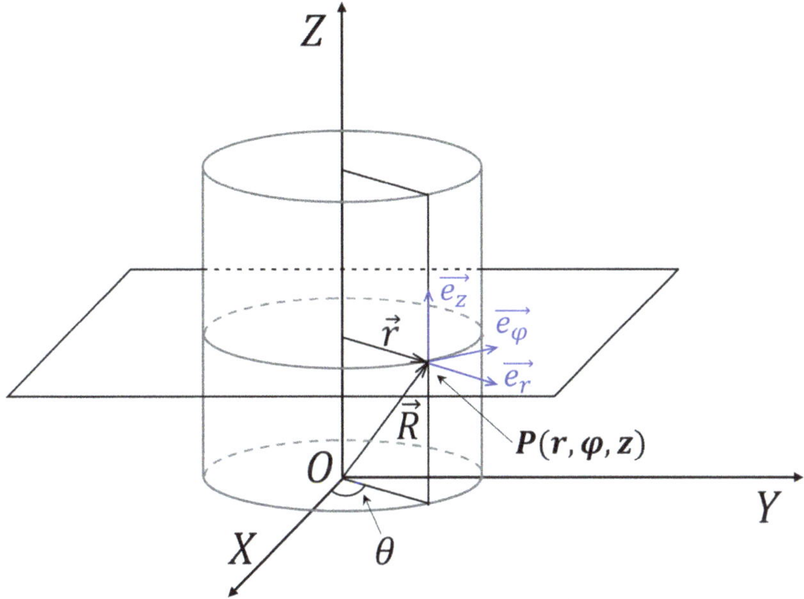

Figure 3.10. *Cylindrical coordinate system for a coaxial cable.*

Exercise 2:

Find the potential in the dielectric of a coaxial cable.
Solution: *in a coaxial cable, for example, there is no free charge between the inner and outer conductors, and equation (3.40) simplifies to:*

$$\Delta V = 0 \tag{3.50}$$

This important special case of Poisson's example equation is called Laplace's equation. Generally, when there is no free charge, Poisson's equation becomes Laplace's equation (3.50). Solving of Poisson's equation or Laplace's equation can be done if the free charge density is known, and then using an appropriate coordinate system required by the symmetry of the problem (see appendix A). So, in this case the appropriate coordinates for the coaxial cable are the cylindrical coordinates (r, θ, z), appendix A.2 (see figure 3.10). In cylindrical polar coordinates the Laplace operator has the form, equation (B.45):

$$\Delta = \frac{1}{r}\frac{\partial}{\partial r}\left(r\frac{\partial}{\partial r}\right) + \frac{1}{r^2}\frac{\partial^2}{\partial\theta^2} + \frac{\partial^2}{\partial z^2} \tag{3.51}$$

Since the coaxial cable has cylindrical symmetry, there is no θ-dependence, and for a long cable we may neglect end effects, so that there is no z-dependence either. Laplace's equation then reduces to:

$$\frac{1}{r}\frac{d}{dr}\left(r\frac{dV}{dr}\right) = 0 \tag{3.52}$$

By integrating this equation, it results in:

$$r\frac{dV}{dr} = A = \text{constant} \tag{3.53}$$

Then, the electric field between the two conductors is:

$$\vec{E} = -\frac{dV}{dr} = -\frac{A}{r} \cdot \frac{\vec{r}}{r} \tag{3.54}$$

Integrating again the equation (3.53), we obtain:

$$\frac{dV}{dr} = \frac{A}{r} \Rightarrow dV = \frac{A}{r}dr \Rightarrow V_{(r)} = A\ln r + B \tag{3.55}$$

where B is another constant of integration. Choosing the zero of potential to be at the radius b of the outer conductor, the boundary conditions when a potential U is maintained across the cable, are: $V_{(a)} = U$ and $V_{(b)} = 0$, where a and b are the radius of inner and outer conductors, respectively. Thus, $B = -A\ln b$. Hence, the values of constants of integration A and B are:

$$U = A(\ln a - \ln b) \Rightarrow A = \frac{U}{\ln\dfrac{a}{b}} = -\frac{U}{\ln\dfrac{b}{a}}; \quad B \geqslant \frac{U}{\ln\dfrac{b}{a}}\ln b$$

$$\Rightarrow V_{(r)} = -\frac{U}{\ln\dfrac{b}{a}}\ln r + \frac{U}{\ln\dfrac{b}{a}}\ln b$$

Consequently:

$$V_{(r)} = -\frac{U}{\ln\dfrac{b}{a}}[\ln r - \ln b] = -\frac{U}{\ln\dfrac{b}{a}}\ln\frac{r}{b} \tag{3.56}$$

The obtained result, i.e., equation (3.56), is in agreement with the result found from the integral form of Gauss' law for dielectrics, as demonstrated later on.

3.4.1 Uniqueness theorem

The potential $V_{(r)}$ given in equation (3.56) obeys Laplace's equation in the region between two conductors $a < r < b$, and also satisfies the conditions imposed at the boundaries of this region. Obviously, $V_{(r)}$ is the only function which has these properties, because we found it by integrating Laplace's equation step by step in a systematic manner. Unfortunately, a systematic integration is not generally possible. However, suppose that in the same region we have managed to find a solution to Poisson's equation or Laplace's equation having specified values on the boundaries region. Then this solution is the **only possible solution** with these boundary values.

It is easy to prove this *uniqueness theorem* by considering first of all the potential inside a cavity in a piece of conducting material. The cavity is a free space, where there is no charge. The potential on the boundary of the cavity **is constant** at a value

V_0. If there is no charge in the cavity, the potential inside it must obey Laplace's equation, $\Delta V = 0$. One solution of this equation, which satisfies the boundary condition, is that $V = V_0$ throughout the cavity. **This is indeed the only solution.** Supposing there was some other solution V_1, then V_1 must have at least one maximum or minimum in cavity, since $V_1 = V_0$ on the walls, but has a different value somewhere inside. At a maximum, for example $\frac{\partial^2 y}{\partial x^2} < 0$, $\frac{\partial^2 y}{\partial y^2} < 0$ and $\frac{\partial^2 y}{\partial z^2} < 0$, which contradicts the original requirement that:

$$\frac{\partial^2 V}{\partial x^2} + \frac{\partial^2 V}{\partial y^2} + \frac{\partial^2 V}{\partial z^2} = 0$$

The sum of the second order derivatives of potential can be zero, only each of them is zero at the same time. Or this requirement which must be satisfied, shows that the potential has no maxima or minima in the free space, thus **it has only one value that equals its value V_0 on the walls.**

3.4.2 Earnshaw's theorem

The Earnshaw's theorem states that a charged particle cannot remain in equilibrium just in one electric field from the free space. In other words, *in free space there are no maxima or minima values of potential.* If there is a maximum potential at one point P then the potential energy of a negative point charge has a minimum value and the particle will be in equilibrium. If there is a minimum for potential, then a positive charge will be in equilibrium at that point.

Suppose that there is in free space a valley in a two-dimensional relief map of V, as shown in figure 3.11. At the point P of the valley, both $\frac{\partial^2 V}{\partial x^2}$ and $\frac{\partial^2 V}{\partial y^2}$ are positive

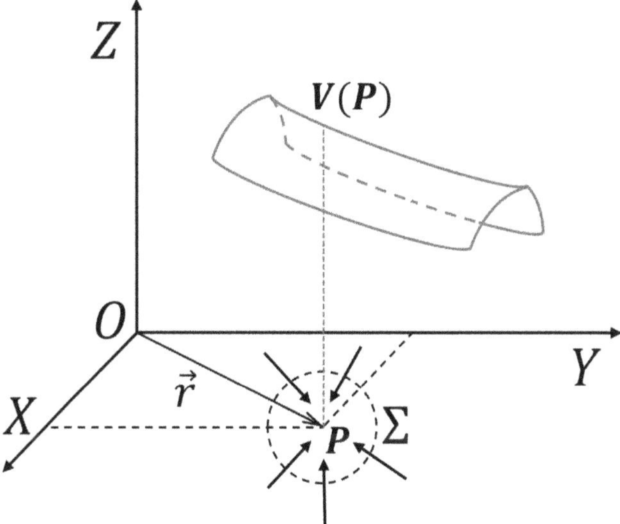

Figure 3.11. Potential surface with a minimum potential at point P.

(the gradient is directed in the direction of maximum increase of potential) in the same direction of x and y axes. Drawing a sphere around point P, one can see that there is an entering flux of the electric field $\vec{E}(P) = -\nabla V(\vec{r})$ (oriented from a higher potential to a lower potential), in opposite directions of the x and y axes. Then, according to the Gauss's Law, $\int_{\Sigma} \vec{E}\,\vec{dS} = -\frac{q}{\varepsilon_0}$, this electric field radially oriented toward point P results from a negative charge, $-q$, present inside the sphere. But we suppose initially that we have a free space where there is no charge, then our supposition that $V(x, y)$ has a minimum is not valid.

Laplace's equation can be used to demonstrate more easily Earnshaw's theorem. Asking the $V(x, y, z)$ surface to have a maximum or minimum it must have all the first order derivatives zero and the second order derivatives must be of the same sign (negative or positive) at that extreme point. On the other hand, their sum must be zero according to Laplace's equation or this is possible just when they are zero simultaneously. If the first order and second order derivatives are simultaneously zero, resulting in $V =$ constant, then there is an equipotential surface without maximum or minimum points.

3.5 Maxwell's equations for electrostatic field

Previously, two fundamental laws for electrostatics were discussed, namely: the electrostatic potential theorem (see section 2.6) and Gauss' law for the electrostatic field (see sections 3.2 and 3.3). Both have been expressed in integral and differential form. The differential form of the two fundamental laws proves Maxwell's equations for electrostatic field. The first equation:

$$\operatorname{div}\vec{E} = \nabla\vec{E} = \frac{\rho}{\varepsilon_0} \tag{3.57}$$

relates locally the electrostatic field to its own sources, while the second equation:

$$\operatorname{curl}\vec{E} = \nabla \times \vec{E} = 0 \tag{3.58}$$

shows that the electrostatic field is a ***conservative field, a potential vector field, an irrotational field*** and ***a field with open lines of field***.

3.6 Electric energy of point charge systems

Imagine an isolated charge q_1. When a second charge q_2 is brought up to a distance $\vec{r_{12}}$ from the first (see figure 3.12), the potential energy of system is:

$$L_{12} = -\int\limits_{\infty}^{r_{12}} \frac{q_1 q_2}{4\pi\varepsilon_0 r_{12}^3}\,\vec{r_{12}}\,\vec{dr} = \frac{q_1 q_2}{4\pi\varepsilon_0 r_{12}} \tag{3.59}$$

To bring up a third charge q_3 requires work to be done against the fields of both q_1 and q_2. If the final position of q_3 is $\vec{r_{13}}$ from q_1 and $\vec{r_{23}}$ from q_2, the additional potential energy is:

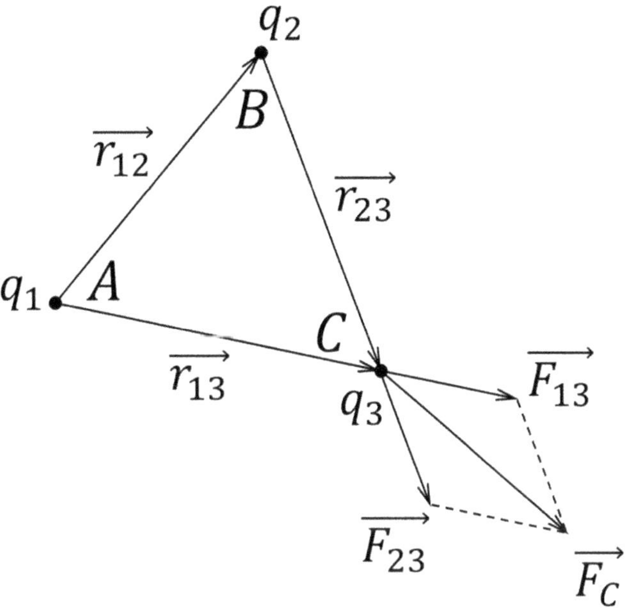

Figure 3.12. Electric energy of point charge assemblies.

$$L_{13} + L_{23} = -\int_{\infty}^{r_{13}} \frac{q_1 q_3}{4\pi\varepsilon_0 r_{13}^3}\, \overrightarrow{r_{13}}\, \overrightarrow{dr} - \int_{\infty}^{r_{23}} \frac{q_2 q_3}{4\pi\varepsilon_0 r_{23}^3}\, \overrightarrow{r_{23}}\, \overrightarrow{dr}$$

$$= \frac{q_3}{4\pi\varepsilon_0}\left(\frac{q_1}{r_{13}} + \frac{q_2}{r_{23}} \right) \tag{3.60}$$

By building up the whole assembly in this way, a single charge at a time, we see that the total potential energy of the assembly is:

$$W = \frac{q_2}{4\pi\varepsilon_0}\left(\frac{q_1}{r_{12}} \right) + \frac{q_3}{4\pi\varepsilon_0}\left(\frac{q_1}{r_{13}} + \frac{q_2}{r_{23}} \right) + \frac{q_4}{4\pi\varepsilon_0}\left(\frac{q_1}{r_{14}} + \frac{q_2}{r_{24}} + \frac{q_3}{r_{34}} \right) + \cdots \tag{3.61}$$

Thus:

$$W = \frac{1}{4\pi\varepsilon_0}\sum_i q_i \left(\sum_{j<i} \frac{q_j}{r_{ij}} \right) \tag{3.62}$$

The restriction $j < i$ in equation (3.62) makes sure that the interaction between each pair of charges is only counted once. We can express the potential energy in a different way by deliberately counting each interaction twice, and then dividing the answer by two, hence writing:

$$W = \frac{1}{2}\sum_{i=1}^{n} q_i \sum_{j \neq i} \frac{q_j}{4\pi\varepsilon_0 r_{ij}} = \frac{1}{2}\sum_{i=1}^{n} q_i V_i \tag{3.63}$$

since, $\sum_{j\neq i}\frac{q_i}{4\pi\varepsilon_0 r_{ij}}$ is the potential V_i at the charge q_i due to all the other charges q_j.

Exercise: electric energy of a point charge assembly

In the sodium chloride (NaCl) crystal, all q_i are either $+e$ or $-e$. If the separation between neighboring sodium and chloride ions is d (lattice constant), then the potential energy of each neighboring pair is $-\frac{e^2}{4\pi\varepsilon_0 d}$. Because there are exactly equal numbers of positive and negative ions, surrounding a particular ion pair, there is a lot of cancellation in the summation, equation (3.63), and the total potential energy per ion pair becomes:

$$U = -\frac{e^2}{4\pi\varepsilon_0 d}x \quad \text{(constant of order unity)}$$

For NaCl, $d = 2.8 \times 10^{-10}$ m, and the constant has the value of 1.75 leading to $U = -8$ eV. Adding the 1.5 eV needed to transfer an electron from sodium atom to chloride atom, the heat of formation of NaCl is found to be 6.5 eV per ion pair, or equivalently 6.5×10^5 joules per mole. This is within formula for the experimenting measured value, and the magnitude of a few electron volts is typical of chemical energy changes. Using the principle of superposition, also the potential energy of the macroscopic charged bodies can be calculated, replacing the sum from equation (3.63) with an appropriate integral:

$$W = \frac{1}{2}\int V dq = \frac{1}{2}\int_{\Gamma} V(r)\lambda(r')dl' + \frac{1}{2}\int_{\substack{\text{all} \\ \text{surface}}} V(r)\sigma(r')dS'$$
$$+ \frac{1}{2}\int_{\substack{\text{all} \\ \text{volume}}} V(r)\rho(r')dv' \tag{3.64}$$

Further reading

[1] Landau L and Lifchitz E 1970 Troisième édition revue et complétée *Physique Théorique— Tome 2, Théorie des Champs* (Moscow: Mir)

[2] Secăreanu I, Ruxandra V, Gherbanovschi N, Logofătu M, Cazan-Corbasca M and Antohe Ş 1984 *Problems of Electricity and Magnetism (Culegere de Probleme de Electricitate şi Magnetism)* (Bucharest: University of Bucharest Publishing House)

[3] Clayton P R, Keith W W and Syed N A 1997 *Introduction to Electromagnetic Fields* (New York: McGraw-Hill)

[4] Jackson J D 1998 *Classical Electrodynamics* 3rd edn (New York: Wiley)

[5] Dávalos A L and Zanette D 1999 *The Poisson and Laplace equations Fundamentals of Electromagnetism* (Berlin: Springer)
[6] Saslow W M 2002 Gauss's law: flux and charge are related *Electricity, Magnetism, and Light* ed W M Saslow and C A San Diego (San Diego, CA: Academic) ch 4 pp 145–83
[7] Feynman R P, Leighton R B and Sands M 2011 *The Feynman Lectures on Physics, Vol II: The New Millennium Edition: Mainly Electromagnetism and Matter* (New York: Basic Books)
[8] Purcell E M and Morin D J 2013 *Electricity and Magnetism* 3rd edn (Cambridge: Cambridge University Press)
[9] Agrawal N 2020 *Electrostatics: Current and Capacitors (for IIT-JEE)* (Independently Published)

IOP Publishing

Electrostatics
Formalism of the electrostatic field in vacuum and matter

Ştefan Antohe and Vlad-Andrei Antohe

Chapter 4

Particular systems of charges

The spherical equipotential surfaces and radial field around a point charge are very simple, as discussed in the previous chapters. However, when more than one charge is present, the field pattern becomes much more complicated. In this chapter, two particular electric charge systems are addressed, namely, the *electric dipole* and *quadrupole*.

4.1 The electric dipole

In a general way, an *electric dipole* deals with the separation of positive and negative charges found. However particularizing, a simple example of this system could be a pair of electric charges of equal magnitude but opposite sign, separated by some typically small distance[1]. Let us then calculate the potential and electric field of such an electric dipole consisting of two equal and opposite point charges, separated by a small distance $\vec{l}$ directed from the negative point charge to the positive point charge, as presented in figure 4.1.

An electric dipole is characterized by a vector quantity called the dipole moment, which represents the vector $\vec{p}$ drawn from the negative to the positive point charge with the magnitude $\left| q\vec{l} \right|$, then:

$$\vec{p} = q\vec{l}; \quad [p]_{SI} = C \cdot m \tag{4.1}$$

In figure 4.1, the dipole moment $\vec{p}$ is directed along Ox axis at an angle θ to the position vector $\vec{r}$. Further, we will calculate the electric potential and the electric field created by the dipole at a point P far away from its middle (O).

---

[1] A permanent electric dipole is called an **electret**.

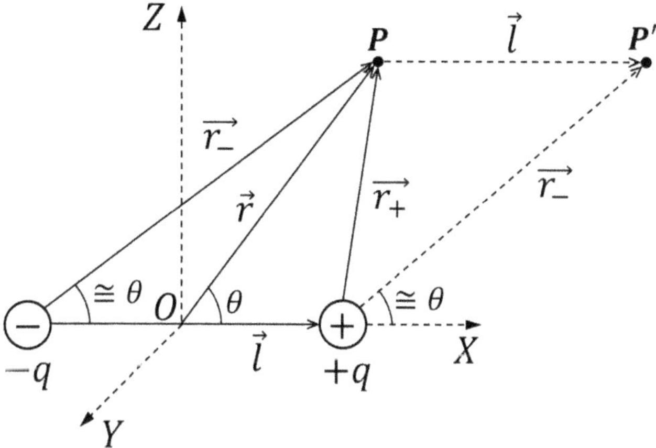

Figure 4.1. The electric dipole.

4.1.1 Potential of an electric dipole

The potential at a point P lying in the X–Z plane, its origin O, with position vector $\vec{r}$ is the sum of the potentials of the electric field generated by each charge separately. Writing $\vec{r_\pm}$ the position vectors joining the point charges $\pm q$ with point P (see figure 4.1), the potential is:

$$V = \frac{q}{4\pi\varepsilon_0}\left(\frac{1}{r_+} - \frac{1}{r_-}\right) \tag{4.2}$$

Equipotential surfaces cutting out the X–Z plane are shown in figure 4.2 together with the field lines, everywhere perpendicular to the equipotential surfaces. What is the magnitude of the potential in different parts of this diagram? Very close to any of the charge, the potential is almost the same as if the charge was on its own. The equipotential surfaces, closely with each charge, are nearly spherical and the potential is reversely proportional to the distance from the closer charge. When $\vec{r_+}$ becomes comparable to $\vec{r_-}$, both terms in equation (4.2) contribute. If θ is the angle between the dipole moment $\vec{p}$ and the relative position vector $\vec{r}$, the module of the vectors $\vec{r_+}$ and $\vec{r_-}$ can be written as:

$$r_\pm = \left(r^2 + \frac{l^2}{4} \mp \frac{2rl}{2}\cos\theta\right)^{1/2} = r\left(1 + \frac{l^2}{4r^2} \mp \frac{l\cos\theta}{r}\right)^{1/2} \tag{4.3}$$

or alternatively:

$$\frac{1}{r_\pm} = \frac{1}{r}\left(1 + \frac{l^2}{4r^2} \mp \frac{l\cos\theta}{r}\right)^{-1/2} \tag{4.4}$$

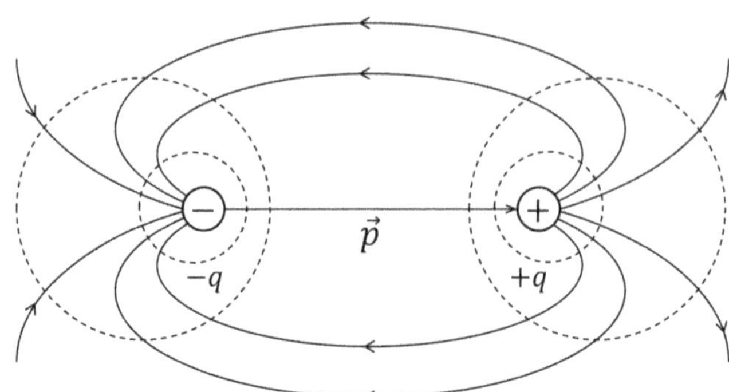

Figure 4.2. Equipotential and field lines near an electric dipole.

Along OZ-axis, where $\theta = \frac{\pi}{2}$, $\cos\theta = 0$, r_+ equals r_-, then the potential is zero (the potential is in fact zero everywhere on the Z–Y plane at $X = 0$). At some distance from dipole, where $r \gg l$, the potential can be developed in powers of $\frac{l}{r}$, becoming:

$$\frac{1}{r_\pm} = \frac{1}{r}\left[1 - \frac{1}{2}\left(\frac{l^2}{4r^2} \mp \frac{l\cos\theta}{r}\right) + \frac{3}{8}\left(\frac{l^2}{4r^2} \mp \frac{l\cos\theta}{r}\right)^2 + \cdots\right] \tag{4.5}$$

Keeping only the term of first order in $\frac{l}{r}$, from equation (4.5), we obtain:

$$\frac{1}{r_\pm} \simeq \frac{1}{r}\left[1 \pm \frac{l\cos\theta}{2r} + \cdots\right] \tag{4.6}$$

Substituting (4.6) in equation (4.2), results in:

$$V = \frac{q}{4\pi\varepsilon_0}\left[\frac{1}{r} + \frac{l\cos\theta}{2r^2} - \frac{1}{r} + \frac{l\cos\theta}{2r^2}\right] = \frac{ql\cos\theta}{4\pi\varepsilon_0 r^2} + \text{terms of higher order} \tag{4.7}$$

The first term by itself is often referred to as a 'dipole potential'. The 'dipole potential' in this sense is exactly proportional to $\frac{1}{r^2}$, but it should be remembered that the potential near a dipole consisting of two separated point charges also includes the higher order terms. The quantity ql is the magnitude of dipole moment and then dipole potential can be written as:

$$V_p(\vec{r}) = \frac{p\cos\theta}{4\pi\varepsilon_0 r^2} = \frac{\vec{p}\cdot\vec{r}}{4\pi\varepsilon_0 r^3} \tag{4.8}$$

Observation: The 'dipole potential' can be obtained very simply by observing that the difference $V_p(\vec{r}) = \frac{q}{4\pi\varepsilon_0 r_+} - \frac{q}{4\pi\varepsilon_0 r_-}$ which appears in equation (4.2) is the variation of the electric potential created by the positive charge along of a displacement vector $\vec{l'} = \vec{l}$ (see figure 4.1) taken with opposite sign:

$$V_p(\vec{r}) = -\left(V_q(P') - V_q(P)\right) = -\left(\frac{q}{4\pi\varepsilon_0 r_-} - \frac{q}{4\pi\varepsilon_0 r_+}\right) = -\Delta V_q = -\nabla V_q \cdot \vec{l}$$

$$V_p(\vec{r}) = -\nabla V_q \vec{l} = -\nabla\left(\frac{q}{4\pi\varepsilon_0 r}\right)\vec{l} = -\frac{q}{4\pi\varepsilon_0}\left(\nabla\frac{1}{r}\right)\vec{l} = -\frac{q}{4\pi\varepsilon_0}\left(\frac{-\vec{r}}{r^3}\right)\vec{l}$$

$$V_p(\vec{r}) = \frac{q\vec{l}\,\vec{r}}{4\pi\varepsilon_0 r^3} = \frac{\vec{p}\,\vec{r}}{4\pi\varepsilon_0 r^3} = \frac{p\cos\theta}{4\pi\varepsilon_0 r^2} \tag{4.9}$$

Notice that this last expression (4.9), which in terms of the vectors $\vec{p}$ and $\vec{r}$ and the magnitude r, makes no reference to a particular choice of a coordinate system. The potential is $V_p = \frac{\vec{p}\,\vec{r}}{4\pi\varepsilon_0 r^3}$, even if $\vec{p}$ is not pointing along the Z-axis and at all points $\vec{r}$, whether or not they lie in the X–Z plane. In all directions the potential at some distance from the dipole is varying as $\frac{1}{r^2}$, a more rapid fall off than the $\frac{1}{r}$ dependence around an isolated point charge (see section 2.1).

4.1.2 Electric field of an electric dipole

Now, having the potential of an electric dipole moment $\vec{p}$, its electric field can be easily calculated taking the gradient of the potential:

$$\vec{E_p} = -\nabla V_p = -\nabla\left(\frac{\vec{p}\,\vec{r}}{4\pi\varepsilon_0 r^3}\right) = -\frac{1}{4\pi\varepsilon_0}\left[\frac{\nabla(\vec{p}\,\vec{r})}{r^3} + \left(\vec{p}\,\vec{r}\right)\nabla\frac{1}{r^3}\right]$$

$$= -\frac{1}{4\pi\varepsilon_0}\left[\frac{\vec{p}}{r^3} - \frac{3(\vec{p}\,\vec{r})\vec{r}}{r^5}\right] = \frac{1}{4\pi\varepsilon_0 r^3}\left[\frac{3(\vec{p}\,\vec{r})\vec{r}}{r^2} - \vec{p}\right] \tag{4.10}$$

As a reminder (see appendix B.2), equation (4.10) was obtained taking the gradient of the product $(\vec{p}\cdot\vec{r})\cdot\left(\frac{1}{r^3}\right)$ and considering $\vec{p}$ a constant vector $\vec{p} = p_x\vec{i} + p_y\vec{j} + p_z\vec{k}$, then: $\vec{p}\,\vec{r} = xp_x + yp_y + zp_z \Rightarrow \nabla(\vec{p}\,\vec{r}) = p_x\vec{i} + p_y\vec{j} + p_z\vec{k} = \vec{p}$.

But the electric field of an electric dipole moment can be also expressed in a polar coordinated system, using the gradient in spherical coordinates (see appendix A.3) in plane (polar coordinates r and θ). Hence:

$$V_p = \frac{p\cos\theta}{4\pi\varepsilon_0 r^2} \Rightarrow \vec{E_p} = -\nabla V_p = -\frac{p}{4\pi\varepsilon_0}\left[\frac{\partial}{\partial r}\left(\frac{\cos\theta}{r^2}\right)\vec{e_r} + \frac{1}{r}\frac{\partial}{\partial\theta}\left(\frac{\cos\theta}{r^2}\right)\vec{e_\theta}\right]$$

Thus:

$$\vec{E_p} = -\frac{p}{4\pi\varepsilon_0}\left[\frac{-2\cos\theta}{r^3}\vec{e_r} - \frac{\sin\theta}{r^3}\vec{e_\theta}\right]$$

and then:

$$\vec{E_p} = \frac{2p\cos\theta}{4\pi\varepsilon_0 r^3}\vec{e_r} + \frac{p\sin\theta}{4\pi\varepsilon_0 r^3}\vec{e_\theta} = E_r\vec{e_r} + E_\theta\vec{e_\theta} \tag{4.11}$$

The radial and azimuthal component of the electric field (E_r and E_θ), respectively, are represented in figure 4.3 and are given by:

$$E_r = \frac{2p \cos \theta}{4\pi\varepsilon_0 r^3}; \quad E_\theta = \frac{p \sin \theta}{4\pi\varepsilon_0 r^3} \tag{4.12}$$

Observation: equations (4.10) and (4.11) are equivalent. Indeed, $\vec{p}$ with respect to polar coordinates, represented in figure 4.3, is:

$$\vec{p} = p \cos \theta \, \vec{e_r} - p \sin \theta \, \vec{e_\theta} \Rightarrow p \sin \theta \, \vec{e_\theta} = p \cos \theta \, \vec{e_r} - \vec{p}$$

and substituting it in equation (4.11), we obtain:

$$\vec{E_p} = \frac{2p \cos \theta}{4\pi\varepsilon_0 r^3} \vec{e_r} + \frac{(p \cos \theta \, \vec{e_r} - \vec{p})}{4\pi\varepsilon_0 r^3} = \frac{3p \cos \theta}{4\pi\varepsilon_0 r^3} \vec{e_r} - \frac{\vec{p}}{4\pi\varepsilon_0 r^3}$$

$$= \frac{1}{4\pi\varepsilon_0 r^3} \left[\frac{3(\vec{p} \cdot \vec{r}) \cdot \vec{r}}{r^2} - \vec{p} \right]$$

which is exactly equation (4.10).

The line of constant field and constant potential are shown in figure 4.2. These equations show that the electric field of a dipole falls off as $\frac{1}{r^3}$ and its potential as $\frac{1}{r^2}$, whereas the corresponding laws for a monopole (a point charge q) are $\frac{1}{r^2}$ and $\frac{1}{r}$, respectively. The significance of these differences is that at large distances, the fields of the two equal and opposite charges which comprise a dipole canceling one with

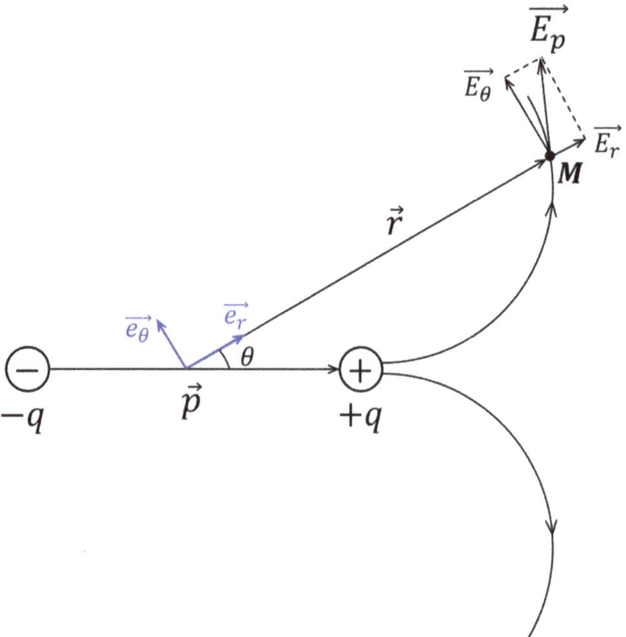

Figure 4.3. Components of the dipole electric field in polar coordinates system.

another in the first approximation (that is, terms varying as $\frac{1}{r^2}$ vanish) but terms in the next order ($\frac{1}{r^3}$ in the field) remain just as in the case of potential when the terms varying as $\frac{1}{r}$ vanish, the terms remaining in the next order ($\frac{1}{r^2}$ in the potential).

4.1.3 Ponderomotive actions experienced by an electric field on an electric dipole moment

The electric couple

If a dipole consisting of two charges $-q$ and $+q$ at distance $\vec{l}$ apart is placed in an uniform electric field $\vec{E}$, there is no nct force on dipole (see figure 4.4(a)), but a couple $qlE \sin \theta$ acts in the sense required for lining up the dipole parallel to the field (see figure 4.4(b)).

The force which acts on the positive charge $+q$ is $\vec{F_+} = q\vec{E}$ and that which acts on $-q$ is $-\vec{F_-} = -q\vec{E}$ being equal as magnitude, parallel and in opposite direction. Their resultant is zero but they act on the dipole with the couple having the moment, with respect to an arbitrary point O given by:

$$\vec{M_c} = \vec{M_{F_+}} + \vec{M_{F_-}} = \left(\vec{r_+} \times q\vec{E}\right) + \left(\vec{r_-} \times -q\vec{E}\right) = \left(\vec{r_+} - \vec{r_-}\right) \times q\vec{E} \Rightarrow$$

$$\vec{M_c} = q\vec{l} \times \vec{E} = \vec{p} \times \vec{E} \tag{4.13}$$

In a uniform electric field $\vec{E}$, there is no translational force acting on the dipole, but there is a *couple* (or '*torque*') which tend to turn the dipole into a position parallel to the field. The elements of the couple are:

- **Module:** $pE \sin \theta$;
- **Line of action:** $\vec{M_c} \perp (\vec{p}, \vec{E})$;
- **Sense:** given by the sense of cross-product between $\vec{p}$ and $\vec{E}$.

The potential energy of a dipole in an electric field

Work must be done against this couple to turn the dipole, whose potential energy thus depends on its orientation. The potential energy of the dipole is:

$$W = \sum_i q_i V_i = q(V_B - V_A) \tag{4.14}$$

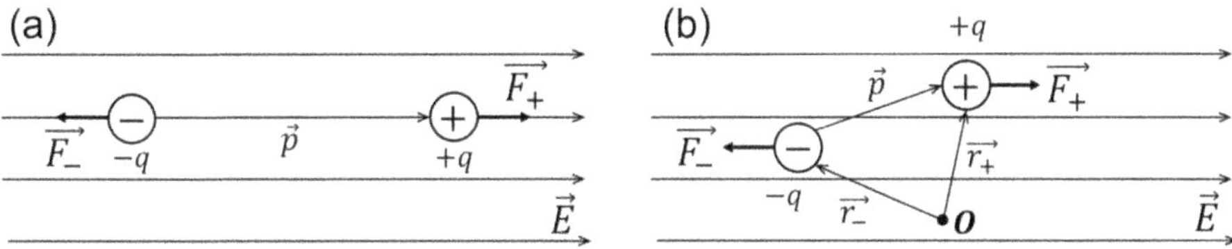

Figure 4.4. Force (a) and couple (b) experienced by an uniform electric field $\vec{E}$ on an electric dipole moment $\vec{p}$.

where V_B and V_A are the potential of the electric field $\vec{E}$ at the position of $+q$ and $-q$, respectively. There should strictly be an additional term to allow for the potential energy of the two charges in one another's fields $\left(\frac{-q^2}{4\pi\varepsilon_0 l}\right)$, but for fixed separation this energy is constant, and we do not have to worry about it, when working out what happens when the dipole turns in the external field. Then, the potential difference $(V_B - V_A)$ is given by:

$$V_B - V_A = -\int_A^B \vec{E}\, \overrightarrow{dl} = -\int_0^l E \cos\theta\, dl = -El \cos\theta \tag{4.15}$$

and then the potential energy will be:

$$W_p = q(V_B - V_A) = -qlE \cos\theta = -pE \cos\theta = -\vec{p}\cdot\vec{E} \tag{4.16}$$

The potential energy is the scalar product of $\vec{p}$ and $\vec{E}$ which shows that the energy depends only on the angle between $\vec{p}$ and $\vec{E}$, and their magnitude.

Another manner to calculate the potential energy presumes the calculus of the work done by the electric field to rotate the dipole with an elementary angle $d\theta$. Knowing the couple acting on the dipole from the field, the work will be:

$$dL_{12} = \int_{\theta_1}^{\theta_2} M_c d\theta = \int_{\theta_1}^{\theta_2} pE \sin\theta\, d\theta = -pE \cos\theta \,\Big|_{\theta_1}^{\theta_2}$$

$$= [-pE \cos\theta_2 - (-pE \cos\theta_1)] = W_B - W_A$$

Then, in the position where the angle between the electric dipole moment $\vec{p}$ and the electric field $\vec{E}$ is θ, the potential energy is:

$$W = -pE \cos\theta = -\vec{p}\cdot\vec{E} \tag{4.17}$$

as was expected, exactly the same relation as that from (4.16).

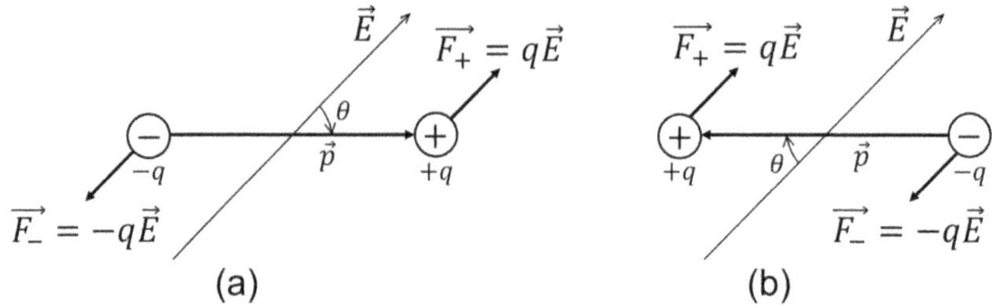

Figure 4.5. Stable (a) and unstable (b) equilibrium of an electric dipole moment in an electric field.

a. The dipole is in **stable equilibrium** for $\theta = 0$, because if it is taken out from its equilibrium position with a small angle θ, just quickly acts on its own a couple which tend to line it up parallel to the field, in other words acting in opposite sense of the displacement from its initial position, see figure 4.5(a). On the other hand, in this case the potential energy has got a minimum value.

$$\text{When } \theta = 0 \quad \Rightarrow \quad \begin{cases} M_c = 0 \\ W_p = -pE = \text{minimum value} \end{cases}$$

b. When $\theta = \pi$ (also $M_c = 0$, but the energy is pE, having the maximum value) the dipole is in **unstable equilibrium** because if it is drawn out from its initial position, a couple acts on it to line up the dipole parallel to the field, acting in the same sense of the displacement from initial position, see figure 4.5(b).

Force acting on a dipole placed in a non-uniform external field

If the electric dipole is placed in a non-uniform electric field, a translational force is exerted on it, a force depending on the variation of the components of the electric field with respect to the ends of the dipole. In the case of a non-uniform electric field, $E(x, y, z)$, the component F_x of the force with which the electric field acts on the dipole along OX axis, will be given by the product of the dipole charge q with the difference between the component E_x of the electric field with respect to the points M and N where are placed the positive and negative charges of the dipole, respectively (see figure 4.6):

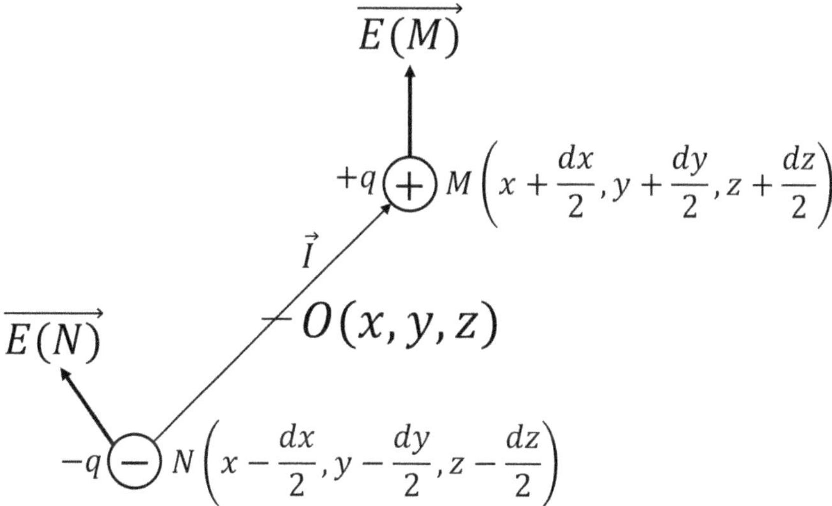

Figure 4.6. Force acting on a dipole placed in a non-uniform electric field.

$$F_x = q\left(E_{x_+} - E_{x_-}\right) \tag{4.18}$$

Supposing that the dipole center is at the point $O(x, y, z)$, the points M and N where are placed the dipole charges $+q$ and $-q$, respectively, will have the coordinates:

$$M\left(x + \frac{dx}{2}, y + \frac{dy}{2}, z + \frac{dz}{2}\right) \text{ and } N\left(x - \frac{dx}{2}, y - \frac{dy}{2}, z - \frac{dz}{2}\right)$$

The x-component of the force acting on the dipole will be:

$$F_x = q[E_x(M) - E_x(N)] \Rightarrow$$

$$F_x = q\left[E_x\left(x + \frac{dx}{2}, y + \frac{dy}{2}, z + \frac{dz}{2}\right) - E_x\left(x - \frac{dx}{2}, y - \frac{dy}{2}, z - \frac{dz}{2}\right)\right] \Rightarrow$$

$$F_x = q\left[E_x(x, y, z) + \frac{\partial E_x}{\partial x}\frac{dx}{2} + \frac{\partial E_x}{\partial y}\frac{dy}{2} + \frac{\partial E_x}{\partial z}\frac{dz}{2} + \cdots - E_x(x, y, z) + \right.$$

$$\left. + \frac{\partial E_x}{\partial x}\frac{dx}{2} + \frac{\partial E_x}{\partial y}\frac{dy}{2} + \frac{\partial E_x}{\partial z}\frac{dz}{2}\right] = q\left(\frac{\partial E_x}{\partial x}dx + \frac{\partial E_x}{\partial y}dy + \frac{\partial E_x}{\partial z}dz\right)$$

or alternatively:

$$F_x = qdx\frac{\partial E_x}{\partial x} + qdy\frac{\partial E_x}{\partial y} + qdz\frac{\partial E_x}{\partial z}$$

$$= p_x\frac{\partial E_x}{\partial x} + p_y\frac{\partial E_x}{\partial y} + p_z\frac{\partial E_x}{\partial z} = (\vec{p}\,\nabla)E_x \tag{4.19}$$

where: $p_x = qdx$, $p_y = qdy$, $p_z = qdz$ are the components along the three coordinate axes of the dipole moment $\vec{p}$. Following the same procedure, the other two components of the force will be:

$$F_y = \left(\vec{p}\,\nabla\right)E_y; \quad F_z = \left(\vec{p}\,\nabla\right)E_z$$

Considering the operator:

$$\left(\vec{p}\cdot\nabla\right) = p_x\frac{\partial}{\partial x} + p_y\frac{\partial}{\partial y} + p_z\frac{\partial}{\partial z}$$

the force acting on the dipole is:

$$\vec{F} = F_x\vec{i} + F_y\vec{j} + F_z\vec{k} = \left(\vec{p}\,\nabla\right)\vec{E} \tag{4.20}$$

The electric field is a conservative field and an irrotational field:

$$\vec{E} = -\nabla V \text{ and } \nabla \times \vec{E} = 0 \Rightarrow$$

$$\frac{\partial E_z}{\partial y} = \frac{\partial E_y}{\partial z}; \quad \frac{\partial E_x}{\partial z} = \frac{\partial E_z}{\partial x}; \quad \frac{\partial E_y}{\partial x} = \frac{\partial E_x}{\partial y} \tag{4.21}$$

Taking into account these relations and replacing them in (4.19) results in:

$$F_x = p_x \frac{\partial E_x}{\partial x} + p_y \frac{\partial E_x}{\partial y} + p_z \frac{\partial E_x}{\partial z} = p_x \frac{\partial E_x}{\partial x} + p_y \frac{\partial E_y}{\partial x} + p_z \frac{\partial E_z}{\partial x} = \frac{\partial \left(\vec{p}\, \vec{E} \right)}{\partial x} \qquad (4.22)$$

and correspondingly:

$$F_y = \frac{\partial \left(\vec{p}\, \vec{E} \right)}{\partial y}; \quad F_z = \frac{\partial \left(\vec{p}\, \vec{E} \right)}{\partial z}$$

Then, generally:

$$\vec{F} = \nabla \left(\vec{p} \cdot \vec{E} \right) = -\nabla W_p \qquad (4.23)$$

Relation (4.23) is very useful in the study of interaction between an electric dipole and an electric field whatever will be its source, then also to study the dipole–dipole interaction.

A very simple example is the force of interaction between an electric charge Q and an electric dipole moment $\vec{p}$, placed at the distance $\vec{r}$ with respect to the point charge Q. If the dipole moment is radially oriented with the negative charge close to the charge Q, the total force of interaction will be attractive (see figure 4.7(a)), having the magnitude:

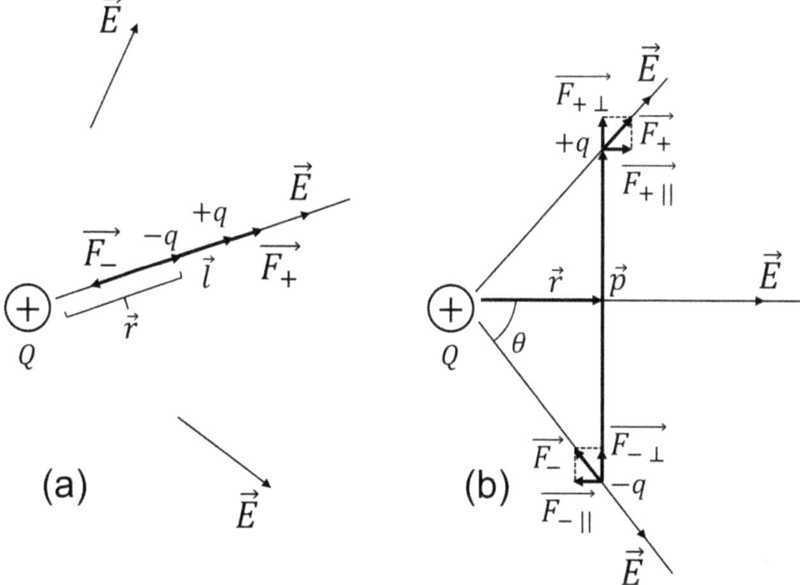

Figure 4.7. (a) Force of interaction between an electric dipole moment placed along the line of the electric field of a point charge Q. (b) Force of interaction between an electric dipole moment placed perpendicular to the line of the electric field of a point charge Q.

$$\vec{F} = -\frac{qQ\,\vec{e_r}}{4\pi\varepsilon r^2} + \frac{qQ\,\vec{e_r}}{4\pi\varepsilon (r+l)^2} = \frac{qQ}{4\pi\varepsilon r^2}\left[-1 + \frac{1}{\left(1+\frac{l}{r}\right)^2}\right]\vec{e_r} \simeq \tag{4.24}$$

$$\simeq \frac{qQ}{4\pi\varepsilon r^2}\left[-1 + 1 - \frac{2l}{r}\right]\vec{e_r} = -\frac{2qlQ\,\vec{e_r}}{4\pi\varepsilon r^3} = -\frac{Q\vec{p}}{2\pi\varepsilon r^3}; \quad \vec{e_r} = \frac{\vec{r}}{r}$$

If the dipole is perpendicular to $\vec{r}$ (see figure 4.7(b)), the forces $\vec{F_+}$ and $\vec{F_-}$ are equal but they are not parallel and in opposite directions, as a result there is the component of force $\vec{F_\perp}$ normal to $\vec{r}$, different to zero acting to the dipole to translate up, unlike the parallel components, $\vec{F_{+\parallel}}$ and $\vec{F_{-\parallel}}$, which are canceled reciprocally.

As concerning the dipole–dipole interaction, it is easy to observe that an electric dipole placed in the electric field of another dipole (see figure 4.8) will interact with it and the relationships for the ponderomotive actions can be re-written as stated below. The couple between two electric dipoles:

$$\vec{M_c} = \vec{p_2} \times \vec{E_1} = \vec{p_2} \times \frac{1}{4\pi\varepsilon_0 r_{12}^3}\left[\frac{3\left(\vec{p_1}\,\vec{r_{12}}\right)\vec{r_{12}}}{r_{12}^2} - \vec{p_1}\right]$$

$$= \frac{1}{4\pi\varepsilon_0 r_{12}^3}\left[\frac{3\left(\vec{p_1}\,\vec{r_{12}}\right)\left(\vec{p_2}\times\vec{r_{12}}\right)}{r_{12}^2} - \vec{p_2}\times\vec{p_1}\right] \tag{4.25}$$

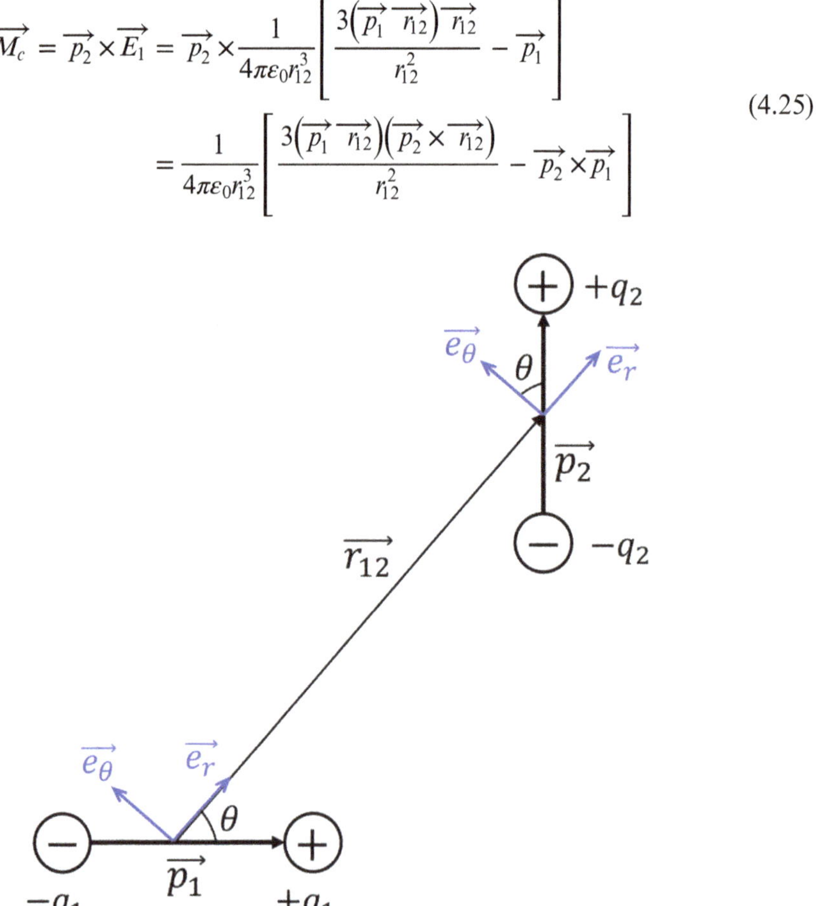

Figure 4.8. The dipole–dipole interaction.

The potential energy of interaction between two electric dipoles:

$$W_p = -\overrightarrow{p_2}\,E_1 = -\frac{1}{4\pi\varepsilon_0 r_{12}^3}\left[\frac{3\left(\overrightarrow{p_1}\,\overrightarrow{n_{12}}\right)\left(\overrightarrow{p_2}\,\overrightarrow{n_{12}}\right)}{r_{12}^2} - \overrightarrow{p_2}\,\overrightarrow{p_1}\right] \tag{4.26}$$

The force of interaction between two electric dipoles:

$$\overrightarrow{F} = -\nabla W_p = \frac{1}{4\pi\varepsilon_0}\nabla\left\{\frac{1}{r_{12}^3}\left[\frac{3\left(\overrightarrow{p_1}\,\overrightarrow{n_{12}}\right)\left(\overrightarrow{p_2}\,\overrightarrow{n_{12}}\right)}{r_{12}^2} - \overrightarrow{p_1}\,\overrightarrow{p_2}\right]\right\} \tag{4.27}$$

We observe that the force between two dipoles is reversely proportional with the fourth power of distance between them, so there *is not a central-type force which characterizes the point charge interaction.*

Exercise: force of interaction for a particular dipole system

Two water molecules having the dipole moments, i.e., $p_1 = p_2 = 6.2 \cdot 10^{-30}$ C m, are placed along of two lines separated by the distance $d = 3 \times 10^{-10}$ m (see figure 4.9). What is the force of interaction between them?
Solution: *Considering figure 4.9, we have:*

$$W_p = -\frac{1}{4\pi\varepsilon_0 d^3}\left[\frac{3\left(\overrightarrow{p_1}\,\overrightarrow{d}\right)\left(\overrightarrow{p_2}\,\overrightarrow{d}\right)}{d^2} - \overrightarrow{p_1}\,\overrightarrow{p_2}\right] = \frac{p_1 p_2}{4\pi\varepsilon_0 d^3}$$

$$\Rightarrow \overrightarrow{F_{12}} = -\nabla W_p = -\nabla\frac{p_1 p_2}{4\pi\varepsilon_0 d^3} = \frac{3 p_1 p_2}{4\pi\varepsilon_0 d^4}\,\overrightarrow{e_d}; \tag{4.28}$$

$$\overrightarrow{e_d} = \frac{\overrightarrow{d}}{d} \quad \text{repulsive}$$

$$W_p = -\frac{1}{4\pi\varepsilon_0 d^3}\left[\frac{3\left(\overrightarrow{p_1}\,\overrightarrow{d}\right)\left(\overrightarrow{p_2}\,\overrightarrow{d}\right)}{d^2} - \overrightarrow{p_1}\,\overrightarrow{p_2}\right] = -\frac{p_1 p_2}{4\pi\varepsilon_0 d^3}$$

$$\Rightarrow \overrightarrow{F_{12}} = -\nabla W_p = \nabla\frac{p_1 p_2}{4\pi\varepsilon_0 d^3} = \frac{p_1 p_2}{4\pi\varepsilon_0}\left(-\frac{3\overrightarrow{d}}{d^5}\right)\overrightarrow{e_d} \tag{4.29}$$

$$= -\frac{3 p_1 p_2}{4\pi\varepsilon_0 d^4}\,\overrightarrow{e_d}; \quad \overrightarrow{e_d} = \frac{\overrightarrow{d}}{d} \quad \text{attractive}$$

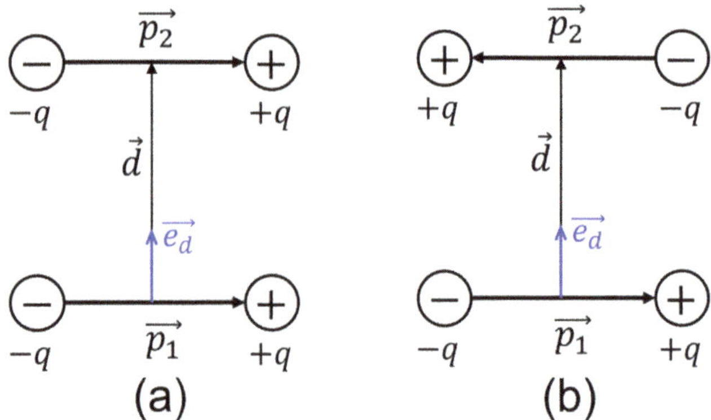

Figure 4.9. *Interactions between two water molecules placed at the distance d between them with the electric dipole moments (a) homo-parallel and (b) anti-parallel.*

4.2 The electric quadrupole

Similarly, if a charge distribution has got a total charge equal to zero, and a total dipole moment zero, the electric field and potential are zero at a small distance, but at large distance they may be different to zero. This is the case of a charge distribution which has got a symmetry axis, called *electric quadrupole*. The potential of a quadrupole falls off as $1/r^3$ and its field as $1/r^4$. A quadrupole is characterized by a physical quantity termed 'quadrupole moment', defined by (see figure 4.10):

$$Q = \frac{1}{2}\sum_j q_j\left(3R_j^2\cos^2\theta_j - R_j^2\right) = \frac{1}{2}\sum_j q_j\left(3z_j^2 - R_j^2\right) \tag{4.30}$$

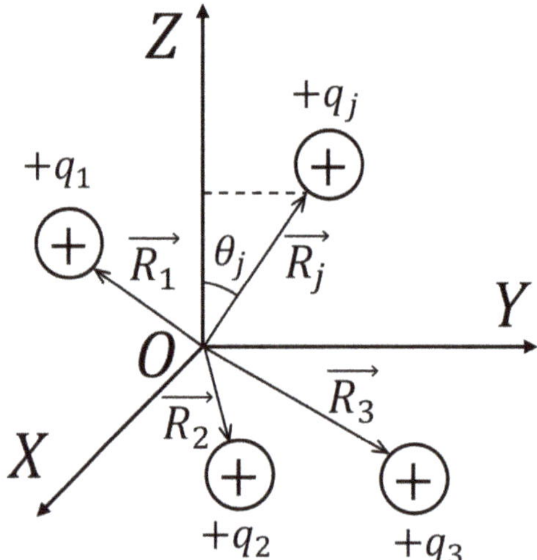

Figure 4.10. The electric quadrupole.

This quantity is very important in the study of the properties of nuclei. The quadrupole moment may be zero, positive or negative (see figure 4.11). A very simple quadrupole is a system of two dipoles which are placed end to end, giving a set of charges as shown in figure 4.12, where their fields cancel one another at large distances, whereas the potential of a quadrupole falls off as $1/r^3$ and its field as $1/r^4$. Indeed, in the case of simple quadrupole formed from a system of two dipoles which are placed end to end (see figure 4.12), the potential and the electric field can be computed at large distance with respect to its center using the superposition principle.

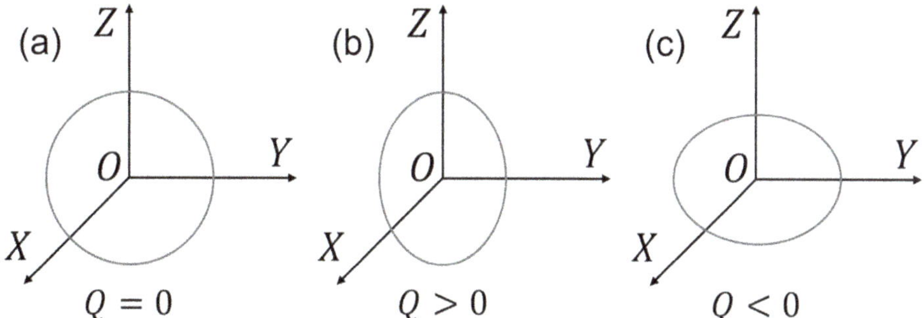

Figure 4.11. Different configurations of an electric quadrupole.

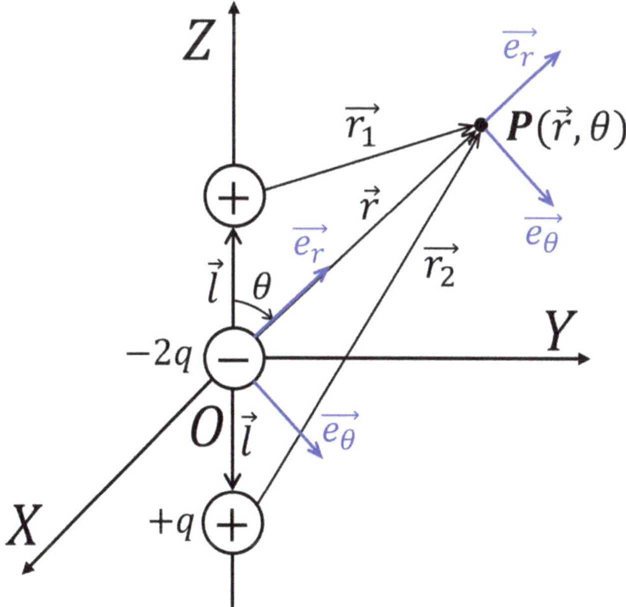

Figure 4.12. Quadrupole formed from two dipoles placed end to end.

According to the superposition principle, the potential of the system formed by the quadrupole charges is:

$$V(r) = \frac{q}{4\pi\varepsilon_0}\left[-\frac{2}{r} + \frac{1}{r_1} + \frac{1}{r_2}\right]$$

(4.31)

where:

$$\begin{cases} r_1 = (r^2 + l^2 - 2rl\cos\theta)^{1/2} \\ r_2 = (r^2 + l^2 + 2rl\cos\theta)^{1/2} \end{cases}$$

Consequently:

$$\frac{1}{r_1} = \frac{1}{r}\left(1 + \frac{l^2}{r^2} - \frac{2l\cos\theta}{r}\right)^{-1/2} = \frac{1}{r}\left[1 - \frac{1}{2}\left(\frac{l^2}{r^2} - \frac{2l\cos\theta}{r}\right)\right.$$

$$\left. + \frac{3}{8}\left(\frac{l^2}{r^2} - \frac{2l\cos\theta}{r}\right)^2 + \cdots\right] = \frac{1}{r}\left[1 - \frac{1}{2}\frac{l^2}{r^2} + \frac{l\cos\theta}{r} + \frac{3}{8}\frac{4l^2\cos^2\theta}{r^2} + \cdots\right]$$

$$= \frac{1}{r} + \frac{l\cos\theta}{r^2} + \frac{1}{2}\frac{3l^2\cos^2\theta - l^2}{r^3}$$

and:

$$\frac{1}{r_2} = \frac{1}{r}\left(1 + \frac{l^2}{r^2} + \frac{2l\cos\theta}{r}\right)^{-1/2} = \frac{1}{r}\left[1 - \frac{1}{2}\left(\frac{l^2}{r^2} + \frac{2l\cos\theta}{r}\right)\right.$$

$$\left. + \frac{3}{8}\left(\frac{l^2}{r^2} + \frac{2l\cos\theta}{r}\right)^2 + \cdots\right] = \frac{1}{r}\left[1 - \frac{l\cos\theta}{r} - \frac{1}{2}\frac{l^2}{r^2} + \frac{3}{8}\frac{4l^2\cos^2\theta}{r^2} + \cdots\right]$$

$$= \frac{1}{r} - \frac{l\cos\theta}{r^2} + \frac{1}{2}\frac{3l^2\cos^2\theta - l^2}{r^3}$$

Hence:

$$V(r) = \frac{q}{4\pi\varepsilon_0}\left[-\frac{2}{r} + \frac{1}{r} + \frac{l\cos\theta}{r^2} + \frac{1}{2}\frac{3l^2\cos^2\theta - l^2}{r^3} + \frac{1}{r} - \frac{l\cos\theta}{r^2}\right.$$

$$\left. + \frac{1}{2}\frac{3l^2\cos^2\theta - l^2}{r^3}\right] = \frac{q}{4\pi\varepsilon_0 r^3}\frac{3l^2\cos^2\theta - l^2}{1} = \frac{ql^2}{4\pi\varepsilon_0}\frac{3\cos^2\theta - 1}{r^3}$$

(4.32)

But according to the equation (4.30), the quadrupole moment of this system is:

$$Q = \frac{1}{2}\sum_i q_i\left(3z_i^2 - R_j^2\right) = \frac{1}{2}\left[q(3l^2 - l^2) + 2q(0) + q(3l^2 - l^2)\right]$$

$$= \frac{1}{2}4ql^2 = 2ql^2$$

(4.33)

Substituting (4.33) in equation (4.32) results in:

$$V = \frac{Q(3\cos^2\theta - 1)}{8\pi\varepsilon_0 r^3} = \frac{Q}{4\pi\varepsilon_0}\frac{\frac{1}{2}(3\cos^2\theta - 1)}{r^3} \tag{4.34}$$

Indeed, the potential of the quadrupole falls off as $1/r^3$.

The electric field of the quadrupole at a point $P(\vec{r}, \theta)$ can be obtained taking the gradient of the potential:

$$
\begin{aligned}
\overrightarrow{E_Q} &= -\nabla V_Q = -\nabla\left[\frac{Q}{4\pi\varepsilon_0}\frac{1}{2}\frac{3\cos^2\theta - 1}{r^3}\right] = -\frac{Q}{4\pi\varepsilon_0}\frac{1}{2}(3\cos^2\theta - 1) \\
&\cdot \frac{\partial}{\partial r}\left(\frac{1}{r^3}\right)\overrightarrow{e_r} - \frac{Q}{4\pi\varepsilon_0 r^3}\frac{1}{2}\frac{1}{r}\frac{\partial}{\partial\theta}(3\cos^2\theta - 1)\overrightarrow{e_\theta} \\
&= \frac{3Q}{8\pi\varepsilon_0 r^4}(3\cos^2\theta - 1)\overrightarrow{e_r} + \frac{6Q}{8\pi\varepsilon_0 r^4}\sin\theta\cos\theta\,\overrightarrow{e_\theta} \\
&= \frac{3Q}{8\pi\varepsilon_0 r^4}[(3\cos^2\theta - 1)\overrightarrow{e_r} + \sin 2\theta\,\overrightarrow{e_\theta}]
\end{aligned}
\tag{4.35}
$$

Then, the electric field of the quadrupole falls off as $1/r^4$.

Synthesizing, we observe that the potential V_q and the electric field E_q of a point charge q falls off as $1/r$ and $1/r^2$, respectively. In contrast, the potential V_p and the field E_p of an electric dipole having the moment $\vec{p}$, falls off as $1/r^2$ and as $1/r^3$, respectively. Ultimately, the potential V_Q and the field E_Q of a quadrupole having the moment Q falls off as $1/r^3$ and as $1/r^4$, respectively. In the case of an arbitrary charge distribution, the potential at a point P, far from the center of the distribution could be written as a sum of three terms:

$$V = V_q + V_p + V_Q \tag{4.36}$$

Each of the terms from equation (4.36) corresponds, respectively, to monopole potential V_q, dipole potential V_p and quadrupoles potential V_Q. For different particular cases, one or two terms from equation (4.36) could be zero ($\sum_i q_i = 0 \Rightarrow V_q = 0$; $\sum_i p_i = 0 \Rightarrow V_p = 0$, and then only $V_Q \neq 0$). If also $Q = 0$ ($V_Q = 0$) we must take the next term from the Taylor expansion of $1/r$ leading to octopole potential $\sim 1/r^4$.

4.3 Multipolar expansion of electrostatic potential

Considering a discrete charge distribution, having the total charge $q = \sum_{i=1}^{n} q_i$, centered at the origin O of a Cartesian coordinate system (see figure 4.13), the potential outside the charge distribution in agreement to the superposition principle will be:

$$V = \sum_{i=1}^{n}\frac{q_i}{4\pi\varepsilon_0|\overrightarrow{R_i}|} = \sum_{i=1}^{n}\frac{q_i}{4\pi\varepsilon_0|\vec{r}-\vec{r_i}|} \tag{4.37}$$

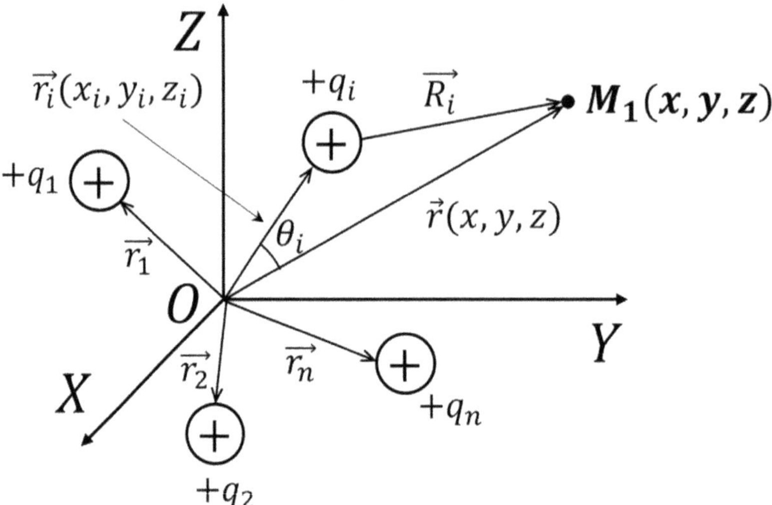

Figure 4.13. Discrete system of charges centered at the origin of a Cartesian coordinate system.

$$|\vec{R_i}| = |\vec{r} - \vec{r_i}| = (r^2 + r_i^2 - 2rr_i \cos\theta_i)^{1/2}$$

$$\frac{1}{|\vec{R_i}|} = \frac{1}{|\vec{r} - \vec{r_i}|} = \frac{1}{r}\left[1 + \frac{r_i^2}{r^2} - \frac{2(\vec{r}\,\vec{r_i})}{r^2}\right]^{-1/2} \tag{4.38}$$

Taking the direct Taylor's series of the ratio $\frac{1}{|\vec{r} - \vec{r_i}|}$, we obtain:

$$\frac{1}{|\vec{r} - \vec{r_i}|} = \frac{1}{r}\left[1 - \frac{1}{2}\left(\frac{r_i^2}{r^2} - \frac{2(\vec{r}\,\vec{r_i})}{r^2}\right) + \frac{3}{8}\left(\frac{r_i^2}{r^2} - \frac{2(\vec{r}\,\vec{r_i})}{r^2}\right)^2 + \cdots\right]$$

$$= \frac{1}{r}\left[1 + \frac{\vec{r}\,\vec{r_i}}{r^2} - \frac{1}{2}\frac{r_i^2}{r^2} + \frac{3}{8}\frac{4(\vec{r}\,\vec{r_i})^2}{r^4} + \cdots\right]$$

$$= \frac{1}{r} + \frac{(\vec{r}\,\vec{r_i})}{r^3} + \frac{\frac{1}{2}\left[3\frac{(\vec{r}\,\vec{r_i})^2}{r^2} - r_i^2\right]}{r^3} \Rightarrow$$

$$\frac{1}{|\vec{r} - \vec{r_i}|} = \frac{1}{r} + \frac{r_i \cos\theta_i}{r^2} + \frac{\frac{1}{2}r_i^2(3\cos^2\theta_i - 1)}{r^3} \tag{4.39}$$

Substituting (4.39) in (4.37) results in:

$$V = \frac{1}{4\pi\varepsilon_0} \frac{\sum\limits_{i=1}^{n} q_i}{r} + \frac{1}{4\pi\varepsilon_0} \frac{\sum\limits_{i=1}^{n} q_i r_i \cos\theta_i}{r^2}$$
$$+ \frac{1}{4\pi\varepsilon_0} \frac{\sum\limits_{i=1}^{n} \frac{1}{2} q_i r_i^2 (3\cos^2\theta_i - 1)}{r^3} = V_q + V_p + V_Q \tag{4.40}$$

The electric potential far from the center of a discrete charge distribution is the sum of three terms representing:

- V_q, the first term from multiple expansion (4.40) is the potential of a point charge (total charge or monopole moment, $q = \sum_{i=1}^{n} q_i$), placed at the center of distribution (the monopole potential which falls as $1/r$);
- V_p, the second term from multiple expansion (4.40) is the potential associated with the total dipole moment $p = \sum_{i=1}^{n} q_i r_i \cos\theta_i$ of charge distribution, and then it is the dipole potential decreasing as $1/r^2$;
- V_Q, the third term from multiple expansion (4.40) is the potential associated with the total quadrupole moment $Q = \frac{1}{2}\sum_{i=1}^{n} q_i r_i^2 (3\cos^2\theta_i - 1)$ of charge distribution, and then it is the quadrupole potential falling as $1/r^3$.

Let's consider now a continuous charge distribution described by the volume charge density $\rho\left(\vec{r'}\right)$, which is non-vanishing only inside a volume V (see figure 4.14).

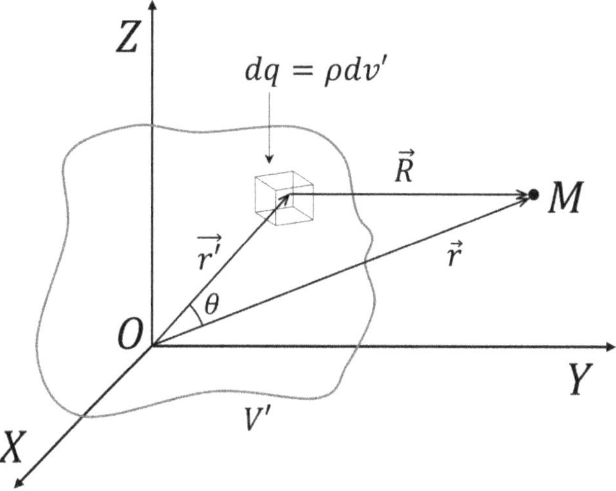

Figure 4.14. The charge uniformly distributed in the volume V.

The potential outside the volume can be written as usual by adding the contributions from all the elementary charges $dq = \rho dv'$:

$$V(r) = \int_{V'} \frac{\rho\left(\vec{r'}\right)dv'}{4\pi\varepsilon_0 R} \tag{4.41}$$

In the integrant dv' is a volume element within the charge distribution, $\rho\left(\vec{r'}\right) = \rho(x', y', z')$ is the charge density there, and R in the denominator is the distance from point M to this particular charge element. The integration is carried out in the coordinates (x', y', z'), of course, and is extended over all regions containing charge. We can express R in terms of $\vec{r}$ (the position vector of point M with respect to the origin) and $\vec{r'}$ (the position vector of charge element with respect to the origin) by:

$$R = (r^2 + r'^2 - 2rr' \cos \theta)^{1/2} \tag{4.42}$$

With this substitution for R, the integral becomes:

$$\begin{aligned}
V_M &= \int_{V'} \frac{\rho(r')dv'}{4\pi\varepsilon_0(r^2 + r'^2 - 2rr' \cos \theta)^{1/2}} \\
&= \int_{V'} \frac{\rho(r')(r^2 + r'^2 - 2rr' \cos \theta)^{-1/2}}{4\pi\varepsilon_0}dv'
\end{aligned} \tag{4.43}$$

Now, we want to take advantage of the fact that for a distant point like M, r' is much smaller than r for all parts of the charge distribution. This suggests that we should expand the square root in equation (4.43) in power of r'/r. Writing:

$$\begin{aligned}
(r^2 + r'^2 - 2rr' \cos \theta)^{-1/2} &= \frac{1}{r}\left(1 + \frac{r'^2}{r^2} - \frac{2r' \cos \theta}{r}\right)^{-1/2} \\
&= \frac{1}{r}\Bigg[1 - \frac{1}{2}\left(\frac{r'^2}{r^2} - \frac{2r' \cos \theta}{r}\right) \\
&\quad + \frac{3}{8}\left(\frac{r'^2}{r^2} - \frac{2r' \cos \theta}{r}\right)^2 + \text{(terms of higher order)}\Bigg]
\end{aligned} \tag{4.44}$$

We get, after collecting together the terms of the same power in r'/r:

$$\begin{aligned}
(r^2 + r'^2 &- 2rr' \cos \theta)^{-1/2} \\
&= \frac{1}{r}\Bigg[1 + \frac{r' \cos \theta}{r} + \frac{1}{2}\frac{r'^2}{r^2}(3 \cos^2 \theta - 1) \\
&\quad + \text{(terms of higher order)}\Bigg]
\end{aligned} \tag{4.45}$$

Now, r is a constant in the integration, so we can take it outside and write the prescription for the potential at point M as follows:

$$V_M = \frac{1}{4\pi\varepsilon_0 r} \int_{V'} \rho(r')dv' + \frac{1}{4\pi\varepsilon_0 r^2} \int_{V'} r' \cos\theta\rho(r')dv'$$
$$+ \frac{1}{4\pi\varepsilon_0 r^3} \int_{V'} \frac{r'^2\rho(r')(3\cos^2\theta - 1)}{2}dv' \tag{4.46}$$

where:

$$\int_{V'} \rho(r')dv' = K_0$$

$$\int_{V'} r' \cos\theta\rho(r')dv' = K_1$$

$$\int_{V'} \frac{r'^2\rho(r')(3\cos^2\theta - 1)}{2}dv' = K_2$$

Each of the integrals above K_0, K_1, K_2 and so on, has a value that depends only on the structure of the charge distribution. Hence the potential at a point M can be written as a power series in $1/r$ with constant coefficients:

$$V_M = V_0 + V_1 + V_2 = \frac{K_0}{4\pi\varepsilon_0 r} + \frac{\overrightarrow{K_1}\overrightarrow{r}}{4\pi\varepsilon_0 r^3} + \frac{K_2}{4\pi\varepsilon_0 r^3} \tag{4.47}$$

To finish the problem we would have to get the electric field. Having the potential, the electric field will be:

$$\overrightarrow{E} = -\nabla V = \frac{K_0\overrightarrow{r}}{4\pi\varepsilon_0 r^3} + \frac{1}{4\pi\varepsilon_0 r^3}\left[\frac{3\left(\overrightarrow{K_1}\overrightarrow{r}\right)\overrightarrow{r}}{r^2} - \overrightarrow{K_1}\right] + \frac{3K_2\overrightarrow{r}}{4\pi\varepsilon_0 r^5} \tag{4.48}$$

Observation:

$$V_2 = \frac{1}{4\pi\varepsilon_0 r^2}\int_V \rho(r')r' \cos\theta dv' = \frac{\overrightarrow{r}}{4\pi\varepsilon_0 r^3}\int_V \rho(r')\overrightarrow{r'}dv = \frac{\overrightarrow{K_1}\overrightarrow{r}}{4\pi\varepsilon_0 r^3}$$

where:

$$\int_V \rho(r')\overrightarrow{r'}dv = \overrightarrow{K_1}$$

We have gone far enough, though, to bring out the essential point: the behavior of the potential at large distances from the source will be dominated by the first term in this series whose coefficient is not zero. Let us look at these coefficients more closely. The coefficient K_0 is $\int_{V'} \rho(r')dv'$, which is the total charge in the distribution. If we have equal amounts of positive and negative charge, as in a neutral molecule, K_0 will be zero. For a single ionized molecule K_0 will have the value e. If K_0 is not zero, then no matter how large K_1, K_2, etc may be, if we go out to sufficiently large distance, the term K_0/r will win out. Beyond that, the potential will approach as that of a point charge placed at origin and so also will be the behavior of the electric field.

Suppose now that we have a neutral molecule, so that K_0 is zero. Our interest now shifts to the second term, with coefficient $K_1 = \int_{V'} r' \cos\theta \rho\left(\vec{r'}\right)dv' = \int_{V'} \left(\vec{r'}\,\vec{e_r}\right)\rho(r')dv'$. This coefficient is a measure of the relative displacement of the positive and negative charge, in the direction toward P (along of position vector $\vec{r}$, having the vector unity $\vec{e_r}$). It has a non-zero value for the water molecule. It is worth pointing out that if the distribution is neutral, the value of K_1 is independent of the position of origin (see figure 4.15). Evidently, if $K_0 = 0$ and $K_1 \neq 0$, the potential will vary asymptotically as $1/r^2$. We expect the electric field strength, then to behave asymptotically like as $1/r^3$, in contrast to the $1/r^2$ dependence of the field from a point charge. If K_0 and K_1 are both zero, and K_2 is not, the potential will behave like $1/r^3$ at large distances, and the field strength will fall off with the inverse fourth power of distance. Figure 4.16 shows a charge distribution for which K_0 and K_1 are both zero, while K_2 is not zero. The quantities K_0, K_1, K_2, ..., are

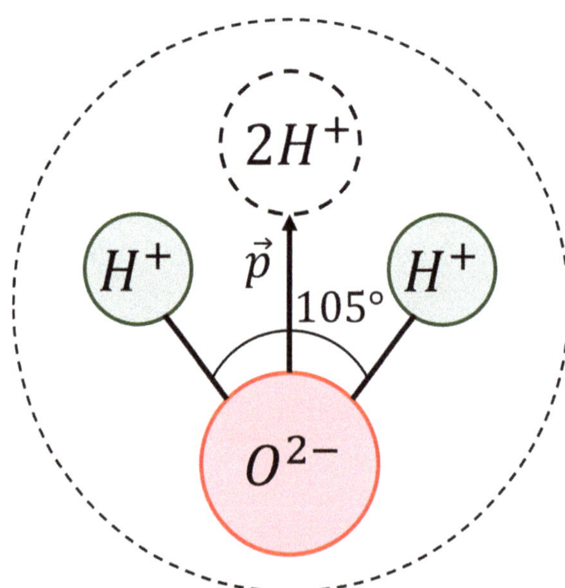

Figure 4.15. Water molecule having net charge zero but non-zero dipole moment.

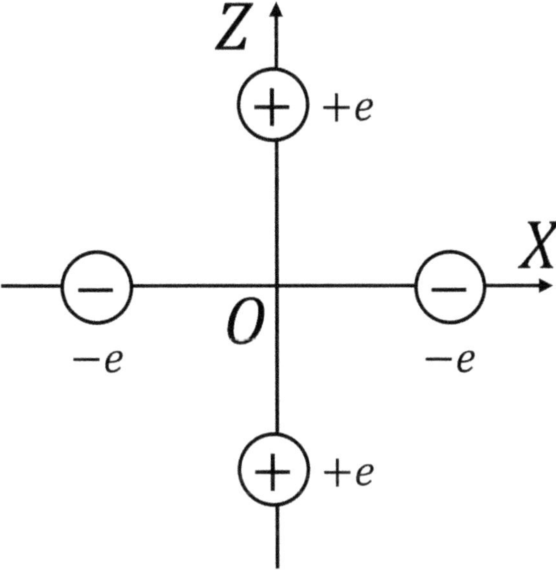

Figure 4.16. A charge distribution with K_0 and K_1 are zero and $K_2 \neq 0$.

called the moments of the charge distribution. Using this language, we call K_0, which is simply the total charge, the **monopole moment** or **monopole strength**, K_1 is the **dipole moment** of the distribution, K_2 is the **quadrupole moment** of the distribution, the next to the **octopole moment**, and so on.

Exercises: calculate the potential of the charge distributions

1. Let's consider a charge distribution like this one from figure 4.17 where eight point charges are placed in the corners apex of a cube having the side a = 1 cm. The values of these charges are given in figure 4.17 in μC. Find the first three terms of potential at a point M located on the OY axis at a distance r = 1 m from the origin of the coordinate system.
Solution:

$$V_p = \frac{K_0}{4\pi\varepsilon_0 r} + \frac{\overrightarrow{K_1}\,\overrightarrow{r}}{4\pi\varepsilon_0 r^3} + \frac{K_2}{4\pi\varepsilon_0 r^3}$$

$$= \frac{3 \cdot 10^{-6}}{4\pi\varepsilon_0 \cdot 1} + \frac{(-3\overrightarrow{i}+6\overrightarrow{j}\,)(\overrightarrow{j}\,)10^{-8}}{4\pi\varepsilon_0 \cdot 1^3} + \frac{3 \cdot 10^{-10}}{4\pi\varepsilon_0 \cdot 1^3}$$

$$= 9 \cdot 10^9[3 \cdot 10^{-6} + 6 \cdot 10^{-8} + 3 \cdot 10^{-10}]$$

$$= 27 \cdot 10^3[1 + 0.02 + 0.0001] = 27 \cdot 1.0201 \cdot 10^3 \ V$$

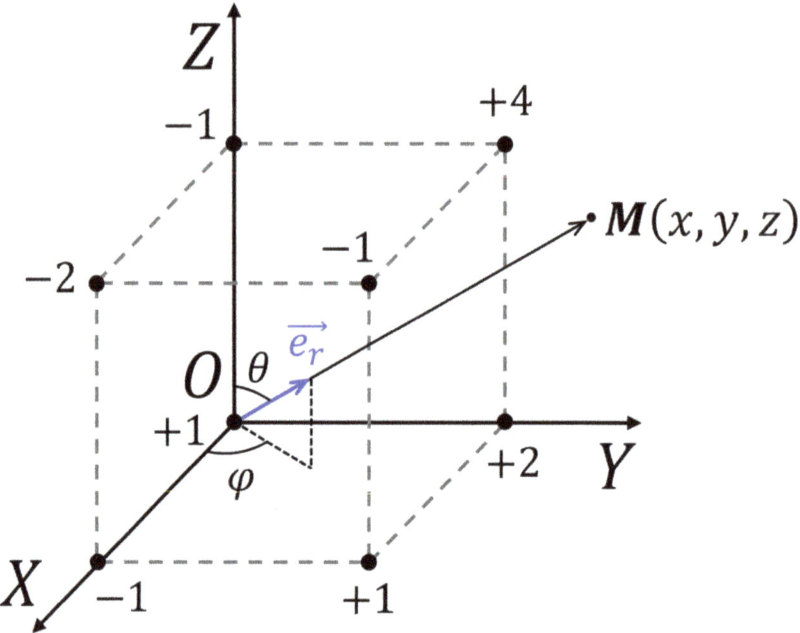

Figure 4.17. System of point charges.

Position	Charge q	$\vec{r'}$	$q\vec{r'}$	$\frac{1}{2}q_i\left[3(\vec{e_r}\vec{r_i})^2 - r_i^2\right]$
(000)	1	0	0	0
(100)	−1	$\vec{i}$	$-\vec{i}$	$\frac{1}{2}(-1)(2) = -1$
(010)	2	$\vec{j}$	$2\vec{j}$	$\frac{1}{2}(2)(2) = 2$
(001)	−1	$\vec{k}$	$-\vec{k}$	$\frac{1}{2}(-1)(2) = -1$
(110)	1	$\vec{i}+\vec{j}$	$\vec{i}+\vec{j}$	$\frac{1}{2}(1)(10) = 5$
(011)	4	$\vec{j}+\vec{k}$	$4\vec{j}+4\vec{k}$	$\frac{1}{2}(4)(10) = 20$
(101)	−2	$\vec{i}+\vec{k}$	$-2\vec{i}-2\vec{k}$	$\frac{1}{2}(-2)(10) = -10$
(111)	−1	$\vec{i}+\vec{j}+\vec{k}$	$-\vec{i}-\vec{j}-\vec{k}$	$\frac{1}{2}(-1)(24) = -12$
	3×10^{-6} C		$(-3\vec{i}+6\vec{j}) \times 10^{-8}$ C m	3×10^{-10} C m^2

2. Let there be a continuous charge distribution with the volume charge density $\rho(x', y', z') = x' + y' + z'$ *C m^{-3} within a cube of side a = 1 cm. Find the potential at a point M located at 1 m from origin along OY axis as in figure 4.18.*

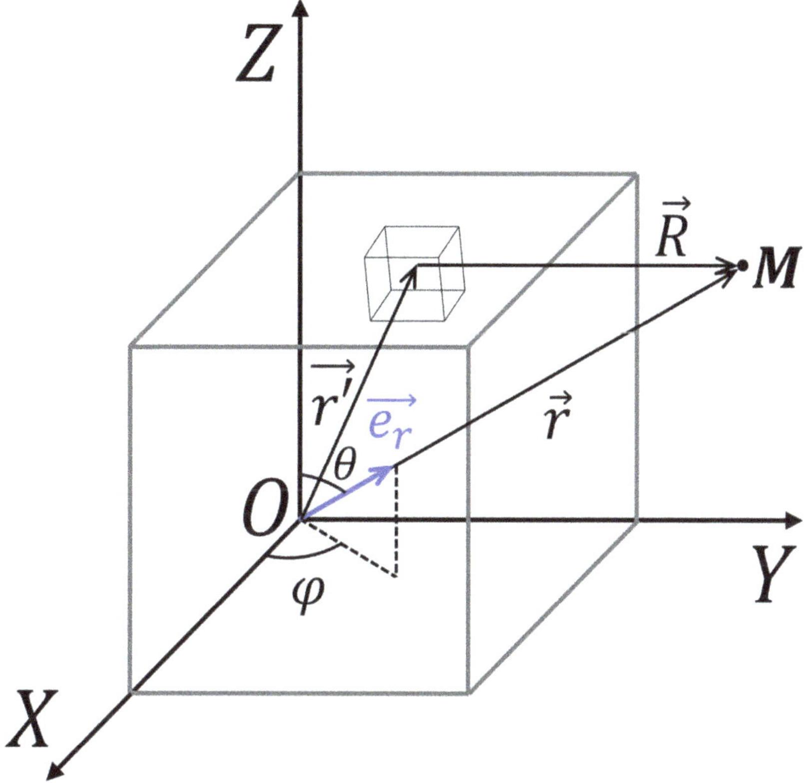

Figure 4.18. *An uniform charge distribution in a cube of side a.*

Solution:

$$K_0 = \int_{V'} \rho \, dv' = \int_0^1 \int_0^1 \int_0^1 (x' + y' + z') dx' dy' dz' = \frac{3}{2}\,\mu\text{C}$$

$$K_1 = \int_{V'} r' \rho \, dv' = \int_0^1 \int_0^1 \int_0^1 (x' + y' + z')\left(x'\vec{i} + y'\vec{j} + z'\vec{k}\right) dx' dy' dz'$$

$$= \int_0^1 \int_0^1 \int_0^1 (x'^2 + y'x' + z'x')\vec{i}\, dx' dy' dz' + \cdots = \frac{5}{6}\left(\vec{i} + \vec{j} + \vec{k}\right)$$

$$K_2 = \int\limits_{V'} \frac{(x' + y' + z')}{2} \left\{ 3\left[\vec{j} \left(x'\vec{i} + y'\vec{j} + z'\vec{k} \right)^2 \right] - \left(x'\vec{i} + y'\vec{j} + z'\vec{k} \right)^2 \right\}$$

$$\cdot \, dx'dy'dz' = \frac{1}{2} \int\limits_{V'} (x' + y' + z')(2y'^2 - x'^2 - z'^2)dx'dy'dz'$$

$$= \frac{1}{2} \int\limits_0^1 \int\limits_0^1 \int\limits_0^1 (x' + y' + z')(2y'^2 - x'^2 - z'^2)dx'dy'dz'$$

$$= \frac{1}{2} \int\limits_0^1 \int\limits_0^1 \int\limits_0^1 [2y'^2 x' - x'^3 - x'z'^2 + 2y'^3 - x'^2 y' - z'^2 y' + 2z'y'^2 - x'^2 z' - z'^3]$$

$$\cdot \, dx'dy'dz' = \frac{1}{2}\left[\frac{1}{3} - \frac{1}{4} - \frac{1}{6} + \frac{1}{2} - \frac{1}{6} - \frac{1}{6} + \frac{1}{3} - \frac{1}{6} - \frac{1}{4} \right] = \frac{1}{2}[0] = 0$$

Thus:

$$V = 9 \cdot 10^9 \left[\frac{\frac{3}{2}10^{-6}}{1} + \frac{\frac{5}{6}\left(\vec{i} + \vec{j} + \vec{k} \right)}{1^2}\vec{j} \, \cdots \right]$$

$$= 9 \cdot 10^9 \cdot 10^{-6} \cdot \left[\frac{3}{2} + \frac{5}{6} \cdot 10^{-2} \right] = 9 \cdot 10^3 \cdot [1.5 + 0.0083] \text{ V}$$

Further reading

[1] Buckingham A D and Longuet-Higgins H C 1968 The quadrupole moments of dipolar molecules *Mol. Phys.* **14** 63–72

[2] Coelho R 1979 *Physics of Dielectrics for the Engineer* 1st edn (Amsterdam: Elsevier)

[3] Savelyev I V 1982 *Fundamentals of theoretical physics* vol 1 (Mir: Moscow)

[4] Clayton P R, Keith W W and Syed N A 1997 *Introduction to Electromagnetic Fields* (New York: McGraw-Hill)

[5] Jackson J D 1998 *Classical Electrodynamics* 3rd edn (New York: Wiley)

[6] Feynman R P, Leighton R B and Sands M 2011 *The Feynman Lectures on Physics, Vol II: The New Millennium Edition: Mainly Electromagnetism and Matter* (New York: Basic Books)

[7] Purcell E M and Morin D J 2013 *Electricity and Magnetism* 3rd edn (Cambridge: Cambridge University Press)

[8] Lacava F 2016 Multipole expansion of the electrostatic potential *Classical Electrodynamics. Undergraduate Lecture Notes in Physics* (Berlin: Springer)

[9] Griffiths D J 2017 *Introduction to Electrodynamics* 4th edn (Cambridge: Cambridge University Press)

Electrostatics
Formalism of the electrostatic field in vacuum and matter
Ştefan Antohe and Vlad-Andrei Antohe

Chapter 5

Conductors within electrostatic fields

Within previous chapters we discussed the properties of the electrostatic field created by stationary charges, in vacuum, in other words we addressed the *formalism of electrostatic field in vacuum*. Now, we are interested in what happens when an electric field is created in matter and how the matter modifies its own properties when exposed to an electric field.

5.1 Conductors at electrostatic equilibrium

It is very well known that a conductor contains free charges which can freely move within the conductor so that, if an electric field is established in a conducting material, charges will move as long as the field exists. If a conductor is isolated, figure 5.1 illustrates what will happen inside the conductor when it is placed in an external electric field, $\overrightarrow{E_{\text{ext}}}$. The total electric field, $\overrightarrow{E_{\text{t}}}$, resulting from the external electric field $\overrightarrow{E_{\text{ext}}}$ applied to the conductor and the electric field $\overrightarrow{E_{\text{int}}}$ induced inside the conductor by the polarization of charge during the transient current flowing in the conductor until the electrostatic equilibrium is installed, is zero in the whole volume of the conductor.

$$\overrightarrow{E_{\text{t}}} = \overrightarrow{E_{\text{ext}}} + \overrightarrow{E_{\text{int}}} = 0 \tag{5.1}$$

As soon as $\overrightarrow{E_{\text{ext}}}$ is established, one part of the conductor is at higher potential than the other, then any free positive charges will move in the direction of $\overrightarrow{E_{\text{ext}}}$, and any negative charges in the opposite direction, respectively. Whatever the sign of the moving charges is, the result is the same: charges reach the surface of the conductor and they can go no further. They collect and produce a field within the conductor, $\overrightarrow{E_{\text{int}}}$, which opposes to the applied external field. This process continues, within the material, until there is no resultant field, the electrostatic equilibrium being installed. Static conditions will then again prevail. We have thus accounted for electrostatic

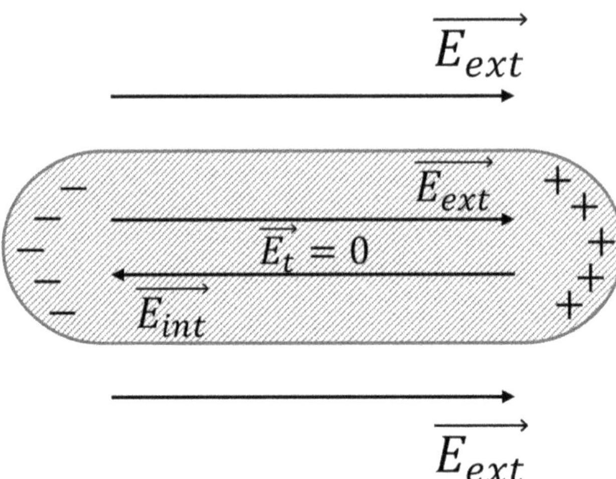

Figure 5.1. An isolated conductor at electrostatic equilibrium in an external electric field.

induction mentioned in the electrization processes. Taking into account this behavior we can underline a few properties of the conductors to electrostatic equilibrium, such as:

1. **A conductor carrying only static charge can have no electric field within its material, and hence throughout the volume of such a conductor there is no potential diffcrence;**

2. **A conductor carrying static charge is an equipotential volume and its surface is an equipotential surface;**

3. **The electric field outside the conductor has no tangential component (parallel to the surface).** If the electric field outside at the conductor surface has a non-zero parallel component, then the charges would move on the surface and there is not an equilibrium state. On the other hand, using the theorem of the electrostatic potential it is easy to demonstrate that the tangential component of the electric field at the surface of the conductor at electrostatic equilibrium is zero. Indeed, considering a very flat closed path (Γ) on the surface of the conductor (see figure 5.2), the line integral of the electric field along this path must be zero:

$$\int_{\Gamma} \overrightarrow{E}\,\overrightarrow{dl} = \int_{A}^{B} \overrightarrow{E_t}\,\overrightarrow{dl} + \int_{B}^{C} \overrightarrow{E_n}\,\overrightarrow{dl} + \int_{C}^{D} \overrightarrow{E_t}\,\overrightarrow{dl} + \int_{D}^{A} \overrightarrow{E_n}\,\overrightarrow{dl} = 0 \tag{5.2}$$

The line integrals along the sides BC and DA are zero, due to the fact that the path Γ is very flat (closely centered on the point P) on the surface of the conductor. The electric field inside the conductor being zero, the line integral along the side CD is also zero. Taking into account these observations, the line integral along the side AB from the right side of equation (5.2) will be zero too, or this is possible just only when the tangential component $\overrightarrow{E_t}$ is zero.

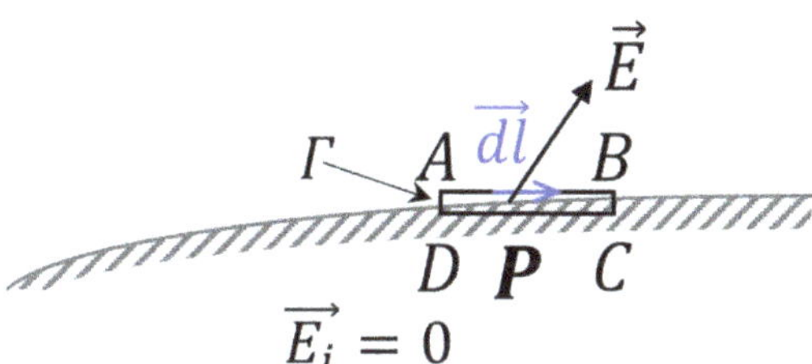

Figure 5.2. Tangential component of the electrostatic field to the surface of an isolated conductor at electrostatic equilibrium is zero at any point on the surface.

4. Because the surface of a conductor is necessarily **a surface of constant potential**, the electric field, which is $-\nabla V$, must be **perpendicular** to the surface at every point on the surface. Proceeding from the interior of the conductor outward, we find at the surface a sudden charge in the electric field; $\vec{E}$ **is not zero outside the surface, and it is zero inside.** The electric field **outside the conductor has only a normal component (perpendicular to the surface)** and has no tangential component (parallel to the surface), as was demonstrated above.

5. The discontinuity in $\vec{E}$ at any point of a conductor surface is accounted for by the presence of a surface charge, of density σ, which we can relate directly to $\vec{E}$ by Gauss's law. Indeed, the free charge within the conductor moves within the material, until there is no resultant field (the electrostatic equilibrium is performed). But the moving charges cannot leave the conductor, **they distribute only onto the surface of the conductor.** Indeed, if we consider a Gaussian surface, Σ' inside of the conductor (see figure 5.3), the flux of the electric field throughout this Gaussian is zero because the electric field inside of the conductor is everywhere zero:

$$\int_{\Sigma'} \vec{E}\,\overrightarrow{dS} = \frac{q_i}{\varepsilon_0} = 0 \Rightarrow q_i = 0 \tag{5.3}$$

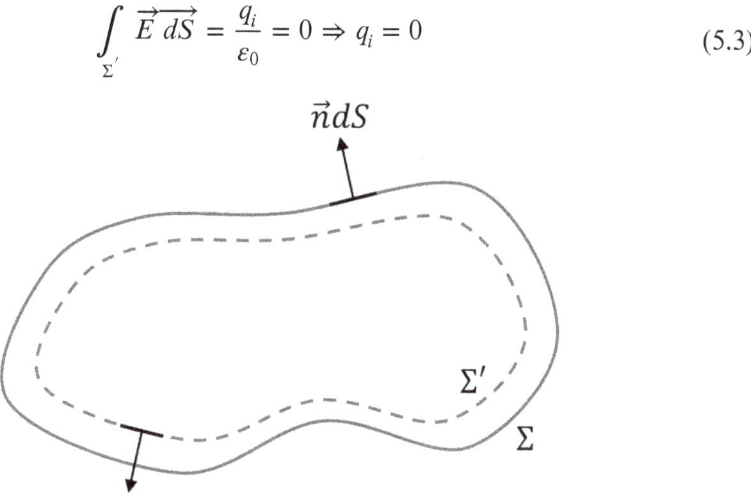

Figure 5.3. The electric charge is distributed only on the surface of a conductor at electrostatic equilibrium.

If we increase the Gaussian Σ', to transform it in another one Σ'', very close to the surface until it touches inside of the surface Σ of the conductor, the result will be the same:

$$q_i = \varepsilon_0 \int_{\Sigma''} \vec{E}\,\vec{dS} = 0 \tag{5.4}$$

due to the fact that the electric field inside of the conductor is everywhere zero. If the charge is zero everywhere inside of the conductor, it could not be any place than the surface, because it cannot leave the conductor.

6. Now, we can relate the electric field outside of the conductor to the charge density σ onto its surface, directly by Gauss's law. We can use a flat box enclosing a part of the surface ΔS, and having the generatrix very small, $h \to 0$ (see figure 5.4). The charge enclosed by this box is $\sigma \Delta S$, and then Gauss's law gives:

$$\int_{\Sigma} \vec{E}\,\vec{dS} = \frac{\sigma \Delta S}{\varepsilon_0} \tag{5.5}$$

Here, there is no flux through the rim and the 'bottom' of the box, but only through the 'upper' base, being equal to the charge which lies inside the conductor, hence we can write:

$$E_n \Delta S = \frac{\sigma}{\varepsilon_0}\Delta S \tag{5.6}$$

Now, we conclude that the normal component of the electric field to the surface of a conductor at electrostatic equilibrium is:

$$\vec{E_n} = \frac{\sigma}{\varepsilon_0}\vec{n} \tag{5.7}$$

where $\vec{n}$ is the normal to the surface directed from inside to outside of the closed surface. We observe here that this field has magnitude double that of the charge distribution on an infinite planar surface, with surface charge

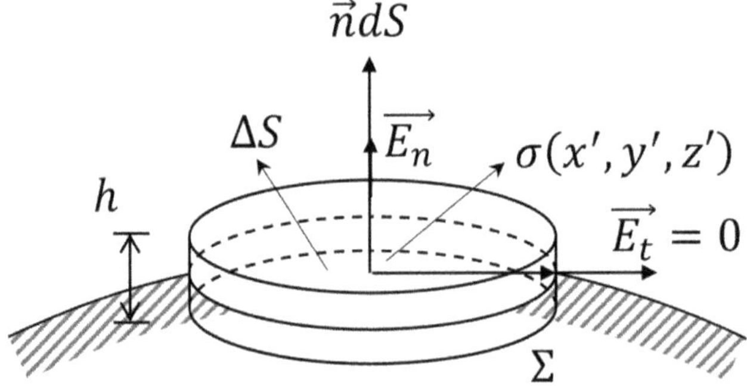

Figure 5.4. The normal component of the electric field to the surface of a conductor at electrostatic equilibrium.

density σ, having the value $\sigma/2\varepsilon_0$. This is easy to demonstrate, as presented further. Choosing a small surface element dS on the surface of the conductor (see figure 5.5), it will carry the charge $dq = \sigma dS$ and will give rise, according to Gauss's Law, to the electric field:

$$\overrightarrow{E_{\text{loc}}} = \pm\frac{\sigma}{2\varepsilon_0}\vec{n} \tag{5.8}$$

$\overrightarrow{E_{\text{loc}}}$ is oriented parallel[1] with the normal to the surface of the conductor outside the conductor, and antiparallel with the normal to the surface of the conductor inside the conductor (see figure 5.5). On the other hand, the rest of the charge on the whole surface of the conductor behaves again as an infinite planar surface charge distribution giving rise to the electric field $\overrightarrow{E_{\text{dist}}} = +\frac{\sigma}{2\varepsilon_0}\vec{n}$, which at point P inside the conductor is in opposite sense than $\overrightarrow{E_{\text{loc}}}$, then giving rise inside to a total electric field that is zero. Outside the conductor, $\overrightarrow{E_{\text{dist}}}$ has the same orientation with $\overrightarrow{E_{\text{loc}}}$ and then the total electric field, at point P on the conductor's surface, will be:

$$\vec{E} = \frac{\sigma}{\varepsilon_0}\vec{n} \tag{5.9}$$

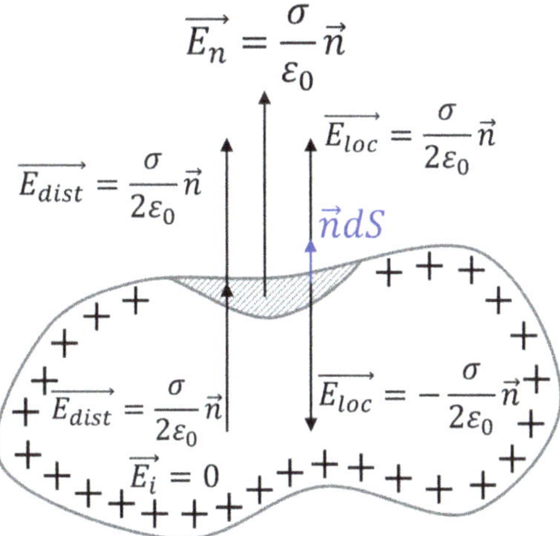

Figure 5.5. The normal component of the electric field to the surface of a conductor at electrostatic equilibrium.

---

[1] *Parallel vectors* have the same sense. *Antiparallel vectors* have opposite sense.

Nevertheless, the electric field outside the conductor, will depend on how the charge is distributed on its own surface. As we have already seen, there is no other component in this case, the field being always perpendicular to the surface. The surface charge must account for the whole charge Q of the conductor. That is, the surface integral of $\sigma_{(\vec{r'})}$ over the whole conductor must equal Q. Summarizing, we can make the following statements about any system of conductors, whatever their shape and arrangement (see figure 5.6):

- $V = V_K$ at all points on the surface of the Kth conductor.

- At any point just outside the conductor, $\vec{E}$ is perpendicular to the surface, and $\vec{E} = \frac{\sigma}{\varepsilon_0}\vec{n}$, where σ is the local density of surface charge.

- The total charge is:

$$Q_K = \int_{S_K} \sigma dS = \varepsilon_0 \int_{S_K} \vec{E}\,\vec{dS}$$

(5.10)

Because the relation (5.9) uniquely relates $\vec{E}$ to σ, the local surface charge density, you may be tempted to think of σ as the source of $\vec{E}$. That would be a mistake, $\vec{E}$ is the total field arising from all the charges in the system, near and far, of which the surface charge is only a part. The surface charge on a conductor is obliged to 'readjust itself' until the relation (5.9) is fulfilled.

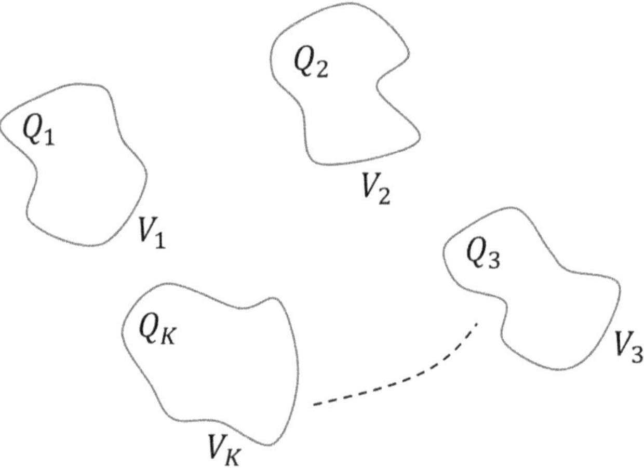

Figure 5.6. System of many conductors: Q_1 is the charge of conductor 1, V_1 is its own potential; Q_2 is the charge of conductor 2, V_2 is its potential; etc.

Summarizing, a charged conductor to electrostatic equilibrium is characterized by the following properties:

    a. The electric field inside of the conductors is zero at any point;
    b. The volume of conductor is an equipotential region;
    c. The charge of a conductor is distributed only on its surface;

   d. The tangential component of the electric field to the surface of a conductor to electrostatic equilibrium is zero at any point on the surface;

   e. The surface of a conductor at electrostatic equilibrium is an equipotential region, having the same potential as any point inside the conductor;

   f. The electric field outside of the conductor has only the normal component to the surface of the conductor, having the value: $\overrightarrow{E} = \dfrac{\sigma}{\varepsilon_0}\overrightarrow{n}$.

5.2 Green's theorem

The electric charge distributes onto the surface of a conductor in such a way that the surface charge density is inversely proportional to the radius of sphericity (curvature) of the conductor. Indeed, if we consider a conductor of an arbitrary shape, its surface having various convex radii of curvature (two in the case presented in figure 5.7), where it may be equivalent with two spherical conductors having the radius R_1 and R_2, bounded by a wire from a conducting material, so both spheres are at the same potential, i.e., V_0, as that on the surface of the conductor:

$$V_1 = V_2 = V_0$$

$$\frac{q_1}{4\pi\varepsilon_0 R_1} = \frac{q_2}{4\pi\varepsilon_0 R_2}$$

$$\frac{\sigma_1 4\pi R_1^2}{4\pi\varepsilon_0 R_1} = \frac{\sigma_2 4\pi R_2^2}{4\pi\varepsilon_0 R_2} \tag{5.11}$$

$$\frac{\sigma_1}{\sigma_2} = \frac{R_2}{R_1} < 1 \Rightarrow \sigma_1 R_1 = \sigma_2 R_2 = \text{const.} \Rightarrow \sigma_1 = \frac{K}{R_1}$$

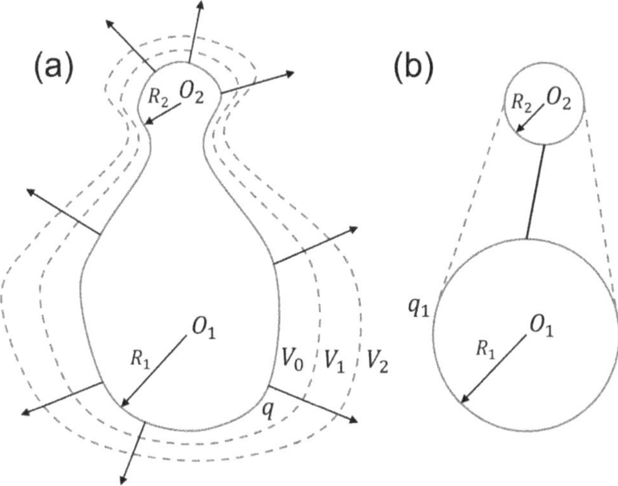

Figure 5.7. Surface of a conductor at electrostatic equilibrium presenting two different radii of curvature.

The electric charge density is higher where the conductor is sharper. Therefore, in general the charge should be denser where the convex radius of curvature is smaller and particularly at points. In other words, Green's theorem states that: *the charge density onto the surface of a conductor is proportional at any point of surface with the derivative of potential in the direction of the normal to the surface at that point.* Indeed, we know that the electric field at a point of the surface is:

$$\vec{E} = -\nabla V = \frac{\sigma}{\varepsilon_0}\vec{n} \Rightarrow \sigma = -\varepsilon_0 \nabla V \vec{n} = -\varepsilon_0 \frac{\partial V}{\partial n} \tag{5.12}$$

This is equivalent to saying that the surface charge density onto a conductor is larger where the equipotential surfaces are dense, or the surface charge density is inversely proportional to the distance from the conductor surface to the next equipotential surface. This property is known as '*the apex power*' and it consists in the fact that the charge density is greater at sharper corners of a conductor where also the electric field strength is large enough to cause the ionization of the air around them: ions of opposite sign are then attracted to the conductor and discharged while those of the same sign are repelled and cause a detectable wind. This property is used for building the **lightning rod** which protects buildings against thunderbolts. The linkage of the charge induced on the metallic roof of a building by a big charged cloud, can equal the potentials of a cloud and then protect the building against a thunderbolt. Even though the current throughout a corona discharge of a lightning rod is very small, i.e., about 10^{-9} A (while the current throughout a thunderbolt is $\sim 10^5$ A), there is a very short time of discharge (10^{-6} s) and so the lightning rod is efficient.

5.3 Force on the surface of a charged conductor. Electrostatic pressure

Let there be a surface element ΔS, centered to the point M placed on the surface of a charged conductor to electrostatic equilibrium, carrying the charge distributed on the surface with surface charge density σ (see figure 5.8(a)). The charge element

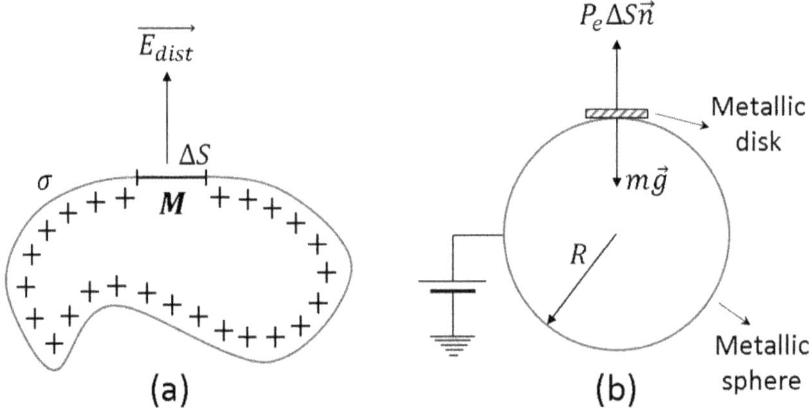

Figure 5.8. Figures showing the presence of the electrostatic pressure.

$\Delta q = \sigma \Delta S$, on the surface element ΔS, is placed in the electric field $\overrightarrow{E_{\text{dist}}} = +\frac{\sigma}{2\varepsilon_0}\overrightarrow{n}$, created by the total charge on the surface of the conductor excepting that on the surface element ΔS. The charge element will be experienced by the electric force:

$$\Delta\overrightarrow{F} = \Delta q \cdot \overrightarrow{E_{\text{dist}}} = \sigma\Delta S\,\overrightarrow{E_{\text{dist}}} \tag{5.13}$$

If outside the conductor there is another electric field, this will act on Δq too, but here, in equation (5.13), we will consider just the force given by the electric field $\overrightarrow{E_{\text{dist}}} = +\frac{\sigma}{2\varepsilon_0}\overrightarrow{n}$, then:

$$\Delta\overrightarrow{F} = \sigma\Delta S\,\overrightarrow{E_{\text{dist}}} = \sigma\Delta S\frac{\sigma}{2\varepsilon_0}\overrightarrow{n} = \frac{\sigma^2}{2\varepsilon_0}\Delta S\overrightarrow{n} \tag{5.14}$$

leading to the electrostatic pressure:

$$P_e = \frac{\Delta F}{\Delta S} = \frac{\sigma^2}{2\varepsilon_0} \tag{5.15}$$

Taking into account the relationship for the normal component of the electrostatic field to the surface of the conductor, $\overrightarrow{E} = \frac{\sigma}{\varepsilon_0}\overrightarrow{n}$, the electrostatic pressure can be rewritten as a function of the electric field on the surface of the conductor:

$$P_e = \frac{\sigma^2}{2\varepsilon_0} = \frac{\varepsilon_0 E^2}{2} \tag{5.16}$$

This force is strictly exerted on the charges on the surface, but because emission of charge does not normally occur, there must be a balancing non-electrostatic force exerted by the conductor on the charges. This causes the force to be communicated to the conductor itself by **Newton's third law**[2]. At sharp points, where σ is very large, **the force can be great enough** to cause charges to leave the surface and **field emission**[3] is then said to occur.

Experimentally, the electrostatic pressure can be measured using a sphere and a disk, both from a conducting material, which are brought in contact, see figure 5.8(b). The disk is very small and makes very close contact with the sphere. Charging the sphere with a *Van de Graaff* electrostatic generator[4], we could observe

---

[2] For every action, there is an equal and opposite reaction. This statement means that in every interaction, there is a pair of forces acting on the two interacting objects. The size of the force on the first object equals the size of the force on the second object. The sense of the force on the first object is opposite to the sense of the force on the second object. In other words, forces always come into pairs—equal and opposite action–reaction force pairs.

[3] Emission of electrons induced by a strong electrostatic field. The most common context is the electron field emission from a solid surface into a vacuum, phenomena exploited for generating the electron beam within the high resolution field emission - scanning electron microscope (FE-SEM).

[4] The device uses a moving belt to accumulate electric charge on a hollow metal globe on the top of an insulated column, creating very high electric potentials. Thus, it produces very high voltage at low direct current (DC) levels. It was invented by American physicist *Robert J Van de Graaff* in 1929.

that at a certain value of charge on the sphere, the disk is going up. In this moment the electric force is greater than or equal to the weight of the disk.

$$P_e \Delta S \geqslant mg \Rightarrow \frac{\sigma^2}{2\varepsilon_0}\Delta S \geqslant mg \tag{5.17}$$

If the radius of disk is r and of sphere R, it results for the surface charge density:

$$\sigma = \sqrt{\frac{2mg\varepsilon_0}{\pi r^2}} \tag{5.18}$$

Now, the total charge of the sphere is:

$$Q = \sigma 4\pi R^2 = \frac{R^2}{r}4\pi\sqrt{\frac{2mg\varepsilon_0}{\pi}} \tag{5.19}$$

and the potential of the sphere is:

$$V = \frac{Q}{4\pi\varepsilon_0 R} = \frac{R}{r}\sqrt{\frac{2mg}{\pi\varepsilon_0}} \tag{5.20}$$

Considering, $R = 2$ cm, $r = 2$ mm, $m = 0.02$ g $\Rightarrow Q = 8.36 \times 10^{-8}$ C and $V = 3.76 \times 10^4$ V, we can see that the electrostatics is in the range of very small charges and very high potential.

5.4 The energy of a charged conductor

The electrostatic pressure expresses the force with which the charge on the charged conductor repel the charge on the unit surface from the conductor. If a ball conductor built from an elastic material is charged, the volume of the conductor increases during the charging process. Reversely, to reduce the volume of the elastic ball conductor, keeping the total charge constant, an external work must be done on the conductor. Supposing we want to compress a charge distribution on a sphere of radius r_0, to one of the radius $r_0 - dr$ (see figure 5.9), the work can be done by the

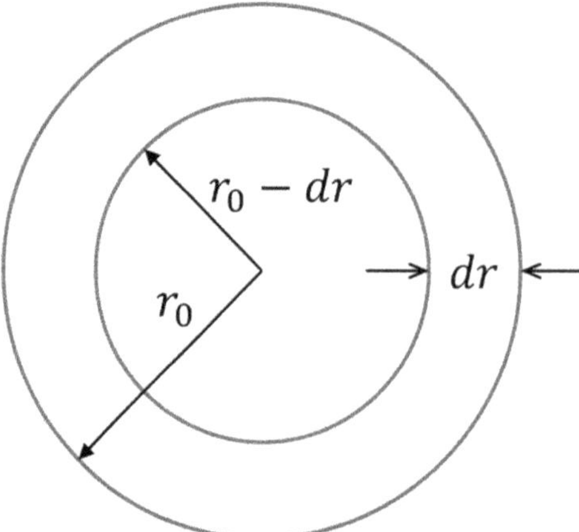

Figure 5.9. Charge distribution compressed on an elastic and conducting sphere.

application on each unit area on the surface sphere of a force $\frac{\sigma^2}{2\varepsilon_0}$, oriented toward the inner conductor. This force moves its own application point along the distance dr, then the work is:

$$dW = \frac{\sigma^2}{2\varepsilon_0}4\pi r_0^2 dr = \frac{\sigma^2}{2\varepsilon_0}dv = \frac{1}{2}\varepsilon_0 E^2 dv \tag{5.21}$$

Defining the energy density as:

$$w = \frac{1}{2}\varepsilon_0 E^2 \tag{5.22}$$

the work done will be:

$$dW = wdv = \frac{1}{2}\varepsilon_0 E^2 dv = \frac{1}{2}\varepsilon_0\frac{\sigma^2}{\varepsilon_0^2}dv = \frac{Q^2 dr}{8\pi\varepsilon_0 r_0^2} \tag{5.23}$$

This work determined the changing of the electric field from the value $E_{\text{int}} = 0$, inside of the sphere, to the value, $E_{\text{ext}} = \sigma/\varepsilon_0$, on its external surface, and this work done to build the charge distribution dW, is stored as energy in all the space occupied by the electric field. According to the equation (5.23), the total energy store in the electrostatic field can be re-written as a function of its own intensity as:

$$W = \frac{1}{2}\int_V \varepsilon_0 E^2 dv \tag{5.24}$$

with the integral extended on the whole volume containing the electric field.

Theorem 5.4.1. *The potential energy W of a system of charges (representing the work done to form the system) can be computed as volume integral of the energy density $w = \frac{1}{2}\varepsilon_0 E^2$ on the whole space where the electric field is.*

Then the work done to charge the initial sphere with the charge Q is:

$$W = \int\limits_{\substack{\text{on the whole}\\\text{space}}} \frac{\varepsilon_0}{2}E^2 dv = \frac{\varepsilon_0}{2}\int\limits_{r_0}^{\infty}\left(\frac{Q}{4\pi\varepsilon_0 r^2}\right)^2 4\pi r^2 dr = \frac{Q^2}{8\pi\varepsilon_0 r_0} \tag{5.25}$$

energy which is found in the whole space around the sphere (from the surface sphere to infinity), which is why sometimes this is called 'the energy stored' in the electric field. Having a conservative system, this energy can be transferred outside of the system. The potential energy can also be expressed as a function of the electrostatic potential $V(x, y, z)$ of a charge distribution as:

$$W = \frac{\varepsilon_0}{2}\int (\nabla V)^2 dv \tag{5.26}$$

5.5 Theorem of the corresponding elements

Theorem 5.5.1. *At electrostatic equilibrium, the charges distributed on the corresponding elements of the surfaces of two conductors in the electrostatic influence, are equal but of opposite sign.*

Let there be two conductors A and B, in reciprocal electrostatic influence (see figure 5.10). Supposing that on conductor A is the total charge q, then the surface element dS bounded by a tube of lines of field will carry the charge $dq = \sigma dS$. Extending the tube of lines starting from A until conductor B, it will determine on B the surface element dS' (called the corresponding element of dS on A), an element which carries the influence charge $dq' = \sigma' dS'$. If we close the line tube with the surfaces drawn with a dashed line in figure 5.10, to generate a close surface Σ which will contain the corresponding elements dS and dS', on conductors A and B, respectively, applying *Gauss's Law* results in:

$$\Phi = \int_\Sigma \vec{E}\,\overrightarrow{dS} = \frac{dq + dq'}{\varepsilon_0} = \frac{\sigma dS + \sigma' dS'}{\varepsilon_0} = 0 \qquad (5.27)$$

The total flux through Σ is zero due to the fact that the flux through the lateral area of the line tube is zero and the flux through each base is zero of course, the electric field inside of each conductor, being everywhere zero. From equation (5.27) results:

$$\sigma dS = -\sigma' dS' \qquad (5.28)$$

representing the mathematical relationship of the enounced theorem. Particularly, if $dS = dS' \Rightarrow \sigma = -\sigma'$. In the case when around conductor A there are many conductors in electrostatic influence, it can be easily demonstrated that:

$$q = -\Sigma q' \qquad (5.29)$$

In other words, an electric charge q induces the total charge $\Sigma q' = -q$, i.e., equal but of opposite sign with the charge of inductor conductor.

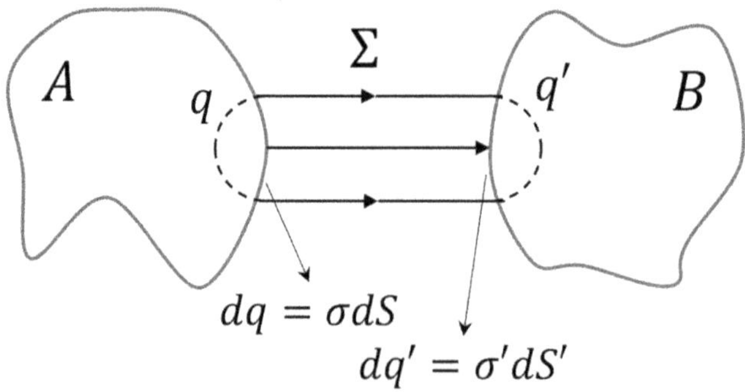

Figure 5.10. Theorem of the corresponding elements.

5.6 Electric field in a hollow conductor. Electrostatic shielding

We know already that in the material of a conductor carrying static charge only, $\vec{E}$ is everywhere zero. Any Gaussian surface drawn entirely within conducting material can therefore have no flux of $\vec{E}$ crossing it and must contain zero charge. A conductor with no interior surface (see figure 5.11(a)) therefore **carries all its excess charge on the surface**, while a hollow one (see figure 5.11(b)) **carries all its excess charge on the outer surface**, unless there is a charge $+Q$ on a body A placed in the hollow (see figure 5.11(c)) when a charge of $-Q$ must be induced on the inner surface. If the conductor in figure 5.11(c) is uncharged as a whole, a charge $+Q$ must be induced on the outer surface whatever the position of A. If A is a conductor and is allowed to touch the inside, the induced $-Q$ and $+Q$ on A neutralize each other leaving the outer $+Q$ unaffected.

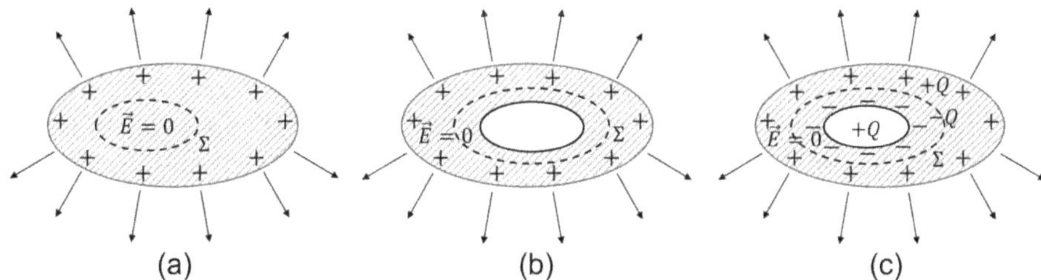

Figure 5.11. Distribution of charge on conductors: (a) a full conductor (with no interior surface); (b) a hollow conductor without any charge inside of hollow; (c) hollow conductor with a charged body A carrying the charge $+Q$, inside the hollow. Σ in all figures is a Gaussian surface.

We can go further. The inner surface S (see figure 5.12) of a charged hollow conductor is an equipotential. Suppose that it is possible for an electric field $\vec{E}$ to exist in the hollow with no charge placed in it. Then there would be points inside at

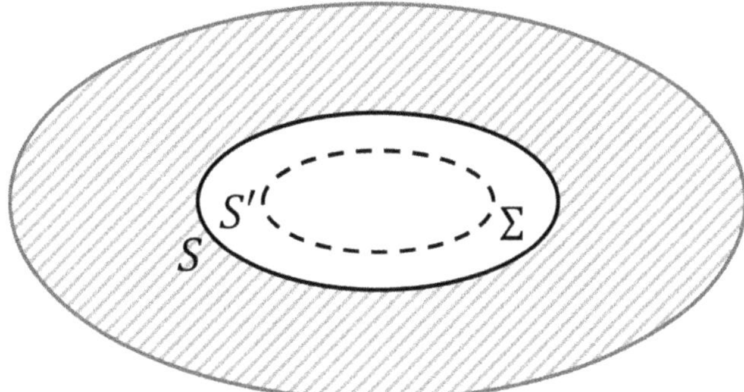

Figure 5.12. Absence of electric field inside a hollow conductor.

potentials different from that of S. If all points just inside S are at higher potential, then an equipotential surface like S' exists and line of force between S and S' are all outwards; the flux of E over S' is not zero and a charge exists within it, which is self-contradictory. If all points just inside S are at lower potential, a similar argument applies. We can only conclude that $\vec{E}$ **is zero at all points within a hollow conductor not containing charge** and that the interior is an equipotential volume at the same potential as the conductor: **all this is irrespective of happenings outside**. That charge residing only on the outer surface of conductors was suggested in 1729 by *Stephen Gray* who showed that a solid and hollow cube of the same size produced the same electrical effects at distant points when charged in the same way: the absence of **the inside** made no difference. The results deduced above have two important applications whose success verifies the validity of our arguments. **Firstly**, a hollow conductor can be given an indefinitely large charge by conveying it to the interior in successive small amounts. **Secondly**, any apparatus placed inside a hollow conductor whose potential is fixed, is unaffected by external electric fields: this is the principle of *electrostatic shielding*. We must take note here, that an isolated hollow conductor whose potential is fixed, is only a unilateral electrostatic shielding, because it shields the inside by the external electrostatic actions but not also the outside by the internal actions. The electric state outside of the hollow conductor is dependent on the electric state inside of the hollow of the conductor, because whatever will be the value and position of the charge q_c inside the hollow, its effect will be transferred outside of the conductor (see figure 5.13(a)), according to Gauss's Law:

$$\int_{\Sigma} \vec{E}\, dS = \frac{q_c - q_c + q_c}{\varepsilon_0} = \frac{q_c}{\varepsilon_0} \neq 0 \tag{5.30}$$

In this case the potential of conductor rests constant and different from zero, even though it is neutral from electrical point of view. If a hollow conductor is **connected to earth (grounded)**, its potential will be zero and its excess charge on the outer

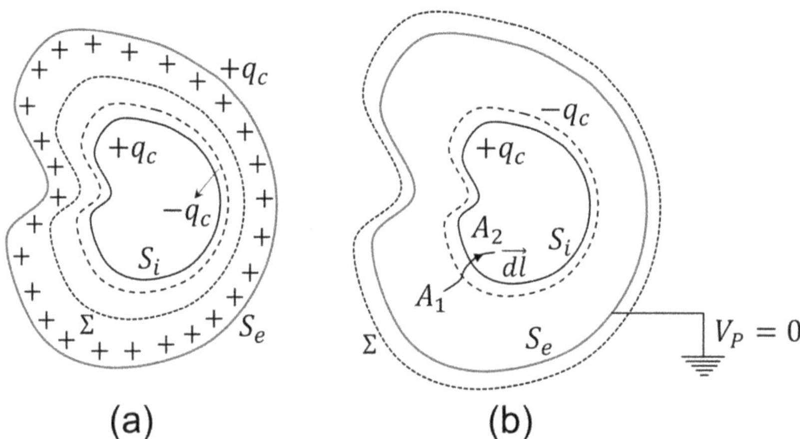

Figure 5.13. (a) An isolated hollow conductor. (b) A grounded hollow conductor.

surface q_c, flows to the Earth. Even though any charge is placed inside of the hollow conductor, the external electric field is unaffected. As shown in figure 5.13(b), the flux of $\vec{E}$ through a Gaussian surface drawn outside of the hollow conductor is zero because it contains zero charge:

$$\int_{\Sigma} \vec{E}\,dS = \frac{q_c - q_c}{\varepsilon_0} = 0 \tag{5.31}$$

Just as in an isolated hollow conductor, the electrostatic state inside of an earth-bounded hollow conductor is unaffected by the external electric fields. That is why an earth-bounded hollow conductor is a bilateral electrostatic shielding. It shields both the electrostatic state outside by the internal electric field and also the electrostatic state inside of the hollow by external electric fields. One of its applications is the *Faraday cage*, very useful in the measurements of very small signals[5].

5.7 Simple systems of conductors

In this section, we shall investigate a few particularly simple arrangements of conductors. In each case, we will compute the electric field outside and inside the conductors, as well as the potential in different regions of the conductors.

Two concentric spheres carrying the charges Q_1 and Q_2

Let's consider two concentric metal spheres, of radius R_1 and R_2, carrying total charges Q_1 and Q_2, respectively (see figure 5.14). The electric field inside of the inner sphere ($\vec{E_I}$) is zero. The electric field between the two spheres is:

$$\int_{\Sigma_{II}} \vec{E_n}\,\vec{dS} = \frac{Q_2}{\varepsilon_0} \Rightarrow \vec{E_{II}} = \frac{Q_2\,\vec{r}}{4\pi\varepsilon_0 r^3} \tag{5.32}$$

The electric field outside the larger sphere is given also by Gauss's law:

$$\int_{\Sigma_{III}} \vec{E_{III}}\,\vec{dS} = \frac{Q_1 + Q_2}{\varepsilon_0} \Rightarrow \vec{E_{III}} = \frac{Q_1 + Q_2}{4\pi\varepsilon_0 r^3}\,\vec{r} \tag{5.33}$$

Hence, the potentials in the third region will be:

$$V_{III} = -\int_{\infty}^{r} \vec{E_{III}}\,\vec{dr} = \frac{Q_1 + Q_2}{4\pi\varepsilon_0 r} \tag{5.34}$$

then the potential on the surface of the outer sphere is:

$$V_1 = \frac{Q_1 + Q_2}{4\pi\varepsilon_0 R_1} \tag{5.35}$$

---

[5] A *Faraday cage* or *Faraday shield* is an enclosure made up of a conducting material, used to block electromagnetic fields. It is named after the scientist *Michael Faraday* who invented it in 1836.

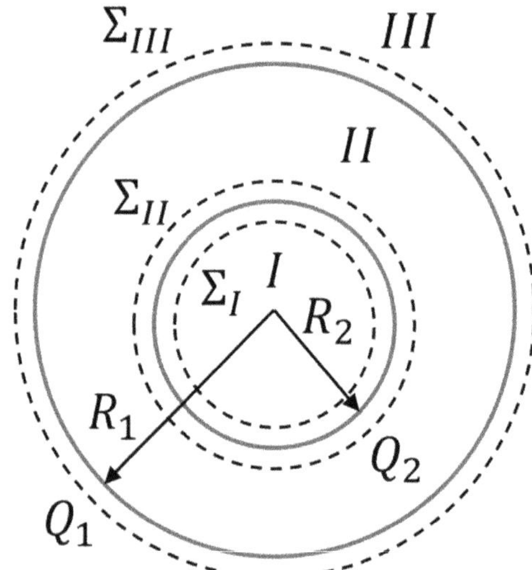

Figure 5.14. Two conducting concentric spheres carrying charges Q_1 and Q_2.

The potential in the second region (between the charged spheres) will be:

$$V_{II} = -\int_{\infty}^{R_1} \overrightarrow{E_{III}}\,\overrightarrow{dr} - \int_{R_1}^{r} \overrightarrow{E_{II}}\,\overrightarrow{dr} = \frac{Q_1 + Q_2}{4\pi\varepsilon_0 R_1} - \int_{R_1}^{r} \frac{Q_2}{4\pi\varepsilon_0 r^2}dr$$

$$= \frac{Q_1 + Q_2}{4\pi\varepsilon_0 R_1} - \frac{Q_2}{4\pi\varepsilon_0}\left(\frac{1}{R_1} - \frac{1}{r}\right) = \frac{Q_1}{4\pi\varepsilon_0 R_1} + \frac{Q_2}{4\pi\varepsilon_0 r}$$

(5.36)

Then the potential on the inner sphere which is also the potential at all points inside it will be:

$$V_2 = \frac{Q_1}{4\pi\varepsilon_0 R_1} + \frac{Q_2}{4\pi\varepsilon_0 R_2}$$

(5.37)

If the spheres carried equal and opposite charges $Q_1 = -Q_2$, only the space between them would have a non-vanishing electric field, generating a spherical capacitor.

A point charge near a conducting plane

The simplest system in which the mobility of the charges in the conductor makes itself evident is the point charge near a conducting plane. Let's consider a point charge Q situated at a perpendicular distance h from an infinite plane conducting surface which we shall take as a zero potential (see figure 5.15(a)). There is axial symmetry about the line QO, but this only means that the surface charge density and the electric field at point P on the conducting plane will be the same as those at any other point on the plane surface at distance r from O (as a result of $\overrightarrow{E} = -\nabla V$ and $V = \text{constant} = 0$ on the conducting plane), see figure 5.15(b). Now we look at the

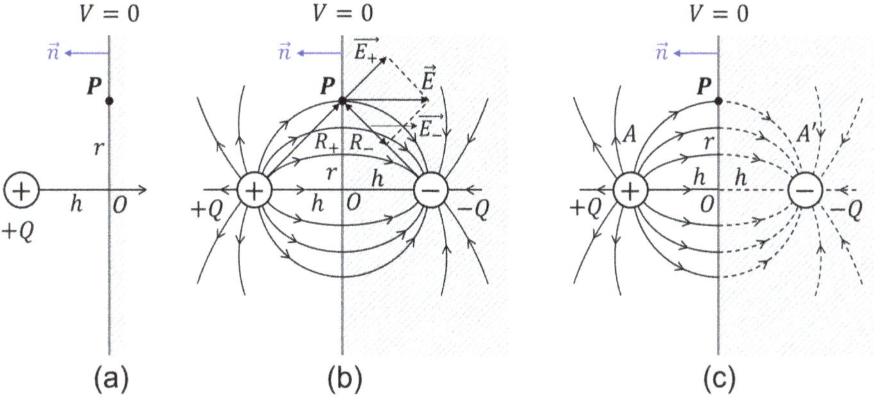

Figure 5.15. (a) A point charge and an infinite conducting plane; (b) a point charge and an image $-Q$ in the plane; (c) lines of $\vec{E}$ (in full) due to a point charge and conducting plane: the dashed lines do not exist.

lines of force and equipotential of a system of two equal and opposite point charges which forms an electric dipole. The equipotential surface with zero potential between the two equal positive and negative charges (see figure 5.15(b)) is the mediator plane of the line $+Q, -Q$, joining the two charges. The field lines near each charge point radially, but at the mediator plane the lines are perpendicular to the mediator plane. So it is just like in the case of a point charge Q located at h distance from the conducting plane (see figure 5.15(c)). In figure 5.15(c) the region to the left of the $V = 0$ equipotential conducting plane is a region of space containing a charge Q situated at perpendicular distance h from an infinite plane of zero potential, exactly the situation to the left of the conducting surface (see figure 5.15(a)). The problem for this region is the same in the two cases and would be expected to give the same lines of force and equipotential. Thus, to solve the given problem we can replace the plane by a charge $-Q$ known as the image charge since it is the same point as the image of Q would be in a plane mirror through OP. At point P in figure 5.15(b), the electric field is:

$$\vec{E} = \vec{E_+} + \vec{E_-} = \frac{Q}{4\pi\varepsilon_0 R^3}\left(\vec{R_+} - \vec{R_-}\right) = \frac{-Q2h}{4\pi\varepsilon_0(h^2 + r^2)^{3/2}}\vec{n} \tag{5.38}$$

and because the electric field on the surface of a conducting plane having at point P the charge density σ is:

$$\vec{E} = \frac{\sigma}{\varepsilon_0}\vec{n} \tag{5.39}$$

resulting that:

$$\sigma_p = -\frac{2hQ}{4\pi(h^2 + r^2)^{3/2}} \tag{5.40}$$

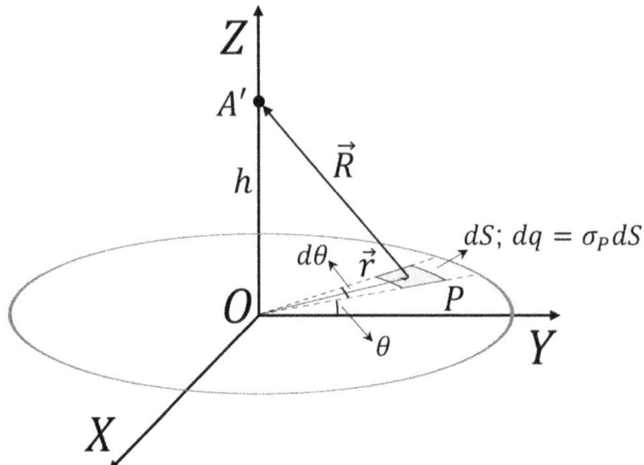

Figure 5.16. The total charge induced by a point charge Q to the surface of an infinite surface conducting plane.

representing the total surface charge ought to amount to $-Q$. Just as a check, we could integrate over the surface and see if it is. Indeed, considering a surface element located at point P in a cylindrical coordinate system (see figure 5.16), it carries the elementary charge:

$$dq = \sigma_p dS \Rightarrow Q_{\substack{\text{all}\\ \text{surface}}} = \int_{\substack{\text{all}\\ \text{surface}}} \sigma_p dS = \int_0^\infty \int_0^{2\pi} -\frac{2Qhr\,dr\,d\theta}{4\pi(h^2+r^2)^{3/2}}$$

$$= -\frac{Qh2\pi}{4\pi}\int_0^\infty \frac{2r\,dr}{(r^2+h^2)^{3/2}} = -\frac{Qh}{2}\int_0^\infty \frac{d(r^2+h^2)}{(r^2+h^2)^{3/2}} \qquad (5.41)$$

$$= -\frac{Qh}{2}\left[-2\frac{1}{\sqrt{r^2+h^2}}\right]\Bigg|_0^\infty = -\frac{Qh}{2}2\left[\frac{1}{h}-0\right] = -Q$$

The method of solution used here has traditionally been called 'the **method of images**'. One thinks of the fictive negative charge located at distance h bellow the plane of a conductor, toward which the field lines appear to plunge, as the 'image' of the point charge Q, something like the virtual image behind a mirror. The electric force which acts on the charge Q, owing to the attraction of the surface charge is equal to the force that a charge $-Q$ in the image position would cause. Note that the actual origin of this force is the surface charge. The force between Q and plane will be:

$$\vec{F} = \frac{-Q^2}{16\pi\varepsilon_0 h^2}\vec{n} \qquad (5.42)$$

The general method of images thus involves replacing conducting surfaces by systems of charges which, together with the given charges, produce equipotential

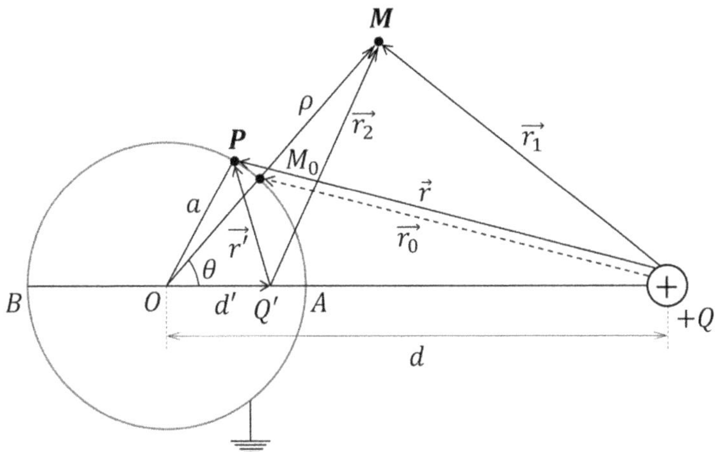

Figure 5.17. Image method in the case of a grounded sphere of radius a, in the presence of a point charge Q, placed at the distance d with respect to the center of the sphere.

surfaces of the correct V identical with those of the original conductors. The field and forces can then be calculated at any point and the surface densities of charge obtained from $\vec{E}$ as above. The 'image' system is not necessarily the same as that in the equivalent optical system, as in the case of a point charge Q at a distance d from the center O of a grounded conducting sphere of radius a, where an image charge $Q' = -\frac{Qa}{d}$ there is at the distance $d' = a^2/d$ from O. Indeed, the potential at a point P on the sphere due to Q charge and its image Q' (see figure 5.17) is:

$$V_p = \frac{Q}{4\pi\varepsilon_0 r} + \frac{Q'}{4\pi\varepsilon_0 r'} = 0 \tag{5.43}$$

At point A, the condition (5.43) leads to the equation:

$$\frac{Q}{d - a} + \frac{Q'}{a - d'} = 0 \tag{5.44}$$

and at point B, to the equation:

$$\frac{Q}{d + a} + \frac{Q'}{d' + a} = 0 \tag{5.45}$$

Consequently:

$$\frac{d + a}{d - a} = \frac{d' + a}{a - d'} \Rightarrow ad + a^2 - dd' - ad' = dd' - ad' + ad - a^2 \Rightarrow d' = \frac{a^2}{d} \tag{5.46}$$

Thus:

$$Q' = -Q\frac{a + d'}{a + d} = -Q\frac{a + \dfrac{a^2}{d}}{a + d} = -Q\frac{a(a + d)}{d(a + d)} = -Q\frac{a}{d} \Rightarrow Q' = -Q\frac{a}{d} \tag{5.47}$$

The potential at any point M outside the conducting sphere will be:

$$V_M = \frac{Q}{4\pi\varepsilon_0 r_1} - \frac{Qa}{d\,4\pi\varepsilon_0 r_2} \Rightarrow \overrightarrow{E_M} = -\nabla V_M = \frac{Q}{4\pi\varepsilon_0 r_1^3}\vec{r_1} - \frac{\frac{a}{d}Q}{4\pi\varepsilon_0 r_2^3}\vec{r_2} \tag{5.48}$$

$$V_M = \frac{Q}{4\pi\varepsilon_0}\left[\frac{1}{(\rho^2 + d^2 - 2d\rho\cos\theta)^{1/2}} - \frac{a}{d(\rho^2 + d'^2 - 2\rho d'\cos\theta)^{1/2}}\right] \tag{5.49}$$

The charge density at the point P on the sphere will be:

$$\sigma = -\varepsilon_0 \frac{\partial V}{\partial\rho}\bigg|_{\rho=a} = -\frac{Q}{4\pi}\frac{d^2 - a^2}{ar_0^3} \quad \text{with} \quad r_0 = \sqrt{d^2 + a^2 - 2ad\cos\theta} \tag{5.50}$$

Further reading

[1] Secăreanu I, Ruxandra V, Gherbanovschi N, Logofătu M, Cazan-Corbasca M and Antohe Ş 1984 *Problems of Electricity and Magnetism (Culegere de Probleme de Electricitate şi Magnetism)* (Bucharest: University of Bucharest Publishing House)

[2] Jackson J D 1998 *Classical Electrodynamics* 3rd edn (New York: Wiley)

[3] Panofsky W K H and Phillips M 2005 *Classical Electricity and Magnetism* 2nd edn (New York: Dover (Dover Books on Physics))

[4] Feynman R P, Leighton R B and Sands M 2011 *The Feynman Lectures on Physics, Vol II: The New Millennium Edition: Mainly Electromagnetism and Matter* (New York: Basic Books)

[5] Purcell E M and Morin D J 2013 *Electricity and Magnetism* 3rd edn (Cambridge: Cambridge University Press)

[6] Peck E R 2013 *Electricity and Magnetism* (New York: Dover)

[7] Pierrus J 2018 The electrostatics of conductors *Solved Problems in Classical Electromagnetism: Analytical and Numerical Solutions with Comments* (Oxford: Oxford University Press)

[8] Agrawal N 2020 *Electrostatics: Current and Capacitors (for IIT-JEE)* (Independently Published)

Electrostatics
Formalism of the electrostatic field in vacuum and matter
Ştefan Antohe and Vlad-Andrei Antohe

Chapter 6

Capacitance and capacitors

The first known practical realization of a ***capacitor***, dates back to 1745 from Germany, when *Ewald Georg von Kleist of Pomerania*[1] found that electric charge could be stored by connecting a high-voltage electrostatic generator through a wire to a volume of water in a hand-held glass jar [5]. The scientist's hand and the water acted as conductors, while the jar was the dielectric medium. Physicists also observed that touching the wire resulted in an electric spark, that proved to be much more powerful than the one obtained from an electrostatic machine. However, the first capacitor in the world is considered to be the '*Leyden jar*' invented one year later by *Pieter van Musschenbroek*, professor at the Leiden University, the Netherlands. Although how it works was not well-understood at the time, physicists discovered that the electric charge could be stored even after the jar was disconnected from the generator. The effect produced by a capacitor is known as ***capacitance***. Initially, there was no practical use for the Leyden jar, except allowing scientists to do various electricity-related experiments. For example, *Benjamin Franklin* proved in 1752 for the first time in his famous kite flying experiment[2] that lightning is electricity, leading further to the invention of the lightning rod.

6.1 Capacitance of an insulated conductor

In a finite system of charged conductors, the algebraic sum of the charges from the system can be zero or different to zero. A system with a finite number of charged bodies having the total charge of the system equal to zero, is called 'complete system' and it is characterized by the fact that all the lines of the electric field starting from

---

[1] Ewald Georg von Kleist of Pomerania (10 June 1700–11 December 1748) was a German jurist, Lutheran cleric, and physicist.
[2] In the kite experiment, a kite with a conductive wire attached to its apex is flown near thunder clouds to collect electricity from the air and conduct it down the wet kite string to the ground. The experiment's purpose was to demonstrate that lightning and electricity were the result of the same phenomenon.

the positive charges end on the negative charges inside of the system and any line of the electric field does not escape outside of the system. In this case the total flux of the electric field through a closed surface, enclosing the system, is zero. If the total charge of the system of charged bodies is different to zero, the system is called 'incomplete', in this case remaining open lines of field ending to infinity. Nevertheless, it can be considered a large enough surface carrying an electric charge equal to that of a finite system but of opposite sign, in such a way that the 'incomplete system' behaves like a 'complete' one.

The most simple 'complete system' is formed from two conductors of arbitrary shape placed at a large distance from each other and carrying equal charges but of opposite sign ($q_1 > 0$ and $q_2 = -q_1 < 0$), see figure 6.1. In this case all the lines of the electric field starting from the conductor charged with positive charge end on the negatively charged conductor, without losses of the lines to infinity. The potentials on the surface of each conductor are constant having the values V_1 and V_2, with $V_1 > V_2$.

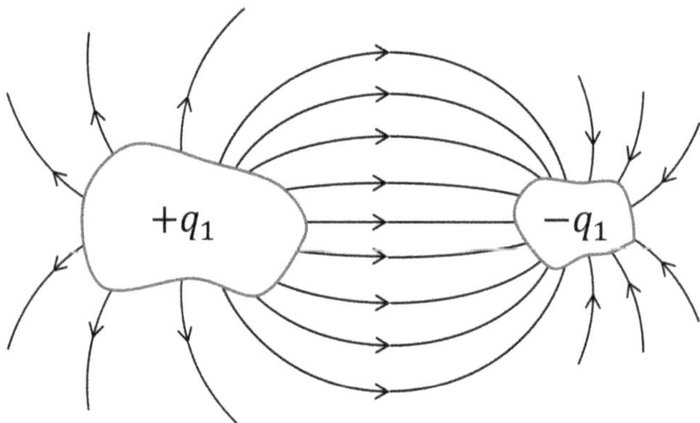

Figure 6.1. Charged conductors with equal but opposite sign charges.

Considering now a charged conductor A carrying the positive charge $+Q$, surrounded by a system of conductors B, C, D, ..., bounded between them by conducting wires and connected to the ground (see figure 6.2), the potential of each of them will be zero and each of them will carry the negative charge: $-q_B$, $-q_C$, $-q_D$. Having a complete system of charged bodies, the charge conservation law can be applied, resulting in:

$$Q - q_B - q_C - q_D = 0 \quad \text{or} \quad Q = q_B + q_C + q_D \tag{6.1}$$

The distribution of charges q_B, q_C, q_D on the surface of each conductor will depend on the geometry of each surface, on the relative positions between the conductors and on the electrical properties of the surrounding medium. The surface of the conductor A is an equipotential surface with the potential V_A which depends on the electric charge present on the surface of each conductor from the system. The surface element dS on the conductor A carries the charge dq placed at distance $\vec{r_i}$ ($i = A$, B, C, D) on the surface of each conductor A, B, C, D, and then, according to the

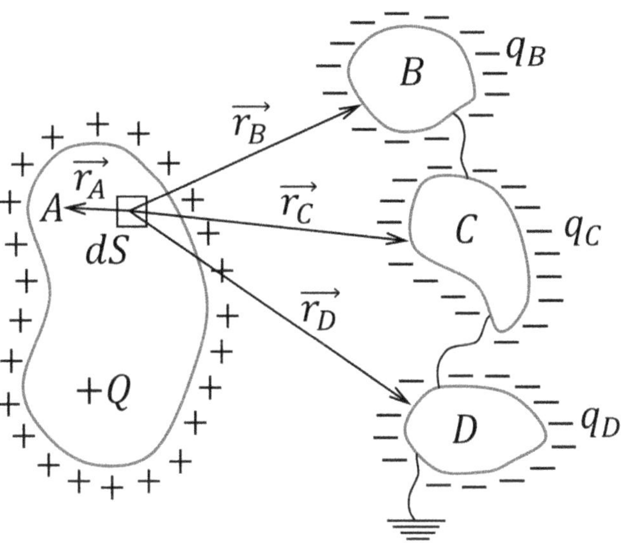

Figure 6.2. Complete system of conductors to maximum electrostatic influence.

superposition principle (see section 2.2) and the theorem of the corresponding elements (see section 5.5), the potential of the conductor A will be:

$$V = \frac{1}{4\pi\varepsilon}\left(\int_{S_A} \frac{dq}{r_A} + \int_{S_B} \frac{dq}{r_B} + \int_{S_C} \frac{dq}{r_C} + \int_{S_D} \frac{dq}{r_D} \right) \tag{6.2}$$

If the charge on the conductor A increases n times, in this new equilibrium state it will have the charge nQ, and then, according to the theorem of the superposition of the equilibrium states, the influence charges on the surfaces of conductors B, C, D, (fixed in the same positions), will be: $-nq_B$, $-nq_C$, $-nq_D$. Keeping the distances $\vec{r_i}$ ($i = A, B, C, D$) constant and using the relation (6.2) the potential of the conductor A in the new equilibrium state will be nV.

Let's try answering the question: '*What is the relation between the charge Q carried by an isolated charged conductor (in our case conductor A) and its potential V at electrostatic equilibrium?*' It is easy to observe that the ratio between the charge Q and its potential V, at any equilibrium state is constant and this ratio is called the '***capacitance of the insulated conductor***':

$$C = \frac{Q}{V} = \frac{nQ}{nV} \tag{6.3}$$

When a finite conductor remote from other bodies is given a charge Q, it's potential V (zero at infinity) increases in magnitude if Q increases, and the ratio is said to be the '***capacitance of the conductor***' (C), so that:

$$C = \frac{Q}{V} \text{ (definition of } C) \tag{6.4}$$

An older name for C is ***capacity***, used because it gives the amount of charge the conductor holds at a certain potential. We can observe that the capacitance is a parameter which characterizes the geometry of a conductor being independent of Q or V, depending only on the size and shape of the conductor. The definition enables C to be calculated for some particular conductors, when we can easily calculate the potential on the surface of a charged conductor as a function of its charge Q.

In the case of a spherical conductor having the radius R, see figure 6.3, the potential on its surface, when the conductor has the charge Q, is:

$$V = -\int_{\infty}^{R} \vec{E}\,\overrightarrow{dr} = -\int_{\infty}^{R} \frac{Q\vec{r}\,\overrightarrow{dr}}{4\pi\varepsilon_0 r^3} = -\int_{\infty}^{R} \frac{Qdr}{4\pi\varepsilon_0 r^2} = \frac{Q}{4\pi\varepsilon_0 R} \tag{6.5}$$

and then its capacitance will be:

$$C = \frac{Q}{\dfrac{Q}{4\pi\varepsilon_0 R}} = 4\pi\varepsilon_0 R \tag{6.6}$$

But for other shapes an accurate calculation is not easy. The SI unit of capacitance is the C V^{-1} or ***Farad*** with the symbol F, although its submultiples are more convenient in practice (1 μF $= 10^{-6}$ F), because F is a very big unit as it results from the calculus of the capacitance of a hypothetical spherical conductor with the radius equal to that of Earth, $R_P = 6370$ km:

$$C = 4\pi\varepsilon_0 R_P = \frac{1}{9 \times 10^9} \times 6370 \times 10^3 = 707.7\,\mu\text{F} = 0.707\,\text{mF}$$

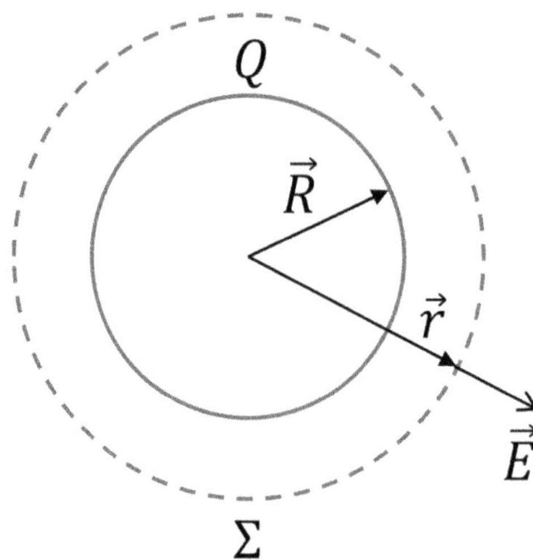

Figure 6.3. Capacitance of an insulated spherical conductor.

6.2 Capacitance and potential coefficients

Most of the practical situations in this chapter have concerned not more than two conductors and the surroundings, but in practice, in different applications, we need to have more complicated systems of conductors, which are in an electrostatic influence until the system reaches electrostatic equilibrium.

6.2.1 The influence coefficients of capacitance

Let there be a system of n static conductors A_1, A_2, ..., A_n, at electrostatic equilibrium, see figure 6.4. The potential of each conductor from the system is: V_1, V_2, ..., V_n and the total charge on each conductor (whatever will be its distribution on the surface of the conductor) is: Q_1, Q_2, ..., Q_n. The electric state of the system at electrostatic equilibrium could be described by the table:

$$
\begin{array}{llllllll}
\text{Conductors} - A: & A_1 & A_2 & \dots & A_i & \dots & A_n \\
\text{Potentials} - V: & V_1 & V_2 & \dots & V_i & \dots & V_n \\
\text{Charges} - Q: & Q_1 & Q_2 & \dots & Q_i & \dots & Q_n
\end{array}
\tag{6.7}
$$

The uniqueness theorem (see section 3.4) guarantees that with V_1, V_2 and V_n given, the electric field is determined throughout the system. It follows that the charges Q_1, Q_2 and Q_n on the individual conductors are likewise uniquely determined. According to the superposition theorem of the equilibrium states (STES), this final equilibrium state described by the parameters from the above can be reached by the superposition of the simplest equilibrium states.

Let there be first, that from the system of n conductors, just the A_1 conductor is kept to the potential $V_1 = 1$ V, in the rest, all the $n - 1$ conductors are kept at potential zero (are grounded). The lines of the electric field starting from the

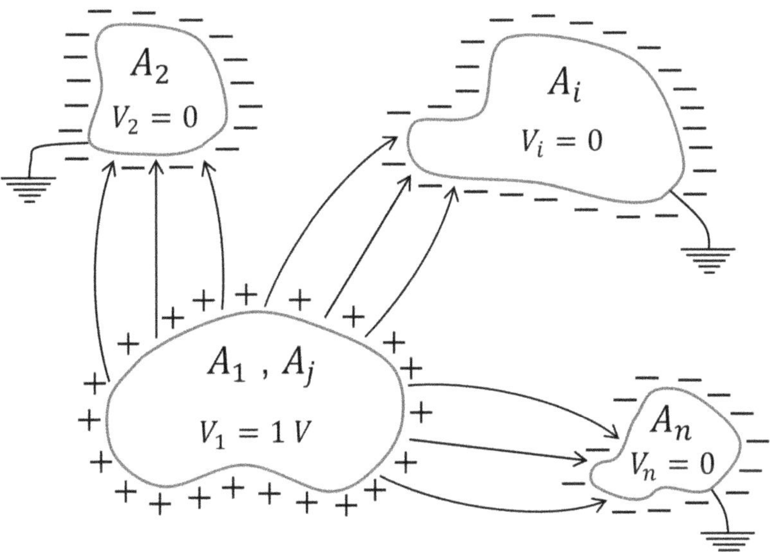

Figure 6.4. System of conductors at electrostatic equilibrium.

conductor A_1 end to the grounded conductors or to infinity, then just the A_1 conductor carries the positive charge, noted as C_{11}, in the rest all the other conductors carry negative charges (the positive influence charges flowing to the Earth), see figure 6.4. These charges will be noted as: C_{21}, C_{31}, C_{i1}, C_{n1} (the first index refers to the conductor influenced and the second one to the conductor having the potential $V_1 = 1$ V, giving rise to the electrostatic influence). The equilibrium state of the system in this case can be described by the parameters from the table below:

$$
\begin{array}{llllll}
A: & A_1 & A_2 & A_i & A_n \\
V: & 1 & 0 & 0 & 0 \\
Q: & C_{11} & C_{21} & C_{i1} & C_{n1}
\end{array}
\tag{6.8}
$$

where: $C_{11} = q_1$, $C_{21} = q_2$,..., $C_{n1} = q_n$ are the electric charges on each conductor when just the conductor A_1 is kept at the potential $V_1 = +1$ V. By the superposition of $m = V_1$ states on this type, will result in a new equilibrium state in which, according to STES, the potential of the conductor A_1 will be increased to the value V_1 and then all the charges from the system will increase V_1 times, then the new equilibrium state will be stated mathematically by:

$$
\left.\begin{array}{l}
\text{State } I \\
V_1 \neq 0 \\
V_2 = V_3... = 0
\end{array}\right\}
\begin{array}{llllllll}
A_1 & A_2 & A_3 & ... & A_j & ... & A_n \\
V_1 & 0 & 0 & ... & 0 & ... & 0 \\
C_{11}V_1 & C_{21}V_1 & C_{31}V_1 & ... & C_{j1}V_1 & ... & C_{n1}V_1
\end{array}
\tag{6.9}
$$

The n constants C_{11}, C_{21} and C_{n1} can depend only on the shape and arrangement of the conducting bodies. Only C_{11} is positive representing the charge on the conductor 1, kept at the potential V_1, and C_{21},..., C_{n1} are negative representing the charges induced on the 2, 3, ..., n conductors. In just the same way we could analyze states in which only $V_2 \neq 0$ and $V_1 = V_3 = ...V_n = 0$, calling such a condition state II. Again, we must find a linear relation between the only non-zero potential, V_2 in this case, and the various charges will be:

$$
\left.\begin{array}{l}
\text{State } II \\
V_2 \neq 0 \\
V_1 = V_3... = 0
\end{array}\right\}
\begin{array}{llllllll}
A_1 & A_2 & A_3 & ... & A_j & ... & A_n \\
0 & V_2 & 0 & ... & 0 & ... & 0 \\
C_{12}V_2 & C_{22}V_2 & C_{32}V_2 & ... & C_{j2}V_2 & ... & C_{n2}V_2
\end{array}
\tag{6.10}
$$

Intermediary, when only non-zero potential is V_j of the conductor j the others $n - 1$ conductors being held at zero, $V_1 = V_2 = ...V_n = 0$, the field and the charges are proportional to V_j and we can write:

$$
\left.\begin{array}{l}
\text{State } j \\
V_j \neq 0 \\
V_{i_{i \neq j}} = 0
\end{array}\right\}
\begin{array}{llllllll}
A_1 & A_2 & A_3 & ... & A_j & ... & A_n \\
0 & 0 & 0 & ... & V_j & ... & 0 \\
C_{1j}V_j & C_{2j}V_j & C_{3j}V_j & ... & C_{jj}V_j & ... & C_{nj}V_j
\end{array}
\tag{6.11}
$$

Finally, when the $n - 1$ conductors $i_{i \neq n}$ are held at zero, $V_{i_{i \neq n}} = 0$, only non-zero potential being V_n, the field and the charges are proportional to V_n.

$$
\left.\begin{array}{l}
\text{State } n \\
V_n \neq 0 \\
V_{i_{i \neq n}} = 0
\end{array}\right\}
\begin{array}{llllllll}
A_1 & A_2 & A_3 & ... & A_j & ... & A_n \\
0 & 0 & 0 & ... & 0 & ... & V_n \\
C_{1n}V_n & C_{2n}V_n & C_{3n}V_n & ... & C_{jn}V_n & ... & C_{nn}V_n
\end{array}
\tag{6.12}
$$

Now, the superposition of the n states like I, II,..., n is also a possible state in which each conductor has the potential which it had in the previous equilibrium state. The electric field at any point is the vector sum of the electric fields at that point in the n cases, while the charge on a conductor is the sum of the charges it carried in the n cases. In this new state the potentials are V_1, V_2,..., V_j and V_n, none of them necessarily zero. In short, we have a completely general state which, in agreement to the uniqueness theorem (see section 3.4) and the superposition principle (see section 2.2), is a new equilibrium state of the system. Relations connecting charges and potentials are obtained simply by adding equations (6.9) through (6.12):

$$\begin{cases} Q_1 = C_{11}V_1 + C_{12}V_2 + ...+C_{1j}V_j + ...+C_{1n}V_n \\ Q_2 = C_{21}V_1 + C_{22}V_2 + ...+C_{2j}V_j + ...+C_{2n}V_n \\ ... \\ Q_j = C_{j1}V_1 + C_{j2}V_2 + ...+C_{jj}V_j + ...+C_{jn}V_n \\ ... \\ Q_n = C_{n1}V_1 + C_{n2}V_2 + ...+C_{nj}V_j + ...+C_{nn}V_n \end{cases} \tag{6.13}$$

It appears that the electrical behavior of this system is characterized by n^2 constants: C_{11}, C_{12},..., C_{nn}. In fact, only $\sum_1^n i = \frac{n(n+1)}{2}$ constants are necessary, for it can be proved that in **any system** $C_{12} = C_{21}$, $C_{13} = C_{31}$,..., $C_{ij} = C_{ji}$. In general, the relations (6.13) can be written as:

$$Q_i = \sum_{j=1}^{n} C_{ij}V_j \quad \text{for } i = 1 \text{ to } n \tag{6.14}$$

in which C_{ii} are known as **coefficients of capacitance** and the C_{ij} as **coefficients of induction**. The coefficients C_{ii} are always positive representing the ratio between the charge of the conductor i when only it has the non-zero potential V_i all the other conductors being grounded ($V_j = 0$). The coefficients C_{ij} are always negative, $j \neq i$ because each of them represents the ratio between the charges Q_i induced on the conductor i when only non-zero potential is V_j (on j conductor) all others being held to zero potential. If only the conductor A_j is held at the non-zero potential $V_j = +1$ V, it will carry the positive charge $C_{jj}V_j = C_{jj} > 0$. According to the theorem of the corresponding elements (see section 5.5), the total charge induced on the surfaces of all the other conductors is at most equal to C_{jj}, then:

$$C_{jj} \geqslant \sum_{j \neq i} C_{ij} \tag{6.15}$$

The equality can only take place when no field line reaches infinity, in other words, when the electrostatic influence is maximum (there is a complete system in which the inductor conductor is enclosed by the other surrounding conductors). The coefficients of capacitance have the unit F because they represent the ratio between the electric charge of a conductor and its potential, in other words, is a capacity. If we define the square matrix of the capacity coefficients $[C]$ and the colon matrix for $[Q]$ and $[V]$, respectively, we shall write:

$$[Q] = [C] \times [V]; \quad [Q] = \begin{bmatrix} Q_1 \\ Q_2 \\ \cdots \\ Q_n \end{bmatrix}, \quad [V] = \begin{bmatrix} V_1 \\ V_2 \\ \cdots \\ V_n \end{bmatrix},$$

$$[C] = \begin{bmatrix} C_{11} & C_{12} & \cdots & C_{1n} \\ C_{21} & C_{22} & \cdots & C_{2n} \\ \cdots & & & \\ C_{n1} & C_{n2} & \cdots & C_{nn} \end{bmatrix}$$

$$(6.16)$$

But we must note here that the coefficient C_{ii} is the capacitance of the conductor i held at the potential V_i when the others have the potential zero, but, C_{ii} is different to the capacitance of the isolated conductor i with respect to the surroundings.

Exercise: coefficients of capacitance of two charged spheres

Let's consider a system of two charged spheres having the radius, R_1 and R_2, at a very large distance $d \gg R_1$, R_2 between their centers, see figure 6.5. The spheres carry the charges q_1 and q_2, respectively. Calculate the corresponding influence coefficients of capacitance.

Solution: *the potential of each of the spheres will be:*

$$V_1 = \frac{q_1}{4\pi\varepsilon_0 R_1} + \frac{q_2}{4\pi\varepsilon_0 d} \tag{6.17}$$

and

$$V_2 = \frac{q_1}{4\pi\varepsilon_0 d} + \frac{q_2}{4\pi\varepsilon_0 R_2} \tag{6.18}$$

respectively. Solving the system formed by equations (6.17) and (6.18):

$$\begin{cases} \dfrac{q_1}{4\pi\varepsilon_0 R_1} + \dfrac{q_2}{4\pi\varepsilon_0 d} = V_1 \\[2ex] q_2 = 4\pi\varepsilon_0 R_2 V_2 - \dfrac{q_1 R_2}{d} \end{cases}$$

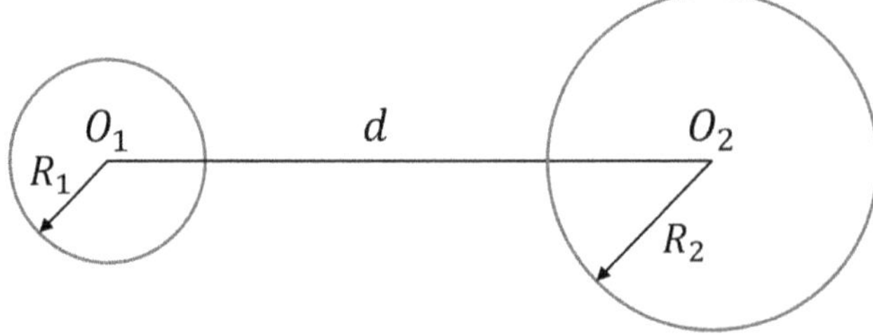

Figure 6.5. *The coefficients of capacitance of a two-spherical-conductors system.*

Consequently:

$$\frac{q_1}{4\pi\varepsilon_0 R_1} + \frac{R_2}{d}V_2 - \frac{q_1 R_2}{4\pi\varepsilon_0 d^2} = V_1$$

$$q_1\left(\frac{1}{4\pi\varepsilon_0 R_1} - \frac{R_2}{4\pi\varepsilon_0 d^2}\right) = V_1 - \frac{R_2}{d}V_2$$

$$q_1 = 4\pi\varepsilon_0 \frac{R_1 d^2}{d^2 - R_1 R_2}V_1 - 4\pi\varepsilon_0 \frac{R_1 R_2 d}{d^2 - R_1 R_2}V_2$$

$$q_1 = \frac{4\pi\varepsilon_0 R_1}{1 - \dfrac{R_1 R_2}{d^2}}V_1 - 4\pi\varepsilon_0 \frac{R_1 R_2}{d\left(1 - \dfrac{R_1 R_2}{d^2}\right)}V_2$$

$$q_2 = 4\pi\varepsilon_0 R_2 V_2 - \frac{R_2}{d}\frac{4\pi\varepsilon_0 R_1}{1 - \dfrac{R_1 R_2}{d^2}}V_1 + 4\pi\varepsilon_0 \frac{R_2}{d}\frac{R_1 R_2}{d\left(1 - \dfrac{R_1 R_2}{d^2}\right)}V_2$$

$$q_2 = \frac{4\pi\varepsilon_0 R_2}{1 - \dfrac{R_1 R_2}{d^2}}V_2 - 4\pi\varepsilon_0 \frac{R_1 R_2}{d\left(1 - \dfrac{R_1 R_2}{d^2}\right)}V_1$$

$$q_1 = \frac{4\pi\varepsilon_0 R_1}{1 - \dfrac{R_1 R_2}{d^2}}V_1 - \frac{4\pi\varepsilon_0 R_1 R_2}{d\left(1 - \dfrac{R_1 R_2}{d^2}\right)}V_2 \simeq 4\pi\varepsilon_0 R_1 V_1 - \frac{4\pi\varepsilon_0 R_1 R_2}{d}V_2 \tag{6.19}$$

$$q_2 = -\frac{4\pi\varepsilon_0 R_1 R_2}{d\left(1 - \dfrac{R_1 R_2}{d^2}\right)}V_1 + \frac{4\pi\varepsilon_0 R_2}{1 - \dfrac{R_1 R_2}{d^2}}V_2 \simeq -\frac{4\pi\varepsilon_0 R_1 R_2}{d}V_1 + 4\pi\varepsilon_0 R_2 V_2 \tag{6.20}$$

$$q_1 = C_{11}V_1 + C_{12}V_2 \Rightarrow C_{11} = \frac{4\pi\varepsilon_0 R_1}{1 - \dfrac{R_1 R_2}{d}}, \quad C_{12} = -\frac{4\pi\varepsilon_0 R_1 R_2}{d\left(1 - \dfrac{R_1 R_2}{d}\right)}$$

$$q_2 = C_{21}V_1 + C_{22}V_2 \Rightarrow C_{21} = -\frac{4\pi\varepsilon_0 R_1 R_2}{d\left(1 - \dfrac{R_1 R_2}{d}\right)}, \quad C_{22} = \frac{4\pi\varepsilon_0 R_2}{1 - \dfrac{R_1 R_2}{d}}$$

Only when $d \gg R_1$, R_2, we shall consider $\frac{R_1 R_2}{d^2} \to 0$ and then $C_{11} = 4\pi\varepsilon_0 R_1$, $C_{12} = C_{21} = -\frac{4\pi\varepsilon_0 R_1 R_2}{d}$, $C_{22} = 4\pi\varepsilon_0 R_2$, and C_{11} and C_{22} seem to be the capacitances of the isolated spheres and $C_{12} = C_{21}$ the influence capacitances.

6.2.2 The influence coefficients of potential

Let's suppose now that the conductors in figure 6.6 possess the charges: Q_1, Q_2,..., Q_n, and are thereby raised to potentials: V_1, V_2,..., V_n with respect to the surrounding shell. The final equilibrium states could be obtained as the super-position of many simplest equilibrium states, shown in figure 6.6. For example, in the state 1, the electric field in the whole system and the potential at every conductor is determined uniquely by the value Q_1 of the charge on the conductor 1, the other charges being zero. Then each of the potentials V_1, V_2,..., V_n must be proportional to Q_1.

$$
\begin{array}{c c c c c c c c c}
 & A & A_1 & A_2 & \dots & A_j & \dots & A_n \\
\text{State 1} & Q & Q_1 & 0 & \dots & 0 & \dots & 0 \\
 & V & S_{11}Q_1 & S_{21}Q_1 & \dots & S_{j1}Q_1 & \dots & S_{n1}Q_1
\end{array}
\tag{6.21}
$$

The constants S_{11}, S_{21},..., S_{n1} depend only on the shape and arrangement of the conducting bodies. In just the same way we could analyze states in which Q_1, Q_2,..., Q_n are zero, only the charge on the 'j' conductor is different to zero (Q_j). Again, we must find a linear dependence between the only charge Q_j in this case, and the various potentials.

$$
\begin{array}{c c c c c c c c c}
 & A & A_1 & A_2 & \dots & A_j & \dots & A_n \\
\text{State } j & Q & 0 & 0 & \dots & Q_j & \dots & 0 \\
 & V & S_{1j}Q_j & S_{2j}Q_j & \dots & S_{jj}Q_j & \dots & S_{nj}Q_j
\end{array}
\tag{6.22}
$$

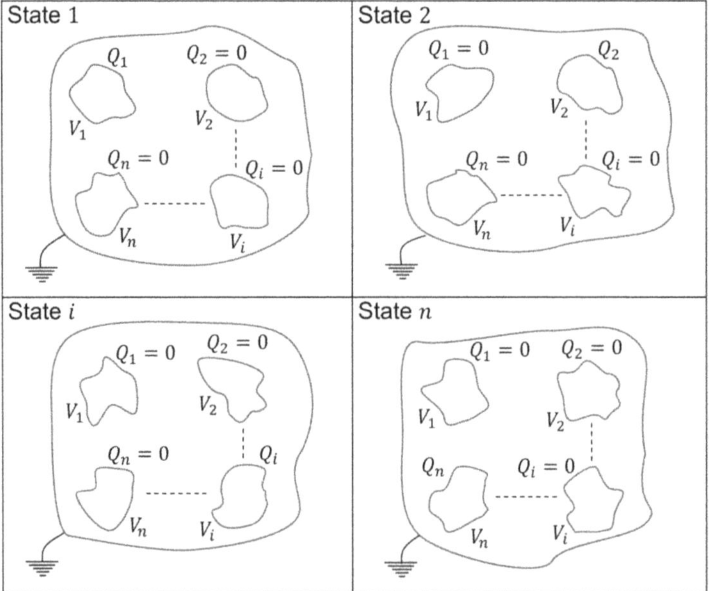

Figure 6.6. Intermediary equilibrium states of a system of charged conductors.

Finally, when $Q_n \neq 0$ and $Q_1 \dots Q_{n-1} = 0$, the field and the potentials are proportional to Q_n.

$$
\begin{array}{ccccccc}
A & A_1 & A_2 & \dots & A_j & \dots & A_n \\
\text{State } n \quad Q & 0 & 0 & \dots & 0 & \dots & Q_n \\
V & S_{1n}Q_n & S_{2n}Q_n & \dots & S_{jn}Q_n & \dots & S_{nn}Q_n
\end{array}
\tag{6.23}
$$

Now the superposition of the states like 1, 'j' and 'n' is also a possible state. The electric field at any point is the vector sum of the electric fields at that point in the 'n' cases, while the potential on a conductor is the sum of its potentials in the 'n' case when Q_1 is placed on 1, Q_2 on 2,..., Q_j on 'j' and Q_n on 'n'.

$$
\begin{cases}
V_1 = S_{11}Q_1 + S_{12}Q_2 + \dots + S_{1j}Q_j + \dots + S_{1n}Q_n \\
V_2 = S_{21}Q_1 + S_{22}Q_2 + \dots + S_{2j}Q_j + \dots + S_{2n}Q_n \\
\dots \\
V_j = S_{j1}Q_1 + S_{j2}Q_2 + \dots + S_{jj}Q_j + \dots + S_{jn}Q_n \\
\dots \\
V_n = S_{n1}Q_1 + S_{n2}Q_2 + \dots + S_{nj}Q_j + \dots + S_{nn}Q_n
\end{cases}
\tag{6.24}
$$

The S_{ij} are known as **coefficients of potential**, or '**elastances**', having the dimension:

$$
[S]_{SI} = \frac{[V]_{SI}}{[Q]_{SI}} = \frac{V}{C} = \frac{1}{F} = F^{-1}
\tag{6.25}
$$

Generalizing for n conductors, we can write:

$$
V_i = \sum_{j=1}^{n} S_{ij}Q_j \quad \text{for } i = 1 \text{ to } n
\tag{6.26}
$$

It appears that the electric behavior of this system is characterized by the n^2 positive constants: $S_{11}, S_{12}, \dots, S_{nn}$. In fact, only $\sum_{i=1}^{n} i = \frac{n(n+1)}{2}$ constants are necessary, for it can be proved that in any system: $S_{12} = S_{21}$, $S_{ij} = S_{ji}$. If we define the square matrix of the potential coefficients $[S]$ and the column matrices for $[Q]$ and $[V]$, respectively, we shall write:

$$
[V] = [S] \times [Q]; \quad [Q] = \begin{bmatrix} Q_1 \\ Q_2 \\ \dots \\ Q_n \end{bmatrix}, \quad [V] = \begin{bmatrix} V_1 \\ V_2 \\ \dots \\ V_n \end{bmatrix},
$$

$$
[S] = \begin{bmatrix} S_{11} & S_{12} & \dots & S_{1n} \\ S_{21} & S_{22} & \dots & S_{2n} \\ \dots & & & \\ S_{n1} & S_{n2} & \dots & S_{nn} \end{bmatrix}
\tag{6.27}
$$

From the system of linear equations (6.24) it results that each conductor from the system has a contribution to the values of the potentials to other conductors from system. This influence depends on the geometrical configuration of the system of conductors and on the charge present on the inductor conductor. The diagonal elements, S_{jj} from (6.26) are called 'self elastances'. The self-elastance S_{jj} of conductor A_j is the ratio between its potential V_j and its own charge Q_j, when the charges on the other conductors from the system are zero. The non-diagonal coefficients S_{ij} are called '**mutual elastances**'. The mutual elastance S_{ij} between the conductors A_i and A_j, S_{ij}, represents the ratio between the potential V_i of the conductor A_i and the charge Q_j, of conductor A_j, when the charges on the other conductors from system are zero. All the potential coefficients are positive because any increasing of a positive charge on a conductor from the system will induce an increase of the positive potential on all conductors. These potential coefficients could also be used to solve different electrostatic problems. An example to demonstrate this will be given below.

Exercise: capacitance of a system of conductors

Let there be a system of two filiform conductors with the same length l, at maximum electrostatic influence, then forming a complete system, see figure 6.7. Considering that the conductors are identical with the radius r and are placed at a very large distance d (d ≫ r) between their axes, find the capacitance of the system.
Solution: *if the charges on the two conductors are Q_1 and Q_2 their potentials are:*

$$\begin{cases} V_1 = S_{11}Q_1 + S_{12}Q_2 \\ V_2 = S_{21}Q_1 + S_{22}Q_2 \end{cases} \tag{6.28}$$

But having a complete system:

$$\left.\begin{array}{l} Q_1 + Q_2 = 0 \\ Q_1 = Q \text{ and } Q_2 = -Q \end{array}\right\} \Rightarrow Q_1 = -Q_2 = Q \Rightarrow$$

$$\implies \begin{cases} V_1 = S_{11}Q - S_{12}Q = (S_{11} - S_{12})Q \\ V_2 = S_{21}Q - S_{22}Q = (S_{21} - S_{22})Q \end{cases} \tag{6.29}$$

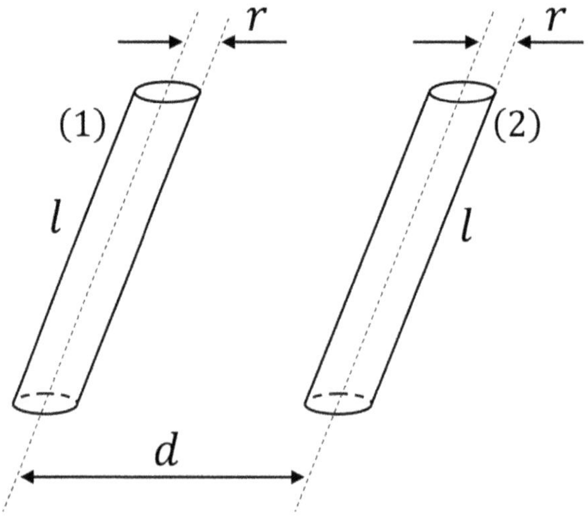

Figure 6.7. *The potential coefficients of a system of two filiform conductors at electrostatic equilibrium.*

Having two identical conductors, $S_{11} = S_{22}$ and $S_{12} = S_{21}$, respectively, resulting in:

$$V_1 - V_2 = Q[(S_{11} - S_{12}) - (S_{21} - S_{22})] = Q[(S_{11} - S_{12}) + (S_{11} - S_{12})]$$
$$= 2Q(S_{11} - S_{12})$$

The capacitance of the system is by definition: $C = Q/(V_1 - V_2)$, then:

$$C = \frac{Q}{V_1 - V_2} = \frac{1}{2(S_{11} - S_{12})} \tag{6.30}$$

Exercise: elastances calculation for a system of two conductors

Compute the elastances of potentials V_1 and V_2, generated by Q on conductor 1 at a point on its surface and on the surface of conductor 2, respectively.
Solution: *the potential of the electric field at the distance r with respect to the axis of a cylindrical conductor on radius R, uniformly charged with the linear charge density λ, see figure 6.8 can be computed using the superposition principle (see section 2.2). Practically, the conductor can be considered like a cylindrical conductor on the radius R carrying the charge Q uniformly distributed on the surface with surface charge density σ, and then Gauss' law permits us to find the electric field at the distance r on the axis of the conductor:*

$$\int \vec{E}\,\overrightarrow{dS} = \frac{q_i}{\varepsilon_0} \Rightarrow E \cdot 2\pi r l = \frac{2\pi R l \sigma}{\varepsilon_0} \Rightarrow \vec{E} = \frac{R\sigma}{\varepsilon_0 r}\frac{\vec{r}}{r}$$

This relation can be re-written as a function of the linear charge density λ of the very long conductor ($Q = 2\pi R \sigma l = l\lambda \Rightarrow \lambda = 2\pi R \sigma$) as: $\vec{E} = \frac{\lambda}{2\pi\varepsilon_0 r}\frac{\vec{r}}{r}$,

$$\vec{E} = \frac{\lambda}{2\pi\varepsilon_0 r}\vec{e_r}; \ \vec{e_r} = \frac{\vec{r}}{r} \tag{6.31}$$

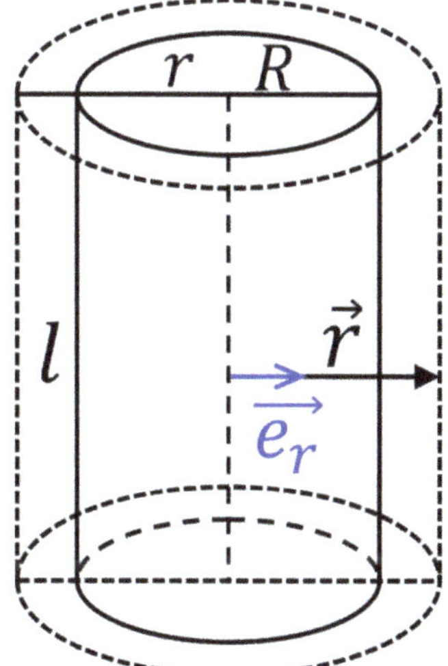

Figure 6.8. *Electric field of a conductor charged with linear charge density λ.*

practically equivalent with the electric field of a linear charge distribution on a filiform conductor placed on the axis of the cylinder. Having the electric field for the uniformly charged conductor, the potential at distance r with respect to its axis will be:

$$V = -\int_{\infty}^{r} \vec{E}\,\vec{dl} \tag{6.32}$$

But due to the fact that the integral in equation (6.32) is divergent, we will use again the superposition principle (see section 2.2). The length element dy from figure 6.9 carries the elementary charge $dq = \lambda dy$ and will give rise at distance R to the elementary potential:

$$dV = \frac{\lambda dy}{4\pi\varepsilon_0 R} = \frac{\lambda dy}{4\pi\varepsilon_0 \sqrt{r^2 + y^2}}$$

Then the potential will be:

$$V = \int_{-l_1}^{l_2} \frac{\lambda dy}{4\pi\varepsilon_0 \sqrt{r^2 + y^2}} = \frac{\lambda}{4\pi\varepsilon_0} \ln\left[y + \sqrt{r^2 + y^2}\right]\Bigg|_{-l_1}^{l_2}$$

$$= \frac{\lambda}{4\pi\varepsilon_0} \ln\left[\frac{\sqrt{r^2 + l_2^2} + l_2}{\sqrt{r^2 + l_1^2} - l_1}\right] \tag{6.33}$$

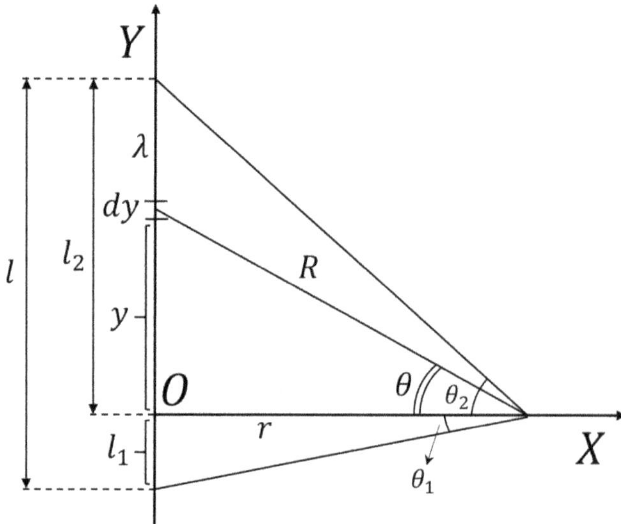

Figure 6.9. *The electric potential of a long conductor charged with a linear charge density λ.*

Having a very long conductor ($l \gg r$), we can suppose the P point placed on the line perpendicular on the middle, and then: $l_2 = l_1 = l/2$. In this case:

$$V = 2 \int_0^{l/2} \frac{\lambda dy}{4\pi\varepsilon_0 \sqrt{r^2 + y^2}} = \frac{2\lambda}{4\pi\varepsilon_0} \ln \frac{\sqrt{r^2 + \frac{l^2}{4}} + \frac{1}{2}}{\sqrt{r^2 + 0} - 0}$$

$$= \frac{\lambda}{2\pi\varepsilon_0} \ln \frac{l}{2r} \left[\left(1 + \frac{4r^2}{l^2} \right)^{1/2} + 1 \right] \cong \frac{\lambda}{2\pi\varepsilon_0} \ln \frac{l}{r} \tag{6.34}$$

The potential of a straight line conductor on finite length l at a distance r on its axis is:

$$V = \frac{\lambda}{2\pi\varepsilon_0} \ln \frac{l}{r} \tag{6.35}$$

The difference of potentials present at two points placed at distances r_1 and r_2 with respect to conductor will be:

$$V_{12} = V_1 - V_2 = \frac{\lambda}{2\pi\varepsilon_0} \left(\ln \frac{l}{r_1} - \ln \frac{l}{r_2} \right) = \frac{\lambda}{2\pi\varepsilon_0} \ln \frac{r_2}{r_1} \tag{6.36}$$

Coming back to the two filiform conductors each having radius r and length l, placed at the distance d between their axes, their potentials can be written using the relation (6.35). The potential of a point on the conductor 1 will be:

$$V_1 = \frac{\lambda_S}{2\pi\varepsilon_0} \ln \frac{l}{r}$$

where $\lambda_S = 2\pi r \sigma$ is the specific linear charge density. Then its self-elastance will be:

$$V_1 = S_{11}\lambda_S = \frac{\lambda_S}{2\pi\varepsilon_0} \ln \frac{l}{r} \Rightarrow S_{11} = \frac{1}{2\pi\varepsilon_0} \ln \frac{l}{r} \tag{6.37}$$

On the other hand, conductor 1 charged with λ_S will determine at a point on the conductor 2 the potential:

$$V_2 = \frac{\lambda_S}{2\pi\varepsilon_0} \ln \frac{l}{d}$$

leading to the mutual elastance $S_{21} = S_{12}$:

$$V_2 = S_{21}\lambda_S = S_{12}\lambda_S = \frac{\lambda_S}{2\pi\varepsilon_0} \ln \frac{l}{d} \Rightarrow S_{12} = \frac{1}{2\pi\varepsilon_0} \ln \frac{l}{d} \tag{6.38}$$

With these, the distributed capacitance of the system is:

$$C = \frac{1}{2(S_{11} - S_{12})} = \frac{1}{2\frac{1}{2\pi\varepsilon_0} \left(\ln \frac{l}{r} - \ln \frac{l}{d} \right)} \Rightarrow C = \frac{\pi\varepsilon_0}{\ln \frac{d}{r}} \; [F/m] \tag{6.39}$$

Similarly, the elastances S_{22} and S_{21} can be computed, the results being:

$$V_2 = -S_{22}\lambda_S = \frac{-\lambda_S}{2\pi\varepsilon_0}\ln\frac{l}{r} \Rightarrow S_{22} = \frac{1}{2\pi\varepsilon_0}\ln\frac{l}{r} = S_{11}$$

$$V_1 = -S_{12}\lambda_S = \frac{-\lambda_S}{2\pi\varepsilon_0}\ln\frac{l}{d} \Rightarrow S_{12} = \frac{1}{2\pi\varepsilon_0}\ln\frac{l}{d} = S_{21}$$

$$(6.40)$$

6.3 Ideal capacitors

In this section, we are going to compute the capacitance of several particular capacitor's configurations.

6.3.1 Capacitance of a conductor placed in the vicinity of another conductor

Let there be a metallic disc A connected to the positive pole of an electric generator placed far from the disc and having the negative earth connected, see figure 6.10. If the potential of the conductor A is V_0, it carries the electric charge:

$$Q_0 = C_0 V_0 \tag{6.41}$$

where C_0 is the capacitance of conductor A considered insulated with respect to other conductors. Approaching conductor A another conductor B connected to earth (at potential $V_B = 0$), on the conductor B will be induced by the electrostatic influence the negative charge $Q_B = C_{BA}V_A$, where the negative constant C_{BA} is the

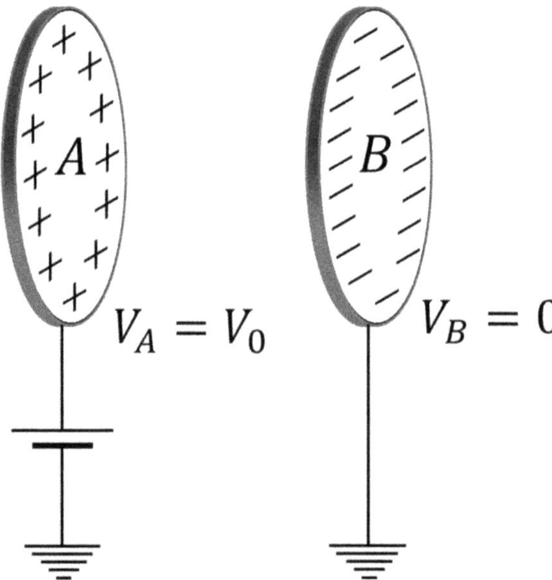

Figure 6.10. System of two conductors at maximum electrostatic influence.

mutual capacitance between conductors A and B, representing the ratio between the influence charge induced on conductor B by conductor A kept to the potential V_A, conductor B being to the potential $V_B = 0$. The negative influence charge on disc B will give rise to the supplementary positive influence charge on conductor A which in turn will generate supplementary negative influence charge on conductor B, the process continues until electrostatic equilibrium is reached. In the final equilibrium state of the system, conductor A carries a charge bigger than that carried when it was insulated (far with respect to conductor B), that is why, at that time physicists said about this experiment that the '*condensation of the charge*' takes place on conductor A when conductor B approaches it, and named the binary system formed by the conductors A and B, at maximum electrostatic influence, '*condenser*'. Effectively, the capacitance of conductor A kept to the potential V_0, in the presence of conductor B earth connected ($V_B = 0$), increased carrying the charge:

$$Q = C_{AA}V_0 > Q_0 = C_0V_0 \tag{6.42}$$

Then the capacitance of a conductor placed in the vicinity of another grounded conductor is bigger than the capacitance of insulated conductor $C_{AA} > C_0$.

6.3.2 The capacitance of a capacitor

In the broad sense of the term, *a capacitor (condenser) is any binary set of conductors separated by a dielectric or vacuum medium, which are at maximum electrostatic influence.* The system of the two conductors called armatures (plates) forms a complete system in the sense specified in section 6.1, equivalent in fact to the sense of maximum or total electrostatic influence. Usually, for the conductor system to be complete, i.e., to be under conditions of total influence (all field lines starting from the conductor with $Q > 0$ to reach the one with $Q < 0$), then one of the conductors here is the outer armature, must completely surround the second, here the inner conductor (armature), see figure 6.11. If the two conductors A and B are kept to the

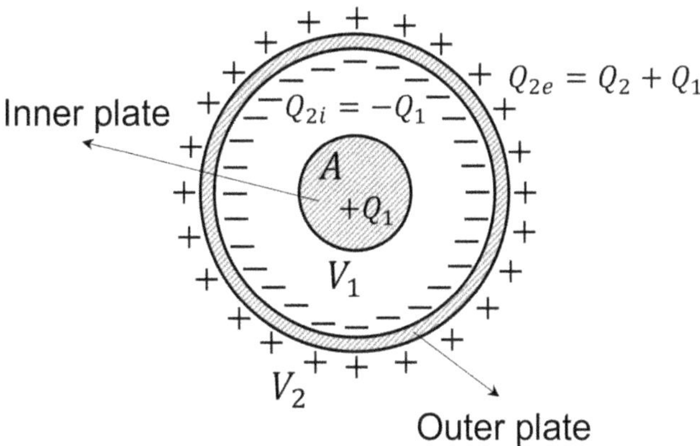

Figure 6.11. Complete system of conductors at electrostatic equilibrium.

potentials V_1 and V_2, then, according to the Maxwell relations generally valid in electrostatic equilibrium conductor systems, they will carry the charges:

$$\begin{cases} Q_1 = C_{11}V_1 + C_{12}V_2 \\ Q_2 = C_{21}V_1 + C_{22}V_2 \end{cases} \tag{6.43}$$

However, it must be taken into account that the electrostatic influence is maximum (total). This means that all the field lines that start from A when $V_1 > 0$, reach conductor B, determining the induction of an influence charge on the inner face of B, a charge which, according to the theorem of the corresponding elements (see section 5.5), will be exactly $-Q_1$. Or, the coefficient of influence between sphere B and A was defined as: $C_{21} = \frac{Q_2}{V_1}$ when $V_1 > 0$ and $V_2 = 0$, so in this case:

$$Q_2 = C_{21}V_1 = -Q_1$$

But: $\quad Q_1 = C_{11}V_1 \Rightarrow C_{11}V_1 = -C_{21}V_1 \Rightarrow C_{11} = -C_{21} = -C_{12}.$ $\qquad$ Noting $\qquad$ that $C_{11} = -C_{21} = -C_{12} = C$, the equations (6.43) will be written in the form:

$$\begin{cases} Q_1 = CV_1 - CV_2 = C(V_1 - V_2) \\ Q_2 = -CV_1 + C_{22}V_2 = -C(V_1 - V_2) + (C_{22} - C)V_2 \end{cases} \tag{6.44}$$

Noting that $V_1 - V_2 = U_{12}$ and $C_{22} - C = C'$ results in:

$$\begin{cases} Q_1 = CU_{12} \\ Q_2 = -CU_{12} + C'V_2 \end{cases} \tag{6.45}$$

Because the electric charge $Q_i = Q_1 = Q = CU_{12}$ is on armature A, the negative influence charge $-Q = -CU_{12}$, will be on the inner surface of armature B and $Q_e = C'V_2$, it will be on the outer face of armature B. The charge Q_e does not depend on the charges on conductor A and on the inner surface of the armature B, for example if they are zero, the charge Q_e is the same as when conductor B alone is brought to the potential V_2. So, C' represents the capacity of the external armature when it is alone (isolated), which is not important when a capacitor is used, because first of all the capacitance of an isolated conductor is very small as compared with that of a capacitor, and on the other hand, the outer armature of a capacitor frequently is earth connected ($V_2 = 0$), then the charge $Q_e = 0$, too. The charge Q on the inner armature of a capacitor is called the charge of capacitor, and neglecting the charge Q_e as in the above comments, the capacitance of a capacitor can be defined by:

$$C = \frac{Q}{V_1 - V_2} = \frac{Q}{U_{12}} \tag{6.46}$$

The positive size defined by the ratio between the charge of one conductor and the potential difference between its potential and that of the other one is called the capacitance of the electric capacitor. Then, the unit for the capacitance, F, can be defined as the capacitance of a capacitor carrying the charge of 1 *Coulomb* when a potential difference of 1 *Volt* is maintained between its armatures.

6.3.3 The capacitance of a simple capacitor

A capacitor is an instrument for storing charge, and a capacitor of large capacity can store correspondingly large quantity of charge for a given potential difference between its armatures. The capacity depends on the geometry of the conductors and the dielectric constant of the medium separating them. In general, calculation of the capacity of a conductor or capacitor is difficult unless simple geometrical shapes are involved. Knowing the geometric shape of the armatures, their relative position and the nature of the dielectric are determined successively: (a) the distribution of the surface charge density σ on each armature; (b) the configuration of the electric field between the armatures, usually using the Gauss theorem; (c) knowing the field, the potential difference $V_1 - V_2$ between the two armatures is calculated starting from the definition relation of the electric voltage between two points:

$$U_{12} = \int_1^2 \vec{E}\,\vec{dl} \tag{6.47}$$

(d) Calculate the charge on the inner armature using:

$$Q = \int_S \sigma dS \tag{6.48}$$

(e) Finally, the capacitance of the capacitor will be:

$$C = \frac{Q}{V_1 - V_2} = \frac{Q}{U_{12}} \tag{6.49}$$

The method indicated here can be applied if the equation of the field lines is known or by making certain simplifying assumptions on the field configuration.

Exercise: spherical capacitor

Compute the capacitance of a spherical capacitor using the protocol described above, in subsection 6.3.3.
Solution: the spherical capacitor is a system formed by two concentric spherical conductors having the radius R_1 and R_2, separated from each other by a dielectric medium of permittivity ε_0 and having between the armatures a potential difference, $U_{12} = V_1 - V_2$, see figure 6.12. Under the action of the potential difference, the system is charged with $+Q$ on the inner armature and $-Q$ on the outer one. As we specified above, the capacity calculation involves the calculation of the potential difference U_{12}, which is possible after finding the intensity of the electric field between the armatures. Considering a spherical surface (Gaussian) Σ, with the radius r concentric with the inner sphere ($R_1 < r < R_2$) and applying Gauss' theorem, we have the electric field between the two spheres:

$$\int EdS = \frac{Q}{\varepsilon_0} \Rightarrow \vec{E} = \frac{Q}{4\pi\varepsilon_0 r^3}\vec{r} \neq 0 \tag{6.50}$$

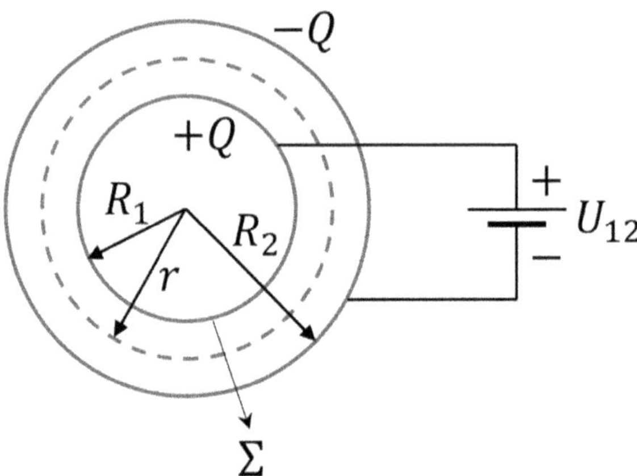

Figure 6.12. *The spherical capacitor.*

while everywhere else $\vec{E}$ is zero. Hence, the potential difference between the two spheres is:

$$U_{12} = \int_{R_1}^{R_2} \frac{Q\,dr}{4\pi\varepsilon_0 r^2} = \frac{Q}{4\pi\varepsilon_0}\left(\frac{1}{R_1} - \frac{1}{R_2}\right)$$

and consequently, the capacity will be:

$$C = \frac{Q}{U_{12}} = \frac{4\pi\varepsilon_0 R_1 R_2}{R_2 - R_1} \tag{6.51}$$

When $R_2 \to \infty$, the equation (6.51) gives the capacity of a conducting sphere remote from others bodies:

$$C_1 = \lim_{R_2 \to \infty} \frac{4\pi\varepsilon_0 R_1 R_2}{R_2 - R_1} = 4\pi\varepsilon_0 R_1 \tag{6.52}$$

If the space between the two sphere is filled with a medium of dielectric constant $\varepsilon = \varepsilon_0 \varepsilon_r$, the capacity increases by a factor ε_r:

$$C = \frac{Q}{U_{12}} = \frac{4\pi\varepsilon_0 \varepsilon_r R_1 R_2}{R_2 - R_1} \quad or \quad C_1 = 4\pi\varepsilon_0 \varepsilon_r R_1$$

By making the ratio between the capacity of an isolated sphere C_1 and the capacity of two concentric spheres, we obtain:

$$\frac{C_1}{C} = \frac{4\pi\varepsilon R_1}{4\pi\varepsilon \dfrac{R_1 R_2}{R_2 - R_1}} = \frac{R_2 - R_1}{R_2} = 1 - \frac{R_1}{R_2} \tag{6.53}$$

a relation that allows us to see what is the influence that a second spherical surface can have on the capacity of an isolated sphere. From relation (6.53) it can be thus assessed whether a spherical surface surrounding the given sphere—for example the walls of a *chamber—can have any influence on the measurement of the sphere's capacity or whether it can be neglected. Note that when $R_2 \gg R_1$, the effect of the second (outer) sphere can be neglected.*

Conversely, if the distance between the armatures is small, the expression of the capacity of the spherical capacitor can be put in another form. Let $R_2 - R_1 = d$ (distance between armatures) and $R = (R_1 + R_2)/2$ the average radius of the two spherical surfaces, then:

$$\begin{cases} R_2 - R_1 = d \\ R_1 + R_2 = 2R \end{cases} \Rightarrow \begin{cases} R_2 = \dfrac{2R + d}{2} = R + \dfrac{d}{2} \\ R_1 = \dfrac{2R - d}{2} = R - \dfrac{d}{2} \end{cases}$$

Resulting in:

$$C = 4\pi\varepsilon \frac{\left(R - \dfrac{d}{2}\right)\left(R + \dfrac{d}{2}\right)}{R + \dfrac{d}{2} - R + \dfrac{d}{2}} = 4\pi\varepsilon \frac{\left(R^2 - \dfrac{d^2}{4}\right)}{d} \cong \frac{4\pi\varepsilon R^2}{d} = \frac{\varepsilon S}{d} \tag{6.54}$$

where $S = 4\pi R^2$ is the area of the spherical surface on the average radius.

If the dielectric of a spherical capacitor is not homogeneous but consists of dielectric spherical layers of different thickness and different electrical permittivity, the electric field is calculated using Gauss' theorem in each medium and then the potential difference between the armatures is calculated by summing the integrals $\int_{R_i}^{R_{i+1}} \overrightarrow{E_i} \, \overrightarrow{dl_i}$ from each dielectric.

Exercise: cylindrical capacitor

Compute the capacitance of a cylindrical capacitor using the protocol described above, in subsection 6.3.3.

Solution: *the cylindrical capacitor consists of two coaxial metallic cylinders of radii R_1 and R_2, their length being very large (practically infinite), in relation to the difference of their diameters, see figure 6.13. Let the potential difference be applied to the two cylinders. The electric field between these two armatures (when they are very long) will have radial symmetry depending only on the distance r from the capacitor axis. Gauss' theorem applied to Gaussian Σ, leads to:*

$$\int_\Sigma \overrightarrow{E} \, \overrightarrow{dS} = \frac{Q}{\varepsilon_0} \Rightarrow \overrightarrow{E} = \frac{Q}{2\pi\varepsilon_0 l r} \overrightarrow{e_r}; \quad \overrightarrow{e_r} = \frac{\overrightarrow{r}}{r}$$

then:

$$U_{12} = \int_{R_1}^{R_2} \frac{Q \overrightarrow{e_r}}{2\pi\varepsilon_0 l r} \overrightarrow{dr} = \frac{Q}{2\pi\varepsilon_0 l} \ln \frac{R_2}{R_1} \Rightarrow C = \frac{2\pi\varepsilon_0 l}{\ln \dfrac{R_2}{R_1}} \tag{6.55}$$

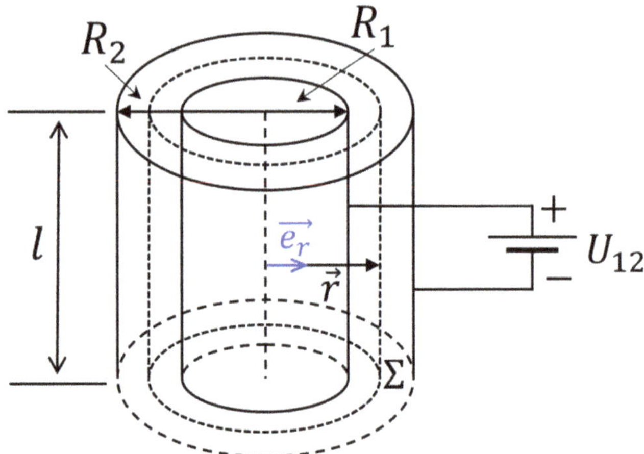

Figure 6.13. *The cylindrical capacitor.*

If the space between the cylinders is filled with a medium of dielectric constant ε_r, the capacity will be:

$$C = \frac{2\pi\varepsilon_0\varepsilon_r l}{\ln \dfrac{R_2}{R_1}} \tag{6.56}$$

Equations (6.55) and (6.56) show that the capacity increases by a factor of ε_r if the space between the armatures or cylinders or spheres is filled with a medium of dielectric constant. This agrees with the original definition of dielectric constant by Faraday, as discussed later.

Distributed capacitance

In a capacitor the capacitance is deliberately localized within a relatively small volume, but in extended conductors, such as coaxial cables or transmission lines used to convey electric currents over large distances, the capacitance is distributed continuously and is an important factor in any electric phenomena which occur. Transmission lines, consist most often of two conductors arranged either as a coaxial cable, see figure 6.14 or as a twin cable (bipolar), see figure 6.15.

The coaxial cable (see figure 6.14) forms an extended cylindrical capacitor and we expect equation (6.55) to apply quite accurately here in the form:

$$C = \frac{2\pi\varepsilon_0}{\ln \dfrac{b}{a}} \quad \text{per unit length} \tag{6.57}$$

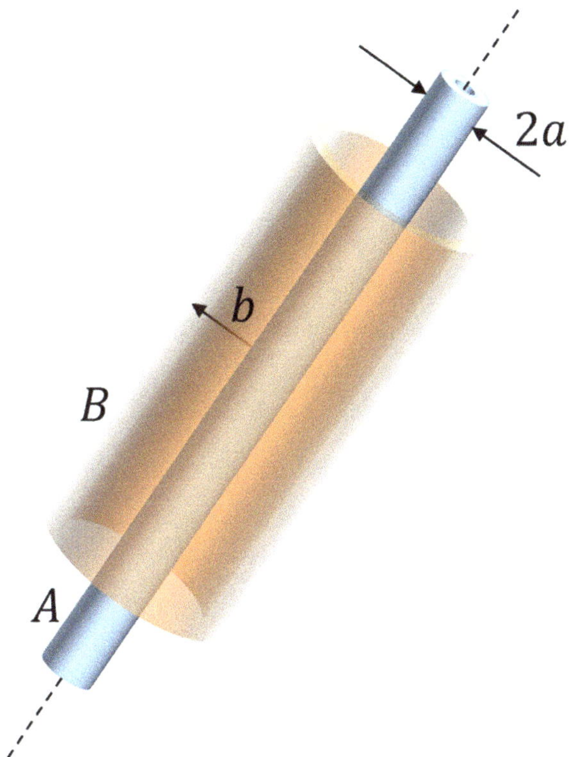

Figure 6.14. *Coaxial cable.*

The capacitance of twin cable (see figure 6.15) can be found most easily if the diameter of wires is small comparing with their distance apart, because the non-uniformity of the distribution of charge over the conductors can be neglected. If a charge λ per unit length is transferred from one to other (by electrostatic induction), the electric field strength at a point placed at distance r from one wire in figure 6.15 is:

$$\vec{E} = \left(\frac{\lambda}{2\pi\varepsilon_0 r} + \frac{\lambda}{2\pi\varepsilon_0(d-r)} \right) \frac{\vec{r}}{r}$$

Indeed, considering the Gaussian surfaces Σ_1, Σ_2, *then:*

$$\int_{\Sigma_1} \overrightarrow{E_1}\,\overrightarrow{dS} = \frac{\lambda l}{\varepsilon_0} \Rightarrow \overrightarrow{E_1} = \frac{\lambda l\,\overrightarrow{e_r}}{2\pi r l\varepsilon_0} = \frac{\lambda}{2\pi\varepsilon_0 r}\frac{\vec{r}}{r}$$

$$\int_{\Sigma_2} \overrightarrow{E_2}\,\overrightarrow{dS} = \frac{-\lambda l}{2\pi\varepsilon_0 l} \Rightarrow -E_2 2\pi(d-r)l = \frac{-\lambda}{\varepsilon_0} \Rightarrow \overrightarrow{E_2} = \frac{\lambda}{2\pi\varepsilon_0(d-r)}\frac{\vec{r}}{r}$$

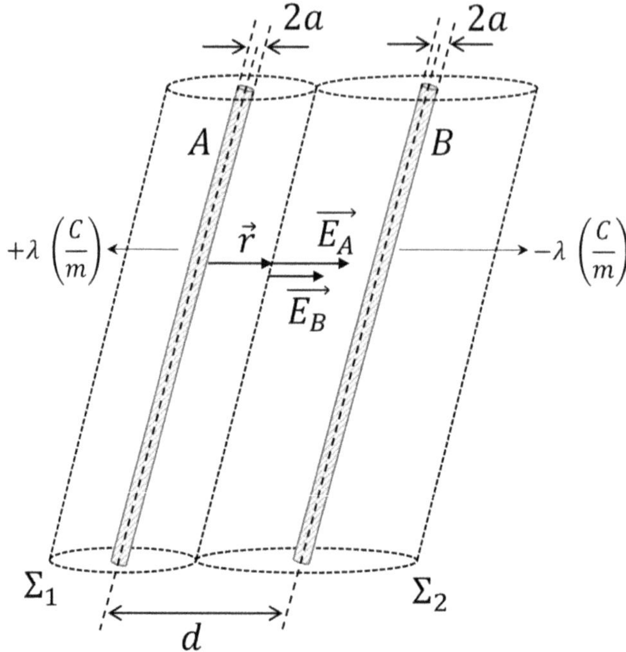

Figure 6.15. *Twin cable.*

Then, the potential difference will be:

$$V_A - V_B = \int_a^{d-a} \left[\frac{\lambda}{2\pi\varepsilon_0 r} + \frac{\lambda}{2\pi\varepsilon_0(d-r)} \right] dr$$

$$= \frac{\lambda}{2\pi\varepsilon_0}\left(\ln\frac{d-a}{a} - \ln\frac{a}{d-a} \right) \tag{6.58}$$

$$= \frac{\lambda}{2\pi\varepsilon_0}\left(\ln\frac{d-a}{a} + \ln\frac{d-a}{a} \right) = \frac{\lambda}{\pi\varepsilon_0}\ln\frac{d-a}{a}$$

Hence, the capacity per unit length will be:

$$C = \frac{\pi\varepsilon_0}{\ln\dfrac{d-a}{a}} \simeq \frac{\pi\varepsilon_0}{\ln\dfrac{d}{a}} \quad per\ unit\ length \tag{6.59}$$

for twin cable in which $d \gg a$.

A single horizontal telegraph or aerial wire at a high h above the Earth's surface would, if charged, have an electric image at a depth h. The capacitance per unit length between such wires of radius a and the Earth will thus be $\pi\varepsilon_0/\ln\frac{2h}{a}$ from equation (6.59).

Estimation of capacitance

Very often a rough estimate of capacitance is sufficient for practical purposes and the applied formulas, as developed are then good enough approximations. They all apply, however, to cases in which the charge is at least approximately uniformly distributed and so, since the ability to estimate capacitance is a useful facility, we look at an example to which they do not apply to all.

A thin conducting disc of radius a, would have greatest charge density at the edges where the curvature is large. The distribution will in fact be intermediate between an uniform one and one in which all the charge is concentrated round the rim. A very simple calculation of the potential at the center of the disc (and therefore of the disc as a whole) for these two extremes can be carried out (see figure 6.16). For a uniform charge density σ, the potential at P point is:

$$V_P = \int_0^a \int_0^{2\pi} \frac{\sigma dS}{4\pi\varepsilon_0 R} = \int_0^a \int_0^{2\pi} \frac{\sigma r dr d\theta}{4\pi\varepsilon_0 \sqrt{r^2 + h^2}} = \frac{\sigma}{2\varepsilon_0} \int_0^a \frac{r dr}{(r^2 + h^2)^{1/2}}$$

$$= \frac{\sigma}{2\varepsilon_0} \sqrt{r^2 + h^2} \Big|_0^a = \frac{\sigma}{2\varepsilon_0}\left(\sqrt{a^2 + h^2} - h\right)$$

$$\text{when:} \quad \begin{cases} h = 0 \Rightarrow V_0 = \dfrac{\sigma a}{2\varepsilon_0} \\[2mm] h \to \infty \Rightarrow V_0 = 0 \end{cases}$$

then:

$$C = \frac{Q}{V_0} = \frac{\sigma\pi a^2 2\varepsilon_0}{\sigma a} = 2\pi\varepsilon_0 a \tag{6.60}$$

If all the charge is round the edge, the potential at P point will be:

$$V = \int_0^{2\pi} \frac{\lambda dl}{4\pi\varepsilon_0(a^2 + h^2)^{1/2}} = \int_0^{2\pi} \frac{\lambda a d\theta}{4\pi\varepsilon_0(a^2 + h^2)^{1/2}} = \frac{\lambda a}{2\varepsilon_0(a^2 + h^2)^{1/2}}$$

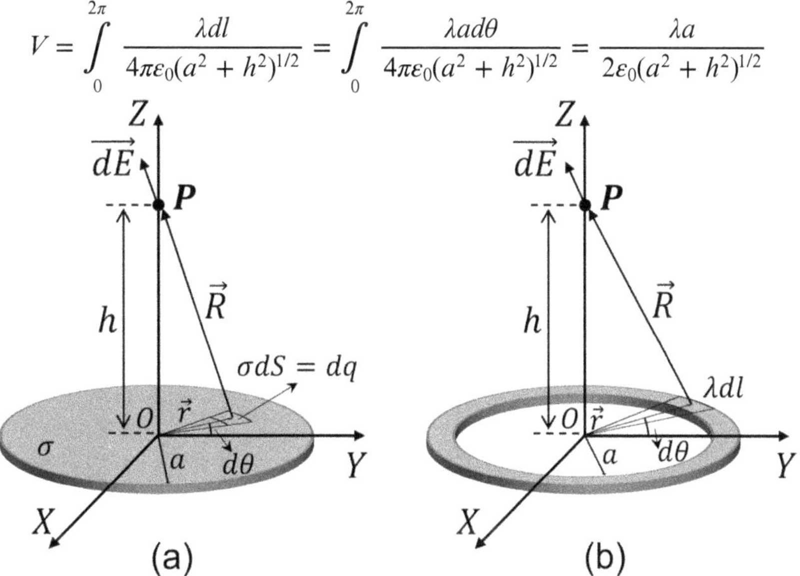

Figure 6.16. *Estimation of the capacitance of a disc. The distribution of charge lies between the uniform distribution (a) and one in which it lies completely round the rim (b).*

For $h = 0 \Rightarrow V_0' = \dfrac{\lambda}{2\varepsilon_0} = \dfrac{Q}{4\pi\varepsilon_0 a}$. Then:

$$C' = \frac{Q}{V_0'} = 4\pi\varepsilon_0 a \tag{6.61}$$

So that the true value should lie between these two $C_a = 2\pi\varepsilon_0 a$ and $C_b = C' = \dfrac{Q}{V_0'} = 4\pi\varepsilon_0 a$, effectively the value $C'' = 8\varepsilon_0 a$ resulted by accurate calculation.

Exercise: planar capacitor

Compute the capacitance of a planar capacitor using the protocol described above, in subsection 6.3.3.

Solution: *an ideal capacitor is best realized by enclosing one conductor within the other as in the spherical capacitor of figure 6.12, (in this case all the lines from one conductor end at the other), but this arrangement is inconvenient because of the inaccessibility of the inner conductor. The shapes in common use are the cylindrical and the parallel-plate capacitor (figures 6.13 and 6.17) and because neither of these can ensure that no lines of force leak to the surroundings, we shall only be able to deal with ideal conditions for the moment and assume no leakage. We shall also assume that the electric fields within capacitors are the same as if they were of infinite extent, which means that we neglect edge effects (sometimes known as fringing). A flat capacitor consists of two flat, parallel, metallic surfaces of area S, placed at a small distance d from the dimensions of the armatures (see figure 6.17). Between the plates there is a homogeneous and isotropic dielectric of permittivity ε. The two plates constitute the capacitor armatures. Suppose that the two plates carry the charges Q_1 and Q_2, respectively. According to the theorem of the corresponding elements (see section 5.5) these charges must be distributed on the two surfaces equally, but their signs are opposite. So:*

$$Q_1 = -Q_2 = Q \Rightarrow Q_1 = Q \text{ and } Q_2 = -Q \tag{6.62}$$

Suppose that these charges are uniformly distributed on the two surfaces with the surface charge density σ. Under these conditions, the capacitor is said to be charged, and on each plate there is a surface charge density: $\sigma_+ = \dfrac{Q}{S} = \sigma$ and $\sigma_- = \dfrac{-Q}{S} = -\sigma$, thus the electric field of each plate will be:

$$\overrightarrow{E_+} = \pm\frac{\sigma}{2\varepsilon_0}\overrightarrow{n}; \ \ \overrightarrow{E_-} = \mp\frac{\sigma}{2\varepsilon_0}\overrightarrow{n} \tag{6.63}$$

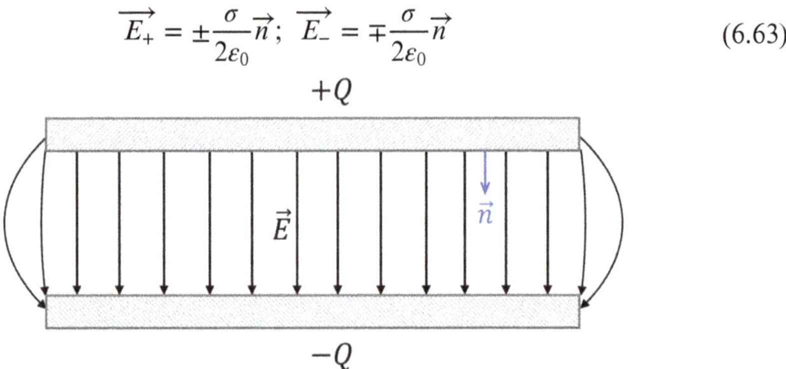

Figure 6.17. *A parallel-plate capacitor showing edge effects.*

The electric field between the two plates is uniform, normal to the planes and is different from zero only between the plates, where it is considered uniform and has the value:

$$\vec{E} = \frac{\sigma}{\varepsilon_0}\vec{n} \tag{6.64}$$

as shown in figure 6.18, the field outside being zero everywhere.

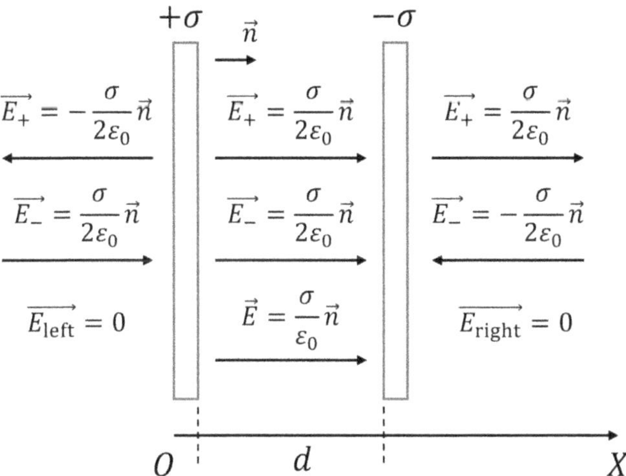

Figure 6.18. *The electric field between the plates of a capacitor.*

The potential difference, between the plates of capacitor will be:

$$U_{12} = V_1 - V_2 = \int_0^d \vec{E}\,\vec{dl} = \int_0^d \frac{\sigma}{\varepsilon_0}\vec{n}\,\vec{dx} = \frac{\sigma}{\varepsilon_0}d \tag{6.65}$$

Then the capacitance of the capacitor is:

$$C = \frac{Q}{U_{12}} = \frac{\sigma S}{\dfrac{\sigma}{\varepsilon_0}d} = \frac{\varepsilon_0 S}{d} \tag{6.66}$$

when a free space there is between the plates. If a material with dielectric constant ε_r fills entirely the space between the plates, the capacity is:

$$C = \frac{\varepsilon_0 \varepsilon_r S}{d} \tag{6.67}$$

A similar relationship was found to that of a spherical capacitor with a large medium radius and a very small spacing between armatures, so we can say that a planar capacitor can be considered as a part of a spherical capacitor with infinite radius, where S represents an element of finite dimensions from the surface of such a sphere.

The edge effect at an electric planar capacitor

Let there be a planar capacitor, formed by the plane armatures A_1 and A_2 assumed to have a length very large compared to their width, see figure 6.19. The A_1 plate is assumed to be charged with the $+Q$ charge which is distributed over its entire surface on both the inner and outer face, with the difference that on the inner face the charge density is much higher than the charge density found on the external face. The situation is similar on plate A_2, with the difference that on this armature the charge density is negative. The distance between the two plates being very small, the intensity of the electrostatic field between them can be considered uniform, the field lines are straight equidistant, normal on the armatures and the equipotential surfaces are flat parallel to the armatures. At the extremities the field lines spread and lengthen. As they are open curves that rest on the two plates, there will be field lines that start from the outer face of plate A_1, reaching the outer face of plate A_2. However, equipotential surfaces are closed surfaces that surround the armatures. The line integral of the electric field between two points, one on plate A_1 and the other on A_2, is the same whatever the position of the two points, on the external or internal faces, and represents the voltage U_{12} between the two plates:

$$U_{12} = V_1 - V_2 = \int_1^2 \vec{E}\,\vec{dl} \tag{6.68}$$

the surface of each plate being an equipotential surface. Previously deduced planar capacitor capacity relation:

$$C = \frac{\varepsilon S}{d} \tag{6.69}$$

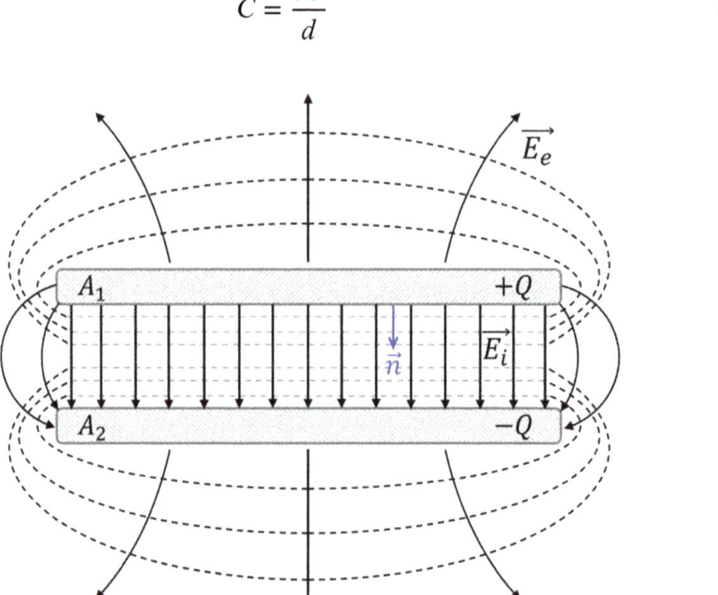

Figure 6.19. *The edge effect at a planar capacitor.*

is valid only for the region of space where the electric field is uniform. In reality, however, the ends of the plates together with their outer surfaces form an electrical capacitor having the capacity C' so that the total capacity of the capacitor is:

$$C_t = C + C' \tag{6.70}$$

The phenomenon of scattering of field lines and equipotential surfaces as well as the appearance of the capacitance C' constitute the edge effect of the capacitor. In practice $C' \ll C$ and it can be neglected, so that:

$$C_t = C \tag{6.71}$$

*However, when it is necessary to rigorously delimit the capacitance C of a planar capacitor, a device is made in which, each of the two plates A_1 and A_2 is surrounded by a circular surface, see figure 6.20. Between these circular surfaces are established the corresponding field lines of the outer surfaces of the plates, thus eliminating the edge effect. This surface system is the **guard ring**. In this assembly, the edge effects are transmitted to the guard ring, between the main armatures A, the electric field remaining perfectly uniform and thus the respective capacity can be determined rigorously. The edge effect occurs especially in the flat capacitor, but can occur in any form of capacitor, usually where the armatures have surfaces without complete symmetry.*

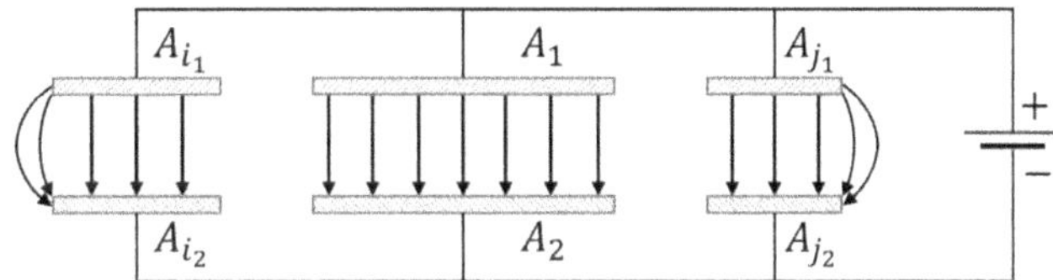

Figure 6.20. *Elimination of edge effects in the case of a flat capacitor, using the guard ring.*

6.4 Combinations of capacitors

The capacitance of the capacitor is limited by construction to a certain value, by the dimensions of the armatures, their shape, and the nature of the dielectric. To obtain a desired value capacity, the capacitors are connected in groups of capacitors. The grouping of the capacitors can be done in series, in parallel and mixed. We now consider more general cases in which several ideal capacitors are connected by conducting wires into a network such as that of figure 6.21. In the case of a network of capacitors with two access terminals A and B, the equivalent capacity is called the size:

$$C_{AB} = \frac{Q_A}{V_A - V_B} = \frac{Q_B}{V_B - V_A} \tag{6.72}$$

wherein $Q_A = -Q_B$ is the total charge absorbed at terminal A, when the voltage $U_{AB} = V_A - V_B$ is applied to the network, initially discharged. The equivalent capacity of a group of capacitors is therefore the capacity of a capacitor which being subjected to

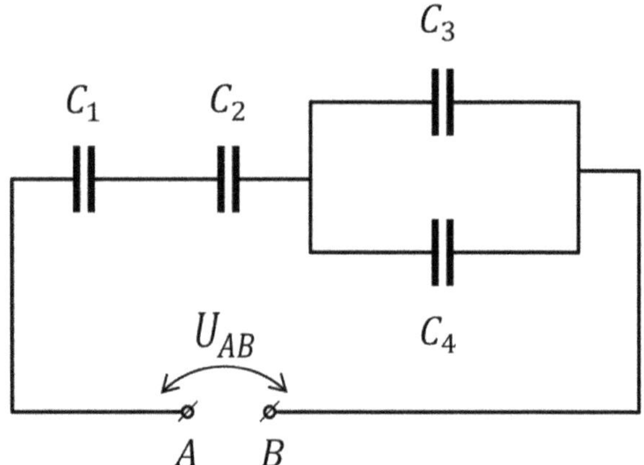

Figure 6.21. A capacitor network.

the same voltage as the capacitor system is charged with the same electrical charge as the given system. Consequently, no change is observed outside the system when replacing the network with the capacitor of equivalent capacity. To obtain the equivalent capacitance, we use two principles already established: *the conservation of charge and the electrostatic potential theorem* (the path independence of potential difference).

6.4.1 Capacitors in series

The series connection of the capacitors, figure 6.22, is made so that between the two external terminals A and B are constituted a succession of elements in which the output terminal of one is connected to the input terminal of the next. The voltage between the external terminals A and B, U_{AB} is equal to the sum of the voltages on each capacitor:

$$U_{AB} = \sum_{i=1}^{n} U_i \tag{6.73}$$

and the electric charge with which each of the capacitors C_1, C_2,..., C_n connected in series is charged, is the same:

$$q_1 = q_A = q_2 = \ldots = q_i = \ldots = q_n = q \tag{6.74}$$

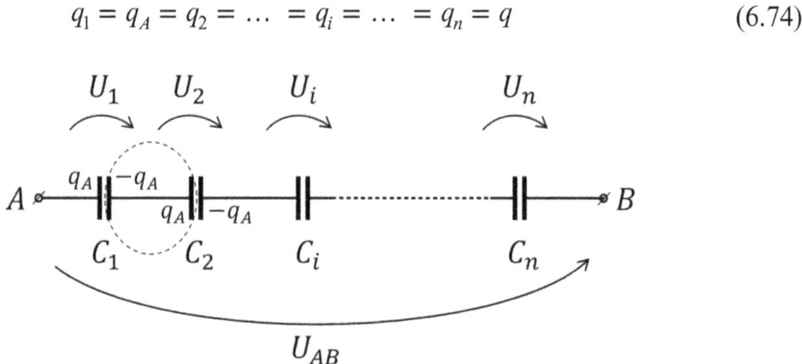

Figure 6.22. Capacitors in series.

Indeed, the charge $q_1 = q_A$, on the positive armature of the first capacitor determines by influence the appearance of the negative charge $-q_A$ on its other armature. Due to the conservation of charge, in the system consisting of the second armature of the first capacitor and the first armature of the second capacitor (see the dotted surface in figure 6.22), the system which must be neutral, results in:

$$-q_A + q_2 = 0 \Rightarrow q_2 = q_A = q \ \ldots \text{ and so on} \tag{6.75}$$

But:

$$U_i = \frac{q_i}{C_i} \Rightarrow U_{AB} = \sum_{i=1}^{n} U_i = \sum_{i=1}^{n} \frac{q_i}{C_i} = q_A \sum_{i=1}^{n} \frac{1}{C_i} \tag{6.76}$$

Then, according with the definition of the equivalent capacitance of the system:

$$C_{es} = \frac{q_A}{U_{AB}} = \frac{q_A}{q_A \sum_{i=1}^{n} \frac{1}{C_i}} \Rightarrow C_{es} = \frac{1}{\sum_{i=1}^{n} \frac{1}{C_i}} \Rightarrow \frac{1}{C_{es}} = \sum_{i=1}^{n} \frac{1}{C_i} \tag{6.77}$$

Or in terms of the elastances, the equivalent relation can be re-written as:

$$S_e = \sum_{i=1}^{n} S_i \tag{6.78}$$

In conclusion:
- The inverse of the equivalent capacity of a series-connected capacitor group is equal to the sum of the inverses of the component capacities;
- The charge with which the system is charged is the same as the charge with which each capacitor is charged;
- The voltage at the terminals of the group is equal to the sum of the voltages on each capacitor;
- Series coupling is indicated when the aim is to obtain an equivalent capacitor that can withstand higher voltages.

6.4.2 Capacitors in parallel

The grouping in parallel of several capacitors of capacity C_i ($i = 1, 2, \ldots, n$), see figure 6.23, is done by connecting together all the armatures of the same name, all the capacitors being charged at the same voltage. In this case, each capacitor is charged with the charge:

$$q_i = C_i U_{AB} \tag{6.79}$$

so that the total charge absorbed by system at terminal A when the U_{AB} voltage is applied is:

$$q_A = \sum_{i=1}^{n} q_i = \sum_{i=1}^{n} C_i U_{AB} = \left(\sum_{i=1}^{n} C_i \right) U_{AB} \tag{6.80}$$

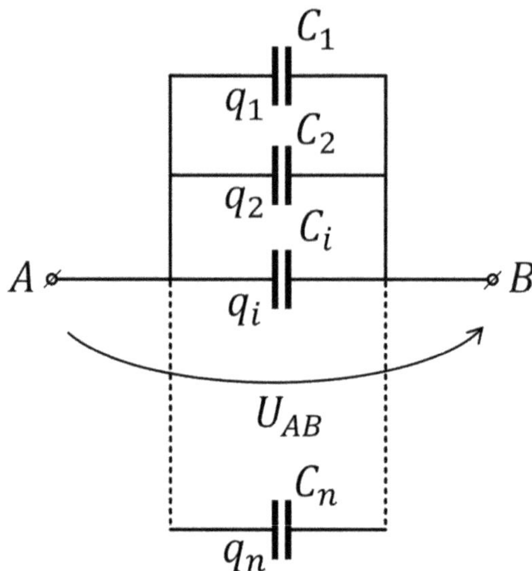

Figure 6.23. Capacitors in parallel.

According to the definition of the equivalent capacitance of the system, results:

$$C_{ep} = \frac{q_A}{U_{AB}} = \sum_{i=1}^{n} C_i \tag{6.81}$$

Consequently:
- The equivalent capacity of a battery of capacitors connected in parallel is equal to the sum of the capacities of the component capacitors.
- The voltage at the terminals of the group is the same as the voltage at the terminals of each capacitor.
- The charge with which the system is charged is equal to the sum of the charges with which each capacitor is charged.
- Parallel grouping is used when aiming to obtain a large capacity.

6.4.3 General networks

Many networks can be treated as a collection of series and parallel combinations and the equivalent capacity found by using equations (6.77) and (6.81) like in the case of the network from figure 6.21 where the equivalent capacitance is:

$$\frac{1}{C_e} = \frac{1}{C_1} + \frac{1}{C_2} + \frac{1}{C_3 + C_4} \tag{6.82}$$

But a network like that of figure 6.24 cannot be treated in this way—neither of the two capacitors are in series (common Q) or in parallel (common U)—so a general method based on the two laws mentioned above (the charge conservation law, and the electrostatic potential theorem) must be used.

Calculation of simple electrostatic networks

An electrostatic network is a set of capacitors and sources of electricity, connected in a somewhat well-defined way, determined by certain situations in practice. The elements of the network are the sides, nodes and meshes or loops of the network. The following rules apply:

- The side of an electrostatic network is called the unbranched element consisting of capacitors and sources, either together or separately;
- The meeting point of at least three sides is called the node;
- Any closed circuit with a certain number of sides that, starting from a certain node, reaches the same node, forms a loop or a network mesh.

For each node of the network, applying the *law of conservation of electric charge*, the relation can be written:

$$\sum_{i=1}^{n} q_i = \text{constant} \tag{6.83}$$

possibly, the constant can be zero. Based on the integral form of the *electrostatic potential theorem* ($\oint \overrightarrow{E} \ \overrightarrow{dl} = 0$), for each mesh of the network we can write the obvious relationship between the potential differences at the terminals of the component elements and the electromotive forces of the present energy sources:

$$\sum_{i=1}^{n} E_i = \sum_{i=1}^{m} U_i \tag{6.84}$$

where E_i is the electromotive voltage at the terminals of the source (positive in the sense of the internal field of the source which is from $-$ to $+$) and $U_i = \frac{q_i}{C_i} = q_i S_i$ is the potential difference between the armatures of a capacitor, considered positive also in the direction of the field between the armatures. So, equation (6.84) can be re-written as:

$$\sum_{i=1}^{n} E_i = \sum_{i=1}^{m} \frac{q_i}{C_i} = \sum_{i=1}^{m} q_i S_i \tag{6.85}$$

Of course, if the mesh of the network does not contain energy sources, it results in:

$$\sum_{i=1}^{m} \frac{q_i}{C_i} = \sum_{i=1}^{m} q_i S_i = 0 \tag{6.86}$$

In solving simple electrostatic circuits, the electromotive voltages of the sources, and the capacitance values of the capacitors are generally given and it is required to determine the charges q_i and drop voltages U_i at their terminals. For this, the relations (6.83) and (6.85) are sufficient (i.e., the law of conservation of charge for nodes and the electrostatic potential theorem for a loop, relations similar to the Kirchhoff theorems in steady-state current networks studied in electrokinetics).

It is shown that if N is the number of nodes in the network, L is the number of sides and B is the number of fundamental meshes (that do not overlap), the law of conservation of charge—relation (6.83)—can be applied by $p = N - 1$ times, and the potential theorem—relation (6.85)—can be applied by $B = L - N + 1$ times, respectively, so that a system of independent equations with L equations and L unknowns, is finally obtained. These relations for p and B, respectively, are called the topological equations of the network. From this general method with the help of special theorems are deduced various methods of calculation of electrostatic networks, methods that generally lead to simpler operations and the possibility of using the computer. They will be presented in detail in the analysis of steady-state current networks.

Application: particular capacitors network

In the network from figure 6.24: $C_1 = 25\ \mu F$, $C_2 = 100\ \mu F$, $C_3 = 50\ \mu F$, $C_4 = 100\ \mu F$, $C_5 = 50\ \mu F$ and $E = 1300\ V$. Calculate the charges with which each capacitor is charged and the voltage at the terminals of each capacitor.

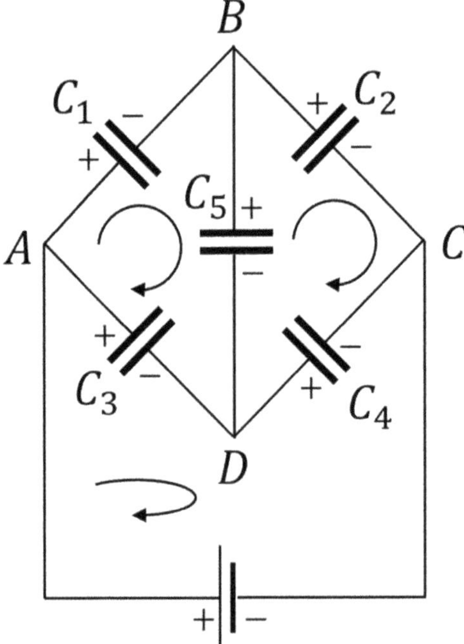

Figure 6.24. *A network which cannot be broken into series and parallel combinations.*

Solution: *the unknown charges q_1, q_2, q_3, q_4, q_5 of the capacitors and the charge q_0 transferred by the source to the network, are considered also unknown. We arbitrarily attribute polarity marks to the capacitor armatures. The circuit having $N = 4$ nodes and $L = 6$ sides, the relations for the nodes will be written three times and the relation for the independent meshes, also three times resulting the system of six equations with six unknowns, written below:*

$$\begin{cases} A: & q_0 = q_1 + q_3 \\ B: & -q_1 + q_2 + q_5 = 0 \\ D: & -q_3 + q_4 - q_5 = 0 \\ ABDA: & \dfrac{q_1}{C_1} + \dfrac{q_5}{C_5} - \dfrac{q_3}{C_3} = 0 \\ BCDB: & \dfrac{q_2}{C_2} - \dfrac{q_4}{C_4} - \dfrac{q_5}{C_5} = 0 \\ ADCA: & E = \dfrac{q_3}{C_3} + \dfrac{q_4}{C_4} \end{cases}$$

Solving the system results: $q_1 = 2.5 \cdot 10^{-2}$ *C,* $q_2 = 3 \cdot 10^{-2}$ *C,* $q_3 = 4.5 \cdot 10^{-2}$ *C,* $q_4 = 4 \cdot 10^{-2}$ *C,* $q_5 = -0.5 \cdot 10^{-2}$ *C,* $q_0 = 7 \cdot 10^{-2}$ *C,* $U_1 = 1000$ *V,* $U_2 = 300$ *V,* $U_3 = 900$ *V and* $U_5 = -100$ *V. The sign '$-$' in q_5 shows that the polarity at the capacitor terminals (5) is reversed from that considered. Knowing the total charge q_0 discharged from the source and absorbed by the terminal A of the circuit and the voltage at the source terminals which is even $U_{AC} = E = 1300$ V, results in the equivalent capacity of the network:*

$$C_e = \frac{q_0}{U_{AB}} = \frac{7 \cdot 10^{-2}}{1300} = 53.85 \cdot 10^{-6} \, F = 53.85 \, \mu F \tag{6.87}$$

The Δ–Y (triangle–star) transformation

The equivalent capacity of the previous network can be easily calculated if the transformation relationship of a part of the network from the triangle (Δ) configuration to the star (*Y*) configuration is known. These relations are very useful in calculating the equivalent capacity of the networks and their deduction means to widen the range of the analysis methods of the electrostatic networks. The transfiguration must be made so that the equilibrium state of the network does not change. In the case of the 'Star' configuration—figure 6.25(a), applying the Kirchhoff theorems, the below system is obtained, which solved leads to the determination of the charge q_1 absorbed by node 1, in the 'Star' configuration.

$$\begin{cases} q_1 + q_2 + q_3 = 0 \\ V_1 - V_3 = \dfrac{q_1}{C_1} - \dfrac{q_3}{C_3} \\ V_2 - V_1 = \dfrac{q_2}{C_2} - \dfrac{q_1}{C_1} \end{cases}$$

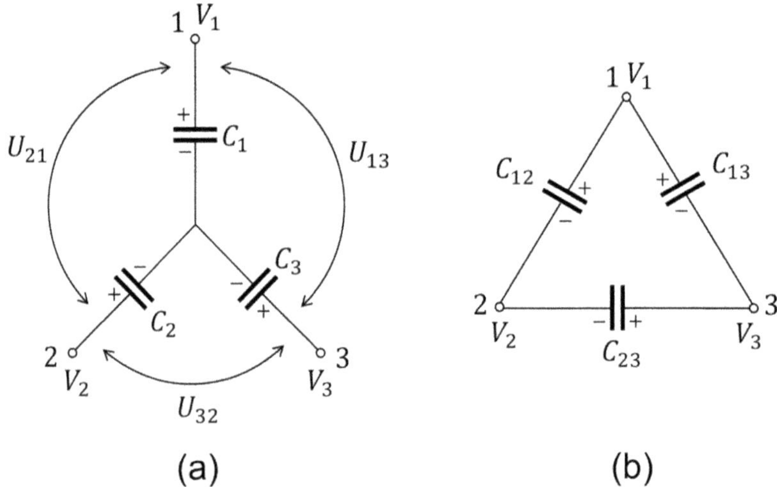

Figure 6.25. The transformation of an electrostatic network from the 'Star' ('Wye') configuration (a) to the 'Triangle' ('Delta') configuration (b).

Thus:

$$\begin{cases} q_3 = \dfrac{C_3}{C_1}q_1 - C_3(V_1 - V_3) \\[2ex] q_2 = \dfrac{C_2}{C_1}q_1 + C_2(V_2 - V_1) \end{cases}$$

$$\Rightarrow q_1\left(1 + \frac{C_3}{C_1} + \frac{C_2}{C_1}\right) = C_3(V_1 - V_3) + C_2(V_1 - V_2)$$

From the 'Star':

$$q_1 = \frac{C_1 C_3}{C_1 + C_2 + C_3}(V_1 - V_3) + \frac{C_1 C_2}{C_1 + C_2 + C_3}(V_1 - V_2) \qquad (6.88)$$

The charge q_1 absorbed by node 1, in 'Triangle' configuration—figure 6.25(b), is from the 'Triangle':

$$q_1 = q_{12} + q_{13} = C_{12}(V_1 - V_2) + C_{13}(V_1 - V_3) \qquad (6.89)$$

Identifying (6.88) with (6.89), we obtain:

$$\text{'Star'} \rightarrow \text{'Triangle'}: \begin{cases} C_{12} = \dfrac{C_1 C_2}{C_1 + C_2 + C_3} \\[2ex] C_{13} = \dfrac{C_1 C_3}{C_1 + C_2 + C_3} \\[2ex] C_{23} = \dfrac{C_2 C_3}{C_1 + C_2 + C_3} \end{cases} \qquad (6.90)$$

Solving the system (6.90) inversely, i.e., having as unknowns the capacities on the arms of the 'Star', and given those on the sides of the 'Triangle', we obtain:

$$\text{'Triangle'} \rightarrow \text{'Star'}: \begin{cases} C_1 = C_{12} + C_{13} + \dfrac{C_{12}C_{13}}{C_{23}} \\[2mm] C_2 = C_{12} + C_{23} + \dfrac{C_{12}C_{23}}{C_{13}} \\[2mm] C_3 = C_{13} + C_{23} + \dfrac{C_{13}C_{23}}{C_{12}} \end{cases} \tag{6.91}$$

6.5 Electric energy of a system of conductors

In this section, we are going to compute the energy (W) that can be 'stored' in the electric field generated by the charged conductors.

6.5.1 The energy stored in a capacitor

Consider a capacitor of capacitance C, with a potential difference V_{12} between the armatures. The charge Q is equal to $C \cdot V_{12}$. There is a charge Q on one armature and $-Q$ on the other. Suppose we increase the charge from Q to $Q + dQ$ by transporting a positive charge dQ from the negative to the positive armature, working against the potential difference V_{12}. The work that has to be done is $dW = V_{12}dQ = \frac{Q}{C}dQ$. Therefore, to charge the capacitor starting from the uncharged state to some final charge Q_f requires the work:

$$W = \frac{1}{C} \int_0^{Q_f} Q dQ = \frac{1}{2}\frac{Q_f^2}{C} \tag{6.92}$$

This is the energy (W) which is 'stored' in the capacitor and it can be also expressed by:

$$W = \frac{1}{2}CV_{12}^2 \tag{6.93}$$

6.5.2 Electric energy of a charged conductor

We can regard a single conductor remote from its surroundings as a special case and use the relations (6.92) and (6.93), replacing V_{12} with V (the potential of the conductor having the capacitance C, placed far away with any other conductor and considering that $V_\infty = 0$). In this case, the energy W which is 'stored' in the electric field of a charged conductor, can be expressed by:

$$W = \frac{1}{2}CV^2 = \frac{1}{2}QV = \frac{1}{2}\frac{Q^2}{C} \tag{6.94}$$

Indeed, let there be a conductor isolated in space, charged with a charge Q and which is at a potential V. To determine the electrostatic energy of this body, first assume the body in a neutral state ($Q = 0$, $V = 0$) then bring the charge Q from infinity on the conductor in infinitely small quantities. The operation is performed very slowly and isothermally so that the electrostatic equilibrium is never disturbed. The total mechanical work that must be performed under these conditions to bring the charge from infinity on the body to be charged is equal to the electrostatic energy of the considered conductor. Let x be a variable parameter ($0 \leqslant x \leqslant 1$) with which the electrization state of the charged body is characterized at any time. At some point, after a certain number of charging processes by bringing elementary charges dQ, the electrization state of the body is characterized by the charge Qx and the potential of the body Vx. To bring from infinity an elementary charge $dQ = Qdx$, on the body which is at the potential Vx, we must perform on the system a mechanical work:

$$-dL = dW = VxdQ = QVxdx \tag{6.95}$$

which is an increase in the electrostatic energy of the body. The total energy of the system is equal to the sum of the elementary mechanical works performed by bringing the elementary charges $dQ = Qdx$ on the charged body, being:

$$-L = W = QV \int_0^1 xdx = \frac{1}{2}QV = \frac{Q^2}{2C} = \frac{1}{2}CV^2 = \frac{1}{2}QV \tag{6.96}$$

6.5.3 Electric energy of a system of charged conductors

If we consider a system of n conductors, having in their final state the charges $Q_1,\ldots,$ Q_n and the potentials $V_1,\ldots,$ V_n, the V_i^{th} is the potential of the conductor carrying its own charge Q_i but also due to all the other charges. The final state of the system was obtained starting from the initial state where all the charges and the potentials are zero and bringing from infinity small charge amounts dq_i, on each conductor. At any moment each of the conductors carries the charge Q_ix and has the potential V_ix, where x is a parameter which characterizes the charging state of the conductor. The parameter x is zero at initial moment when it is uncharged and its potential is zero, respectively, and x is equal to unity at final state when the conductor has the final charge Q_i and the potential V_i. Then the small amount dq_i is also proportional to Q_i (i.e., $dq_i = Q_idx$). So, the work which must be done in charging the system with the elementary charges dq_i on each conductor is:

$$-dL = dW = \sum_{i=1}^{n} V_ixQ_idx = \sum_{i=1}^{n} Q_iV_ixdx \tag{6.97}$$

And then the total work done for charging the system in the final state will be:

$$-L = W = \int_0^1 \sum_{i=1}^{n} Q_iV_ixdx = \sum_{i=1}^{n} Q_iV_i \int_0^1 xdx = \frac{1}{2}\sum_{i=1}^{n} Q_iV_i \tag{6.98}$$

So, for any number of charged conductors the **total potential energy shared in the system** because of the positions of the charges and the electric forces between them is:

$$W_E = \frac{1}{2}\sum_{i=1}^{n} Q_i V_i \tag{6.99}$$

representing the **internal energy** of the system.

Observations:

- The potential energy of Q in an external field is QV, not $\frac{1}{2}QV$;
- When the forces are allowed to move the charges (i.e., through conducting wires), the electric energy decreases at the expense (dissipation) of heat energy and becomes zero when the conductors equalize their potentials;
- Alternatively, the forces may move the conductors as a whole when the electric energy decreases due to the external work done, as discussed later;
- In all the cases the zero of W is taken when all the charges are remote from each other at infinity or when all conductors are uncharged.

If we use Maxwell's equations which relate the conductor charges Q_i to their potentials V_i or reversely, we obtain other shapes for the total potential energy of the conductor system:

$$W_E = \frac{1}{2}\sum_{i,j} C_{ij} V_i V_j \tag{6.100}$$

or:

$$W_E = \frac{1}{2}\sum_{i,j} S_{ij} Q_j Q_i \tag{6.101}$$

The relations (6.99), (6.100) and (6.101) can be used to calculate the potential energy of various conductor systems. For example, in the case of a capacitor, its energy will be:

$$W = \frac{1}{2}\sum_{i} Q_i V_i = \frac{1}{2}(QV_1 - QV_2) = \frac{Q}{2}(V_1 - V_2) \tag{6.102}$$

or:

$$W = \frac{1}{2}\sum_{i,j} C_{ij} V_i V_j = \frac{1}{2}[C_{11}V_1^2 + 2C_{12}V_1V_2 + C_{22}V_2^2]$$

$$= \frac{1}{2}C(V_1 - V_2)^2 = \frac{Q}{2}(V_1 - V_2)$$

but:

$$C_{11} = C_{22} = -C_{12} = C \Rightarrow W = \frac{1}{2}C(V_1 - V_2)^2 = \frac{1}{2}CU_{12}^2 \tag{6.103}$$

or:

$$W = \frac{1}{2}\sum_{i,j} S_{ij}Q_iQ_j = \frac{1}{2}\left[S_{11}Q_1^2 + 2S_{12}Q_1Q_2 + S_{22}Q_2^2\right] =$$

$$= \frac{1}{2}[S_{11}Q^2 - 2S_{12}Q^2 + S_{22}Q^2] =$$

$$= \frac{1}{2}Q^2(S_{11} + S_{22} - 2S_{12})$$

But:

$$V_1 = S_{11}Q - S_{12}Q = Q(S_{11} - S_{12})$$
$$V_2 = -S_{22}Q + S_{21}Q = Q(S_{21} - S_{22})$$
$$V_1 - V_2 = [(S_{11} - S_{12}) - (S_{21} - S_{22})]Q = 2(S_{11} - S_{12})Q$$

Because $S_{11} = S_{22}$ and $S_{12} = S_{21}$, then:

$$C = \frac{Q}{V_1 - V_2} = \frac{1}{2(S_{11} - S_{12})} \Rightarrow 2(S_{11} - S_{12}) = \frac{1}{C}$$

Thus:

$$W = \frac{1}{2}Q^2(S_{11} + S_{22} - 2S_{12}) = 2(S_{11} - S_{12})\frac{1}{2}Q^2 = \frac{1}{2}\frac{Q^2}{C} \tag{6.104}$$

6.5.4 The electric energy in terms of $\vec{E}$

We now transfer our attention to the electric field rather than the charge and consider a single charged conductor at a potential V (zero at infinity) carrying a charge Q. The electric energy of the system is internal since Q is brought to the conductor against its own field and thus $W_E = \frac{1}{2}QV$ as in the last section. A small part of the surface carries a charge dQ and has energy $dW_E = \frac{1}{2}VdQ$ associated with it. The tube of lines of force generated by dQ is shown in figure 6.26, intersecting two typical equipotential surfaces differing in potential by dV and carrying out a small volume of area dS and thickness dr. If the electric field at the volume element is denoted by $\vec{E}$, then Gauss' law, applied to the volume bounded by the conducting surface, dS, and the sides of the tube, gives $EdS = \frac{dQ}{\varepsilon_0}$, so that:

$$dW_E = \frac{1}{2}VdQ = \frac{1}{2}\varepsilon_0 VEdS \tag{6.105}$$

But $V = -\int_{\infty}^{S'} Edr = \int_{S'}^{\infty} Edr$ and is the sum of elementary potential differences spreading from S' to infinity. Because $dQ = \varepsilon_0 EdS$ applies to all the equipotentials cut by the tube:

$$dW_E = \frac{1}{2}\varepsilon_0 EdS \int_{S'}^{\infty} Edr = \int_{S'}^{\infty} \frac{1}{2}\varepsilon_0 E^2 dSdr \tag{6.106}$$

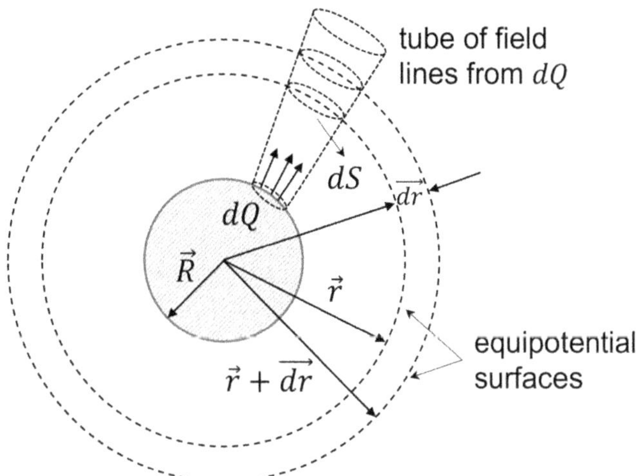

Figure 6.26. The tube of lines of force generated by Q, intersecting two typical equipotential surfaces differing in potential by dV.

The same expression applies to every tube starting from a dQ and the total energy is obtained by integrating over the other two coordinates represented by dS. Thus, if the volume $dSdr$ is denoted by dv:

$$W_E = \int_V \frac{1}{2}\varepsilon_0 E^2 dv \tag{6.107}$$

where the volume integration extends over the whole of space, including now the inside of the conductor where $\vec{E} = 0$ and no contribution to W_E is made. We have, in short, assumed that the every energy $\frac{1}{2}QV$ can be divided up so that $\frac{1}{2}dQdV$ is associated with the volume element dv; shown that $\frac{1}{2}dQdV$ can be written as $\frac{1}{2}\varepsilon_0 E^2$; and integrated over space to obtain the equation (6.107). This derivation can be extended to an ideal capacitor in which all the tubes of lines end on another conductor instead of going to infinity, and theme to sets of charged conductors by considering all tubes started from dQ on one conductor and ending at $-dQ$ on another. That the two expressions (6.99) and (6.107), give the same result for the total energy can be exemplified by a parallel-plate capacitor with a charge Q and capacitance $C = \frac{\varepsilon_0 S}{d}$. Equation (6.99) gives: $W = \frac{1}{2}CV_{12}^2 = \frac{1}{2}\frac{Q^2}{C}$ such as we have already shown. The equation (6.107) leads to:

$$W_E = \frac{1}{2}\int_0^d \varepsilon_0 \frac{\sigma^2}{\varepsilon_0^2} S dx = \frac{1}{2}\frac{\sigma^2}{\varepsilon_0} S d = \frac{Q^2}{2\varepsilon_0 S^2} S d = \frac{Q^2}{2\frac{\varepsilon_0 S}{d}} = \frac{Q^2}{2C}$$

i.e., the same as $\frac{1}{2}QU_{12}$ or $\frac{1}{2}CU_{12}^2 = \frac{1}{2}\frac{Q^2}{C}$. Equation (6.107) can be integrated by saying that the **energy per unit volume**, or the **energy density**, is $W_E = \frac{W_E}{V} = \frac{1}{2}\varepsilon_0 E^2$

but we shall not take the view that this implies only storage of energy in space since we must always use the integrated expression giving the total energy of the system **as a whole**; there is a little point in arguing the location of the energy.

6.5.5 The electrostatic energy

In the case of a point charge system, the electrostatic energy of the system is given, as we have also seen, from the general relation:

$$W = \frac{1}{2}\sum_{i=1}^{n} q_i V_i \tag{6.108}$$

where V_i is the potential of the point where the charge q_i ($i = 1, 2, \dots , n$) is placed, potential established without the contribution of this charge. In the case of continuous charge distributions we show that electrostatic energy is expressed by the relations:

$$W_\lambda = \frac{1}{2}\int_L \lambda V dl \quad \text{for linear charge distribution} \tag{6.109}$$

$$W_\sigma = \frac{1}{2}\int_S \sigma V dS \quad \text{for surface charge distribution} \tag{6.110}$$

$$W_\rho = \frac{1}{2}\int_V \rho V dV \quad \text{for volume charge distribution} \tag{6.111}$$

With these, the electrostatic energy of a region of space in which there are both conductors carrying electric charges, as well as point charges or charged bodies with linear, surface and volume charge distribution will be, according to the principle of superposition, given by:

$$W = \frac{1}{2}\sum_{i=1}^{n} q_i V_i + \frac{1}{2}\sum_{i=1}^{n}\sum_{j=1}^{n}\frac{q_i q_j}{4\pi\varepsilon r_{ij}} + \frac{1}{2}\int_L \lambda V dl + \frac{1}{2}\int_S \sigma V dS \\ + \frac{1}{2}\int_V \rho V dV \tag{6.112}$$

If for one or more conductors in the system, either the potential or the charge are zero, the corresponding terms in the sums in the relations (6.112) are null. At first glance, it would seem that the presence of these conductors is no longer taken into account in the calculation of electrostatic energy, but it must be remembered that those conductors cannot be removed from the system because this would change the equilibrium state of the system. Indeed, the potentials and charges at the surface of the bodies would change, which would lead to another value of the electrostatic energy of the system.

Further reading

[1] Secăreanu I, Ruxandra V, Gherbanovschi N, Logofătu M, Cazan-Corbasca M and Antohe Ş 1984 *Problems of Electricity and Magnetism (Culegere de Probleme de Electricitate şi Magnetism)* (Bucharest: University of Bucharest Publishing House)

[2] Landsberg G 1987 *Cours élémentaire de physique—Tome 2, Electricité et magnétisme* (Moscow: Mir)

[3] Jackson J D 1998 *Classical Electrodynamics* 3rd edn (New York: Wiley)

[4] Feynman R P, Leighton R B and Sands M 2011 *The Feynman Lectures on Physics, Vol II: The New Millennium Edition: Mainly Electromagnetism and Matter* (New York: Basic Books)

[5] Williams H S 2013 *A History of Science: The Leyden Jar Discovered* vol II part VI (Online) (https://www.gutenberg.org/files/1706/1706-h/1706-h.htm)

[6] Purcell E M and Morin D J 2013 *Electricity and Magnetism* 3rd edn (Cambridge: Cambridge University Press)

[7] Cross M and Plunkett E 2014 Capacitors and capacitance *Physics, Pharmacology and Physiology for Anaesthetists: Key Concepts for the FRCA* (Cambridge: Cambridge University Press), pp 60–2

[8] Griffiths D J 2017 *Introduction to Electrodynamics* 4th edn (Cambridge: Cambridge University Press)

[9] Agrawal N 2020 *Electrostatics: Current and Capacitors (for IIT-JEE)* (Independently Published)

IOP Publishing

Electrostatics
Formalism of the electrostatic field in vacuum and matter
Ştefan Antohe and Vlad-Andrei Antohe

Chapter 7

Dielectric materials

Dielectric materials are electric insulators that can be polarized by an applied electric field. When a dielectric material is placed in an electric field, electric charges do not flow through the material as in a conductor, but they only slightly shift from their average equilibrium positions inside the constituent atoms or molecules, causing the dielectric polarization. Studying the properties of dielectrics is important for understanding the storage and dissipation of electric energy in materials, as well as for explaining various phenomena in electronics, optics, solid-state physics and biophysics.

7.1 Bound charge and dielectrics

As opposed to conductors, *dielectrics* ('perfect dielectrics') have no free (mobile) charge. A 'good dielectric' has some free charge in the form of free electrons, as in a conductor. However, it is a very small amount relative to the amount of free charge in a 'good conductor'. The charges in a 'perfect dielectric' are strongly bound to the parent atoms. Even though a dielectric is electrically neutral, an externally applied electric field may cause microscopic separations of the centers of positive and negative charges which thus behave like dipoles of charges described in chapter 4. The charge separation distances are on the order of atomic dimensions, but a large number of such dipoles may provide a significant effect. This phenomenon is referred to as the **polarization of dielectric**. The polarization of a dielectric due to some externally applied electric field may occur as a result of three effects: *electronic polarization, ionic polarization,* or *orientational polarization.*

Electronic polarization occurs when the externally applied electric field causes a shift in the atom's positive and negative charges. Equilibrium is attained when the internal Coulomb' force produced by the charge separation balances the applied force. When the charge separation occurs, we essentially have a microscopic electric dipole.

doi:10.1088/978-0-7503-5859-0ch7　　　　7-1　　　　© IOP Publishing Ltd 2023

Ionic polarization occurs in molecules composed of positively and negatively charged ions. An externally applied electric field again results in a microscopic separation of charge centers thus resembling a dipole of charge.

Orientational polarization occurs in materials that possess permanent, microscopic separations of charge centers. A solid dielectric possessing persistent polarization is called an *electret* which is the analog of a magnet. In the absence of an applied electric field, these permanent dipoles are randomly oriented. In the presence of an applied electric field, these permanent dipoles tend to rotate to align with the applied field, as discussed when we referred to the ponderomotive actions exerted by an electric field on an electric dipole moment (see section 4.1). Materials (such as water) possessing these permanent dipoles in the absence of a field are said to be *polar substances* (*polar dielectrics*), unlike the materials having non-polar molecules which are called *non-polar dielectrics*.

In many substances, the polarization occurs as a combination of electronic, ionic and orientational polarization effects. However, each of these effects results in production of a dipole of charge, and we may view the result as a large number of dipoles in free space, each composed of charge $-q_b$ and q_b. The charge q_b is referred to as **bound charge**, since it is not normally available for conduction except under the application of very large electric fields. Each of these microscopic dipoles has a dipole moment:

$$\vec{p} = q_b \vec{l} \tag{7.1}$$

where each dipole is composed of bound charges $-q_b$ and q_b. This is illustrated for the case of orientational polarization in figure 7.1(a). The material can be characterized by a total electric moment $\vec{p}$ which is the vector sum of all individual microscopic dipole moments:

$$\vec{p} = \sum_{i=1}^{n} \vec{p_i} \tag{7.2}$$

A polarized dielectric is characterized by the vector quantity $\vec{p}$ at macroscopic level. This moment may have two components: the *temporal moment* $(\vec{p_t})$ dependent

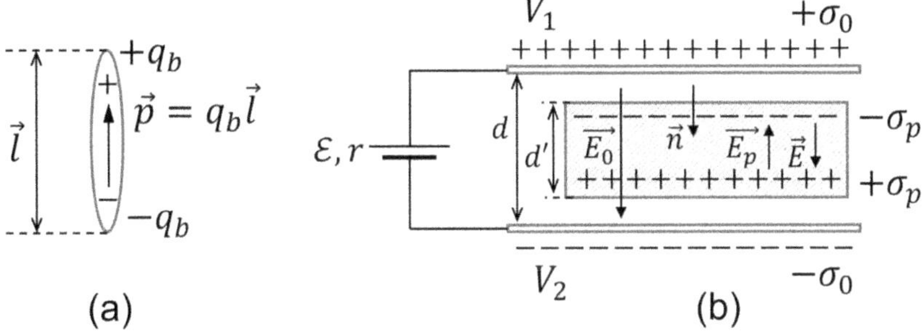

Figure 7.1. (a) Configuration of an electric dipole of moment $\vec{p} = q_b\vec{l}$. (b) Orientational polarization of a dielectric material.

on the external electric field and the *permanent moment* $(\overrightarrow{p_p})$ independent of the electric field, therefore:

$$\overrightarrow{p} = \overrightarrow{p_t} + \overrightarrow{p_p} \tag{7.3}$$

To characterize a polarized body at microscopic scale, the polarization vector $\overrightarrow{P}$ can be used, which is defined as the electric moment per unit volume, that is:

$$\overrightarrow{P} = \lim_{\Delta v \to 0} \frac{\sum_i \overrightarrow{p_i}}{\Delta v} = \frac{\overrightarrow{dp}}{dv} \tag{7.4}$$

where we sum (vectorially) all of the individual dipole moments in the volume Δv. In the absence of an externally applied electric field, the polarization vector would be zero: $\overrightarrow{P} = 0$.

For non-polar substances, in the absence of an electric field, there would be no shift of charge centers and no resultant dipoles. For polar substances, the permanent dipoles would be oriented randomly resulting in no net polarization, too. Note that the individual dipole moments are, by definition, directed from $-q_{\mathrm{b}}$ to q_{b} (in opposition to the direction of the electric field between these charges produced by the charge separation). For the block of material immersed in an electric field shown in figure 7.1(b), the tendency of the dipoles to align with the field causes charges to appear at the surfaces of the material. Thus, it appears that overall separation of charge has been achieved between the surfaces of the material as shown in figure 7.1(b). This is one of the important results of the polarization process.

7.2 Polarization state

We already know that two different electrization states there are named the electrization state of a charged body and the electrization state of a polarized body, are also called the 'polarization state'. Now, we have the opportunity to characterize the polarization state specific to the polarized dielectrics. *By definition, a body is in a 'polarized state' when an externally applied electric field acts on it with supplementary ponderomotive actions (couples and forces) and also it creates an electric field outside its own volume, even though it carries no true electric charge (in other words is uncharged).* The situation is like that when we have introduced the electric charge (q) to characterize the electrization state of a charged body (see section 1.1), now we will introduce the vector quantity $\overrightarrow{p}$ (electric moment) to characterize the polarization state. Obviously, if an external electric field acts on a polarized body with a couple, its polarized state must be characterized by a vector quantity such as $\overrightarrow{p}$. Indeed, the experiments showed that an external electric field $\overrightarrow{E}$ acts on a 'small polarized body' (see figure 7.2) having the moment $\overrightarrow{p}$, both with a couple:

$$\overrightarrow{M_p} = \overrightarrow{p} \times \overrightarrow{E} \tag{7.5}$$

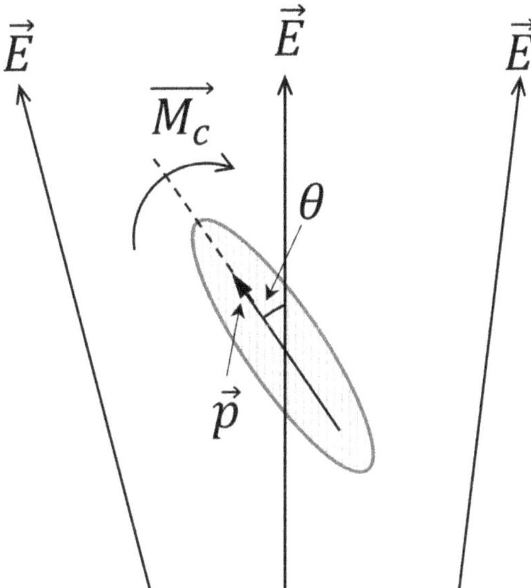

Figure 7.2. A small polarized body in an external non-uniform electric field.

and a force:

$$\vec{F} = \nabla\left(\vec{p}\cdot\vec{E}\right) = \left(p_x\frac{\partial E_x}{\partial x} + p_y\frac{\partial E_y}{\partial x} + p_z\frac{\partial E_z}{\partial x}\right)\vec{i}$$
$$+ \left(p_x\frac{\partial E_x}{\partial y} + p_y\frac{\partial E_y}{\partial y} + p_z\frac{\partial E_z}{\partial y}\right)\vec{j} \qquad (7.6)$$
$$+ \left(p_x\frac{\partial E_x}{\partial z} + p_y\frac{\partial E_y}{\partial z} + p_z\frac{\partial E_z}{\partial z}\right)\vec{k}$$

As shown by equation (7.6), measuring the components of the force experienced by a non-uniform electric field, and knowing the components of field gradient $\left(\frac{\partial E_x}{\partial x}, \frac{\partial E_x}{\partial y}, \ldots\right)$, we can experimentally determine the components p_x, p_y, p_z. So, the electric moment $\vec{p}$ of a polarized body is a **primitive physical quantity**. At microscopic level, a polarized body is characterized by the polarization vector $\vec{P}$ which will have the same two components: temporal and permanent, hence:

$$\vec{P} = \vec{P_t} + \vec{P_p} \qquad (7.7)$$

The temporal polarization is dependent on the electric field and it is different to zero as long as the applied electric field is non-zero. The permanent polarization may be different to zero even though there is no applied electric field. But it may be dependent on the various external conditions such as: the mechanical stress (in piezoelectricity), the temperature (in pyroelectricity), a previous temporal polarization of a liquid dielectric followed by its solidification (in ferroelectricity), etc. Some materials, such as quartz, Seignett salt and others are ferroelectric materials.

A dielectric can have a uniform polarization $\vec{P}$ or a non-uniform polarization $\vec{P}\left(\vec{r'}\right)$. If we know the polarization at a point $\left(\vec{r'}\right)$ inside the dielectric, the electric moment will be given by:

$$\vec{p} = \int_{V'} \vec{P}\left(\vec{r'}\right) dv' \tag{7.8}$$

the integral being extended over all the volume V' of the polarized dielectric.

Electric dipole–small polarized body equivalence theorem

In order to understand the effects of the presence of polarized dielectrics outside, let us see how the polarization vector $\vec{P}$ is related to the potential and field created around. But, for this purpose we need to know the mathematical relations which allow us to do it. Such a way results from the equivalence theorem between an electric dipole having the moment $\vec{p_d}$ and a small polarized body having the moment $\vec{p}$. *This theorem says that an electric dipole and a small polarized body are equivalent both as the ponderomotive actions they experience by an external electric field as well as the electric field created by them, if their moments are equal.* Indeed, we already know ponderomotive actions experienced on an electric dipole by an external field:

$$\begin{cases} \vec{M_d} = \vec{p_d} \times \vec{E} \\ \vec{F_d} = \nabla\left(\vec{p_d} \cdot \vec{E}\right) \end{cases} \tag{7.9}$$

which when $\vec{p_d} = \vec{p}$ are the same as those experienced on a small polarized body, see equations (7.5) and (7.6). So, the first part of the theorem is demonstrated. For the second part, we will consider an electric dipole (having the moment $\vec{p_d}$) and a small polarized body (moment $\vec{p}$) in the electric field $\vec{E_{V'}}$ of a point charge q', located at distance $\vec{r}$ with respect to the middle of dipole and small polarized body, respectively (see figure 7.3). The electric field $\vec{E_{V'}}$ acts both on the dipole with the force $\vec{F_d'}$ and on the small polarized body with the force $\vec{F'}$, respectively. According to 'Newton's Third Law of Motion'[1], the dipole acts on the point charge with the force $\vec{F_d} = q'\,\vec{E_{V_d}} = -\vec{F_d'}$ and the small polarized body acts on q' with the force $\vec{F} = q'\,\vec{E_V} = -\vec{F'}$, where $\vec{E_{V_d}}$ and $\vec{E_V}$ are the electric field of dipole at the point where the charge q' is located, and the electric field of the small polarized body at the same point, respectively. But, from the first part of the theorem $\vec{F_d'} = \vec{F'}$ and then $\vec{F_d} = \vec{F}$, resulting in $\vec{E_{V_d}} = \vec{E_V}$. So, we demonstrated the equivalence theorem, and as a result of its existence, we can use the expressions of the dipole potential or dipole electric field to calculate the potential or the electric field of a small polarized body.

---

[1] Also called the '*Law of Action and Reaction*', states that when two bodies interact, they apply forces to one another that are equal in magnitude and with opposite sense.

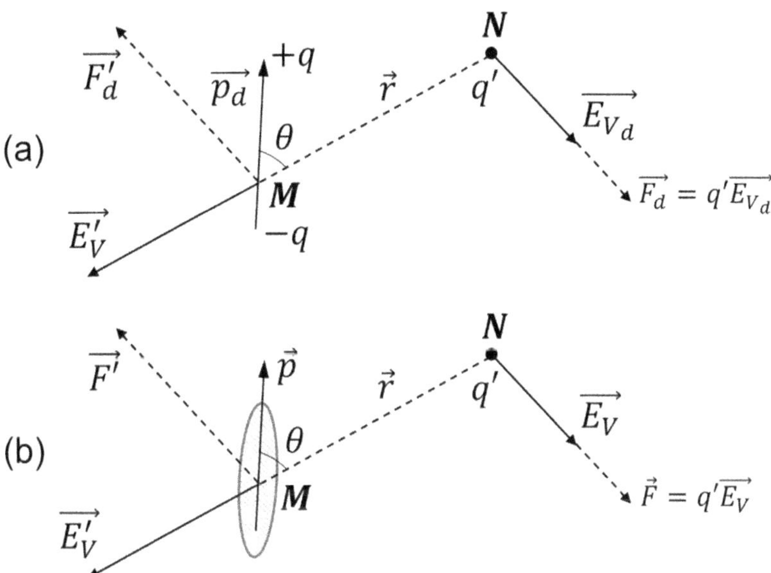

Figure 7.3. An electric dipole (a) and a small polarized body (b) in the electric field $\overrightarrow{E_V'}$ created in vacuum by the point charge q'.

7.3 Potential and electric field of the polarized dielectrics

In order to understand the effects of the presence of polarized dielectrics, let us see how the polarization vector $\overrightarrow{P}$ is related to potentials and charge distributions. Notice that the polarization vector $\overrightarrow{P}$ depends on the dipole moment $\overrightarrow{p}$, which in turn is related to the potential V. According to the equivalence theorem, we may use the expression of dipole potential as:

$$V_p = \frac{\overrightarrow{p}\,\overrightarrow{r}}{4\pi\varepsilon_0 r^3} \tag{7.10}$$

to calculate the potential of a polarized dielectric (see figure 7.4), where $\overrightarrow{r}$ is the vector directed from the origin toward point M. Now let's consider a volume element dv within a polarized dielectric of volume V, as shown in figure 7.4. The vector $\overrightarrow{r'}$ is directed from the origin to dv (the 'source point') and is given by:

$$\overrightarrow{r'} = x'\overrightarrow{i} + y'\overrightarrow{j} + z'\overrightarrow{k} \tag{7.11}$$

We wish to determine the potential at a point M located by the vector $\overrightarrow{r}$ where:

$$\overrightarrow{r} = x\overrightarrow{i} + y\overrightarrow{j} + z\overrightarrow{k} \tag{7.12}$$

From equation (7.10) we have:

$$dV_M = \frac{\overrightarrow{dp}\,\overrightarrow{R}}{4\pi\varepsilon_0 R^3} \tag{7.13}$$

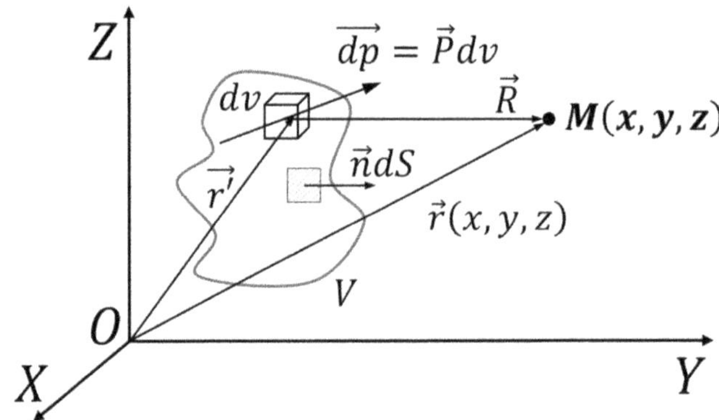

Figure 7.4. A dielectric with a non-uniform polarization.

where $\overrightarrow{dp}$ is the electric moment of the volume dv and $\vec{R}$ is the vector joining the volume element dv with point M, as shown in figure 7.4. Now, from the definition of polarization $\vec{P}$, we have $\overrightarrow{dp} = \vec{P}\,dv$. Hence, equation (7.13) may be written as:

$$dV_M = \frac{\vec{P}\vec{R}}{4\pi\varepsilon_0 R^3}dv \tag{7.14}$$

Since $\vec{R} = \vec{r} - \vec{r}' = (x - x')\vec{i} + (y - y')\vec{j} + (z - z')\vec{k}$, we can show that $\nabla\left(\frac{1}{R}\right) = -\frac{\vec{R}}{R^3}$ and $\nabla'\left(\frac{1}{R}\right) = \frac{\vec{R}}{R^3}$, where ∇' is the gradient with respect to the primed coordinates. Therefore, equation (7.14) becomes:

$$dV_M = \frac{1}{4\pi\varepsilon_0}\left[\vec{P}\nabla'\left(\frac{1}{R}\right)dv\right] \tag{7.15}$$

If we use the vector identity $\nabla(u\vec{v}) = \vec{v}\nabla u + u\nabla\vec{v}$, here $u = \frac{1}{R}$ and $\vec{v} = \vec{P}$, equation (7.15) becomes:

$$dV_M = \frac{1}{4\pi\varepsilon_0}\left[\nabla'\left(\frac{\vec{P}}{R}\right) - \frac{\nabla'\vec{P}}{R}\right]dv \tag{7.16}$$

Hence, the potential at point M is given by:

$$V_M = \frac{1}{4\pi\varepsilon_0}\int_V \left[\nabla'\left(\frac{\vec{P}}{R}\right) - \frac{1}{R}\nabla'\vec{P}\right]dv$$

$$= \frac{1}{4\pi\varepsilon_0}\int_V \nabla'\left(\frac{\vec{P}}{R}\right)dv + \frac{1}{4\pi\varepsilon_0}\int_V \frac{-\nabla'\vec{P}}{R}dv \tag{7.17}$$

The first term on the right side of equation (7.17) may be transformed into a surface integral by the divergence theorem (*Gauss–Ostrogradsky Theorem*—see appendix B.5.1), so that:

$$V_M = \int_S \frac{\overrightarrow{P}\,\overrightarrow{dS}}{4\pi\varepsilon_0 R} + \int_V \frac{-\nabla'\overrightarrow{P}}{4\pi\varepsilon_0 R}dv \tag{7.18}$$

Comparing equation (7.18) with the potential of a surface and volume charge distribution, we observe that the potential at point M arises from equivalent bound surface and volume charge densities, σ_p and ρ_p, respectively, such that:

$$V_M = \int_S \frac{\sigma_p dS}{4\pi\varepsilon_0 R} + \int_V \frac{\rho_p dv}{4\pi\varepsilon_0 R} \tag{7.19}$$

For equations (7.18) and (7.19) to be equivalent, we must have:

$$\sigma_p = \overrightarrow{P}\cdot\overrightarrow{n} = P_n \tag{7.20}$$

where P_n is the component of $\overrightarrow{P}$ normal to the surface, while:

$$\rho_p = -\nabla\overrightarrow{P} = -\mathrm{div}\,\overrightarrow{P} \tag{7.21}$$

Equation (7.19) shows that a polarized dielectric can be replaced by an equivalent surface and volume charge distributions for the purpose of determining the resultant electric field and electric potential. The potential at point M, outside of the polarized dielectric, will be given by these polarization charge distributions:

$$V_M = \frac{1}{4\pi\varepsilon_0}\int_S \frac{\sigma_p dS}{R} + \frac{1}{4\pi\varepsilon_0}\int_V \frac{\rho_p dv}{R} \tag{7.22}$$

and the electric field at the same point will be:

$$\overrightarrow{E_M} = -\nabla V_M = \frac{1}{4\pi\varepsilon_0}\int_S \frac{\sigma_p dS}{R^2}\overrightarrow{e_R} + \frac{1}{4\pi\varepsilon_0}\int_V \frac{\rho_p dv}{R^2}\overrightarrow{e_R} \tag{7.23}$$

Clearly, **the source of the polarization vector $\overrightarrow{P}$ is bound charge**. The units of $\overrightarrow{P}$ are '*Coulombs per square meter*', as is clear from the definitions of $\overrightarrow{p}$ and $\overrightarrow{P}$, and it is logical to expect that this charge per unit surface area would be related to $\overrightarrow{P}$ at this surface. Note the similarity of equation (7.21), i.e., $\rho_{vp} = -\nabla\overrightarrow{P}$, to differential form of Gauss' law, i.e., $\mathrm{div}\,\overrightarrow{E} = \frac{\rho}{\varepsilon_0}$. The negative sign in equation (7.21) is a result of the dipole moment $\overrightarrow{p}$ being directed from the negative charge of the dipole to the positive charge. This is opposite to the direction of an electric field produced by these two charges, field lines of the vector $\overrightarrow{P}$ terminate on polarization charges. The appearance of the polarization charge density is consistent with the requirement that the total polarization charge carried by a piece of dielectric should be zero. The surface charge on an element dS on the surface of the dielectric is $\overrightarrow{P}\,\overrightarrow{dS}$, and summing up surface and volume charges, the total polarization charge is:

$$\int_S \sigma_p dS + \int_V \rho_p dv = \int_S \overrightarrow{P}\,\overrightarrow{dS} - \int_V \nabla\overrightarrow{P}\,dv = \int_V \operatorname{div}\overrightarrow{P}\,dv$$
$$- \int_V \operatorname{div}\overrightarrow{P}\,dv = 0$$

(7.24)

The divergence theorem tells us that the right side is identically zero.

7.4 Polarization charge densities

The fictive charges defined by the relations (7.20) and (7.21), whose distribution in the volume or on the surface of a body, is equivalent to the state of real polarization of the respective body, are called polarization charges. If the volume of a polarized body divided into volume elements along the polarization lines is considered, each volume element can be replaced by an electric dipole, see figure 7.5(a). The excess dipole charge of a name represents the polarization charge q_P. To determine how much polarization charge there is, locally in a body, consider a closed surface Σ, in the electrically polarized body and draw the field lines of the polarization vector. If the polarization intensity $\overrightarrow{P}$ is lower in the part where the field lines of the polarization enter inside the surface than in the part through which these lines exit, then inside the surface has excess of negative dipole charges. If $\overrightarrow{P}$ were more intense at the entrance of Σ than $\overrightarrow{P}$ at the exit of Σ, then it would contain an excess of positive dipole charges. It is observed that the excess charge of a name over the opposite name depends only on the value of the electric polarization $\overrightarrow{P}$, on the surface Σ because the dipoles equivalent to the elements of the body away from Σ, both inside and outside it, do not contribute to the charge of polarization from Σ. So, only dipoles that are intersected by the surface Σ contribute to the polarization charge q_P inside the surface. Consider such a dipole intersected by the surface Σ equivalent to a cylindrical volume element with length l and the base area $dS \cos\alpha$, figure 7.5(b), where dS is the

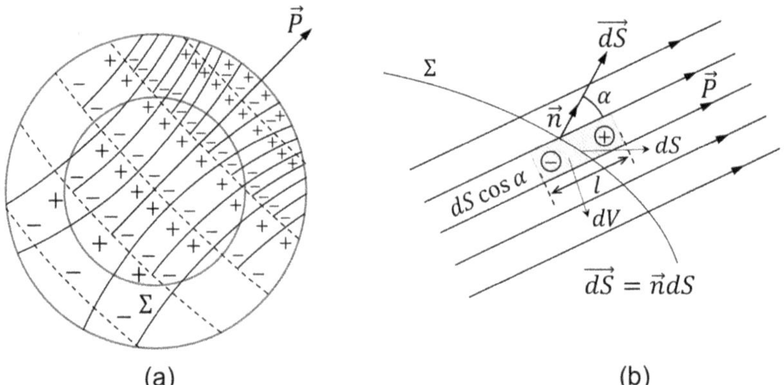

Figure 7.5. A polarized body divided in small volume elements (a) and a polarized charge (b).

area intercepted by the volume element on the surface Σ, and α the angle between the normal at Σ and $\vec{P}$. The electric moment of this element of volume is $dp = Pdv = PldS\cos\alpha$, therefore the dipole charge left outside the surface Σ is:

$$dq_p = \frac{dp}{l} = PdS\cos\alpha = \vec{P}\,\vec{dS} \qquad (7.25)$$

Obviously, inside the dipole charge $-dq_p = -\vec{P}\,\vec{dS}$ remained and this contributes to the total polarization charge q_P of Σ. So the polarization charge inside Σ is:

$$q_P = -\int_\Sigma \vec{P}\,\vec{dS} \qquad (7.26)$$

The total polarization charge located inside a closed surface Σ is equal and of opposite sign to the flux of the electric polarization intensity vector $\vec{P}$ through that surface.

7.4.1 Bulk polarization charge density

By analogy with the effective electric charge density, the density of the polarization charge in volume ρ_P can be defined by the relation:

$$q_{P_{\text{inside}}} = \int_{V_\Sigma} \rho_P dv \qquad (7.27)$$

and taking into account the Gauss–Ostrogradsky theorem (see appendix B.5.1):

$$q_{P_{\text{inside}}} = -\int_\Sigma \vec{P}\,\vec{dS} = \int_{V_\Sigma} -\operatorname{div}\vec{P}\,dv = \int_{V_\Sigma} \rho_P dv \qquad (7.28)$$

From equations (7.26), (7.27) and (7.28), results:

$$\rho_P = -\operatorname{div}\vec{P} \qquad (7.29)$$

So, indeed the second term in the expression of the potential equation (7.22) and the field equation (7.23), corresponds to a volume charge distribution described by ρ_P.

7.4.2 Surface polarization charge density

In the case of a polarized dielectric placed in vacuum, on its surface there is a polarization charge given by the dipole charges of a certain sign of the oriented dipoles. In the case of the dielectric inserted between the capacitor armatures in *Faraday's experiment*, it was the polarizing charge on its faces that caused the electric field to change. The evaluation of the charge density is done, using the representation from figure 7.6(a), as follows:

$$\vec{P} = \frac{\vec{dp}}{dV} = \frac{dQ_P\vec{l}}{dV} = \frac{dQ_P\vec{l}}{dS \cdot l} = \sigma_P \cdot \vec{n} \Rightarrow \sigma_P = \vec{P}\cdot\vec{n} \qquad (7.30)$$

So the surface polarization charge density is given in the case of the dielectric placed in vacuum by the normal component of the polarization intensity to the surface of

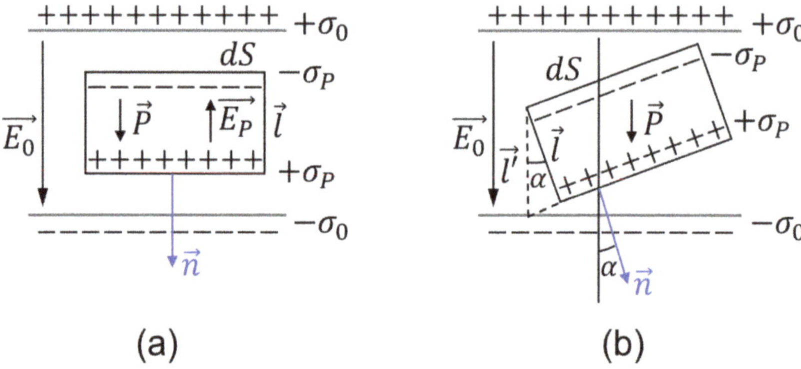

Figure 7.6. A dielectric placed in an electric field in the position: (a) the normal to the dielectric surface is parallel with the electric field $\overrightarrow{E_0}$; (b) the normal to the dielectric surface make the angle α with the electric field $\overrightarrow{E_0}$.

dielectric, whatever will be the orientation of the dielectric surface with respect to the field. Indeed, for a certain angle α made by the normal at the dielectric surface and $\overrightarrow{E_0}$, figure 7.6(b), the same reasoning leads to:

$$\overrightarrow{P} = \frac{\overrightarrow{dp}}{dV} = \frac{dQ_P \cdot \overrightarrow{l}}{dSl} = \sigma_P \frac{\overrightarrow{l}}{l'\cos\alpha} \Rightarrow \overrightarrow{P} \cdot \overrightarrow{n} = \sigma_P \frac{\overrightarrow{l}\,\overrightarrow{n}}{l'\cos\alpha} = \sigma_P \tag{7.31}$$

Basically, the surface charge density is due to the discontinuity of the polarization $\overrightarrow{P}$ at the surface S of the dielectric. Inside the polarized dielectric, it has a certain direction and a finite size, while outside it is zero. In general, when there are discontinuity surfaces of polarization when passing from one medium to another, the surface polarization charge density is given by the surface divergence of the polarization intensity vector $\overrightarrow{P}$:

$$\sigma_P = -\lim_{dS\to 0} \frac{\int_\Sigma \overrightarrow{P}\,\overrightarrow{dS}}{dS} = -\mathrm{div}_S\,\overrightarrow{P} \tag{7.32}$$

In order to calculate the surface divergence of $\overrightarrow{P}$, as a function of $\overrightarrow{P_1}$ and $\overrightarrow{P_2}$ values of vector $\overrightarrow{P}$ on the two sides 1 and 2 of the discontinuity surface, respectively, the surface of a very flat parallelepiped is chosen as closed surface Σ, figure 7.7 and using equation (7.32), we obtain:

$$\mathrm{div}_S\,\overrightarrow{P} = \frac{\overrightarrow{P_1}\,\overrightarrow{n_1}dS + \overrightarrow{P_2}\,\overrightarrow{n_2}dS}{dS} = \overrightarrow{P_1}\,\overrightarrow{n_1} + \overrightarrow{P_2}\,\overrightarrow{n_2}$$

If the normal $\overrightarrow{n_{12}}$ is considered positive when it is oriented from medium (1) toward medium (2), then:

$$\overrightarrow{n_{12}} = \overrightarrow{n_2} \quad \text{and} \quad \overrightarrow{n_{12}} = -\overrightarrow{n_1} \tag{7.33}$$

$$\mathrm{div}_S\,\overrightarrow{P} = \overrightarrow{n_{12}}\left(\overrightarrow{P_2} - \overrightarrow{P_1}\right) \tag{7.34}$$

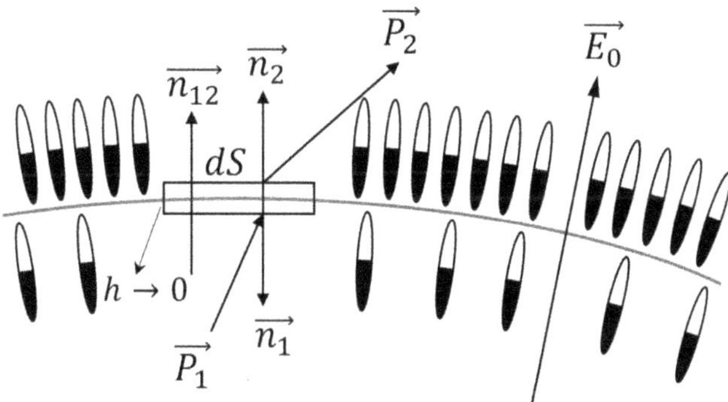

Figure 7.7. Polarization charge density at the interface between two different dielectric media (black ends carry negative charge).

With this, equation (7.32) becomes:

$$\sigma_P = -\left.\overrightarrow{n_{12}}\left(\overrightarrow{P_2} - \overrightarrow{P_1}\right)\right|_{\Sigma} = -\left.(P_{2n} - P_{1n})\right|_{\Sigma} \tag{7.35}$$

If a dielectric body is placed in vacuum, $\overrightarrow{P_2} = 0 \Rightarrow \sigma_P = \overrightarrow{P} \cdot \overrightarrow{n}$, then relation (7.30) is obtained again. If $\overrightarrow{P_1} \neq \overrightarrow{P_2}$, of course $\sigma_P = \sigma_{P_1} + \sigma_{P_2}$. But $\sigma_{P_1} = \overrightarrow{P_1}\ \overrightarrow{n_{12}}$ are positive charges formed from the positive ends of the electric dipoles, unlike $\sigma_{P_2} = \overrightarrow{P_2}\ \overrightarrow{n_{12}}$ representing the negative charge given by the negative ends of the dipoles, then:

$$\sigma_P = \overrightarrow{P_1}\ \overrightarrow{n_{12}} - \overrightarrow{P_2}\ \overrightarrow{n_{12}} = -\overrightarrow{n_{12}}\left(\overrightarrow{P_2} - \overrightarrow{P_1}\right)$$

Having the polarization charge densities ρ_P and σ_P, the total polarization charge can be computed by:

$$q_P = \int_{V_2} \rho_P dv + \int_{\Sigma} \sigma_P dS \tag{7.36}$$

ρ_P and σ_P can be different than zero but $q_P = 0$, because the dielectric which initially was neutral from an electrical point of view, it rests neutral after polarization, too. In the case of a dielectric polarized in a radial electric field (as a dielectric sphere carrying inside a point charge Q, see figure 7.8(a)), at its center there is an accumulation of negative charge due to the fact that divergence of $\overrightarrow{P}$ at the center is positive. On the surface of the dielectric there is an accumulation of positive charge. In the case of polarization of a homogeneous and isotropic dielectric in a uniform electric field, figure 7.8(b), the bulk charge density is zero and the polarization charges appear on the faces of the dielectric parallel to the armatures $+\sigma_P$ and $-\sigma_P$, so that the total polarization charge is null. Usually the calculation of bulk polarization charge density, $+\sigma_P$ and $-\sigma_P$ is done using the relations (7.29), (7.32) and (7.35), after calculating the polarization as a function of the polarizing electric field.

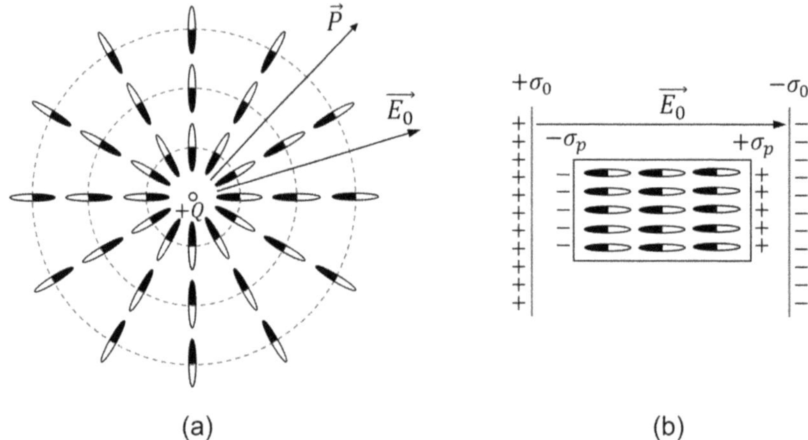

(a)   (b)

Figure 7.8. A polarized dielectric in an electric field: (a) non-uniform and (b) uniform.

7.5 Electric displacement vector

In the general case, dielectric materials may not be electrically neutral even when they are non-polarized. If the dielectric carries a charge density ρ_f of 'free' charges, representing a net surplus or deficit of electrons in the atoms of the dielectric, then the total charge density is:

$$\rho = \rho_f + \rho_p \tag{7.37}$$

It must be emphasized that both terms in equation (7.37) represent real physical charge. Nevertheless, we shall find the distinction between free charge and polarization charge to be a useful one, particularly when we consider the energy associated with fields in a dielectric material. The macroscopic electric field $\vec{E}$ is related to the total charges density, and in matter Gauss' law becomes:

$$\operatorname{div} \vec{E} = \frac{\rho}{\varepsilon_0} = \frac{\rho_f + \rho_p}{\varepsilon_0} = \frac{\rho_f - \operatorname{div} \vec{P}}{\varepsilon_0} \tag{7.38}$$

Re-arranging equation (7.38), we can write:

$$\varepsilon_0 \operatorname{div} \vec{E} + \operatorname{div} \vec{P} = \operatorname{div}\left(\varepsilon_0 \vec{E} + \vec{P}\right) = \operatorname{div} \vec{D} = \rho_f \tag{7.39}$$

Here we have introduced the new vector field:

$$\vec{D}(\vec{r}) = \varepsilon_0 \vec{E}(\vec{r}) + \vec{P}(\vec{r}) \tag{7.40}$$

With which the **differential form of Gauss' law** in dielectrics becomes:

$$\operatorname{div} \vec{D} = \rho_f \tag{7.41}$$

The vector field $\vec{D}$ has the same dimensions as $\vec{P}$, namely dipole moment per unit volume and it is called the **electric displacement**. Equation (7.41) is **still really the differential form of Gauss' law**, now modified in such a way that the effects of polarization charge are automatically included. Gauss' law for the electric field looks a bit different in that it contains a factor $1/\varepsilon_0$ on the right side. This is simply because we have chosen to give $\vec{D}$ and $\vec{E}$ different dimensions. Gauss' law for the electric field can be expressed alternatively in the integral form $\int_S \vec{E}\,\vec{dS} = \frac{Q_f}{\varepsilon_0}$, and similarly Gauss' law for the displacement vector $\vec{D}$, equivalent to equation (7.41), will be:

$$\int_S \vec{D}\,\vec{dS} = \int_V \operatorname{div}\vec{D} = \int_V \rho_f dV = Q_f \tag{7.42}$$

or in other words: **flux of $\vec{D}$ out of a closed surface S is equal to the total free charge enclosed within S**. Unlike the electric field $\vec{E}$ (which is the force acting on unit charge) or the polarization $\vec{P}$ (the dipole moment per unit volume), the electric displacement $\vec{D}$ has no clear physical meaning. The only reason for introducing it is that it enables one to calculate fields in the presence of dielectrics without first having to know the distribution of polarization charges such as we can see in the following example.

Exercise: distributed capacitance of a coaxial cable

A coaxial cable has the dielectric formed from two coaxial cylindrical shells with the dielectric constants ε_{r_1} and ε_{r_2}. The radius of the interface between the two cylindrical shells is b (see figure 7.9). The central conductor having the radius a and the external cylindrical conductor on radius c, both of length l, represents the armatures of a cylindrical capacitor. Find the distributed capacitance of the coaxial cable.

__Solution:__ when a potential difference $V_1 - V_2$ is applied between the central and outer conductors (armatures), on the inner conductor, there is the free charge Q uniformly distributed with the surface charge density σ, so that $Q = 2\pi a l \sigma$. Using the integral

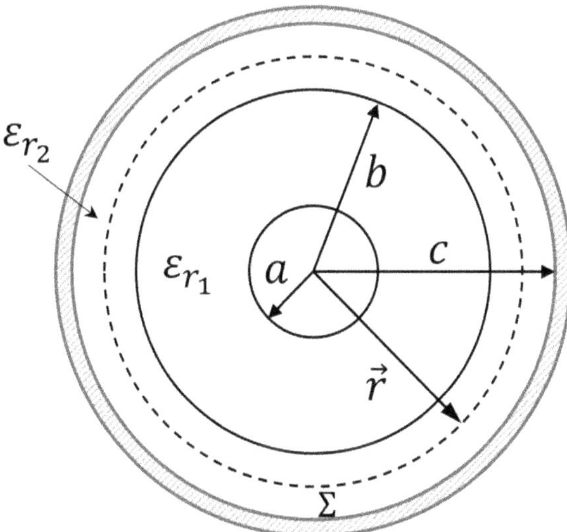

Figure 7.9. *Cross-section through a coaxial cable with two dielectric shells.*

form of Gauss' law, at the distance r from the axis of the capacitor, the displacement vector $\vec{D}$ will be:

$$\int_{\Sigma} \vec{D}\,\overrightarrow{dS} = Q \Rightarrow 2D\pi rl = 2\pi a\sigma l \Rightarrow D(r) = \frac{\sigma a}{r} \tag{7.43}$$

with radial symmetry. If the displacement vector D(r) has as source just the free charge σ, for the electric field in the two dielectric cylindrical shells, this results in:

$$\varepsilon_0\varepsilon_{r_1}E_1 = \varepsilon_0\varepsilon_{r_2}E_2 = D(r) = \frac{\sigma a}{r} \Rightarrow$$

$$\Rightarrow E_1(r) = \frac{\sigma a}{\varepsilon_0\varepsilon_{r_1}r} \quad \text{and} \quad E_2(r) = \frac{\sigma a}{\varepsilon_0\varepsilon_{r_2}r} \tag{7.44}$$

In other words, without knowledge about the polarization charge distributions in the dielectric shells, we can compute the electric field inside them and the potential difference between the armatures can be computed as:

$$V_1 - V_2 = \int_a^c \vec{E}\,\overrightarrow{dr} = \int_a^b \vec{E_1}\,\overrightarrow{dr} + \int_b^c \vec{E_2}\,\overrightarrow{dr} = \frac{\sigma a}{\varepsilon_0}\left[\frac{1}{\varepsilon_{r_1}}\ln\frac{b}{a} + \frac{1}{\varepsilon_{r_2}}\ln\frac{c}{b}\right] \tag{7.45}$$

Then the distributed capacitance of the coaxial cable becomes:

$$C = \frac{Q}{V_1 - V_2} = \frac{2\pi\sigma a\varepsilon_0}{\sigma a\left[\dfrac{1}{\varepsilon_{r_2}}\ln\dfrac{b}{a} + \dfrac{1}{\varepsilon_{r_1}}\ln\dfrac{c}{b}\right]} = \frac{2\pi\varepsilon_0\varepsilon_{r_1}\varepsilon_{r_2}}{\varepsilon_{r_2}\ln\dfrac{b}{a} + \varepsilon_{r_1}\ln\dfrac{c}{b}} \tag{7.46}$$

the parameter very important when the performances of the cable are tested with respect to the transmission of the high frequency signals.

We notice therefore that in the dielectric, the electric field varies, depending on $\frac{1}{r}$, as if there were a vacuum between the armatures. Everywhere the field is non-zero and the permittivity is constant, the Maxwell equation for dielectric:

$$\text{div}\,\vec{D} = \rho_f \tag{7.47}$$

becomes:

$$\text{div}\,\vec{E} = \frac{\rho_f}{\varepsilon_0\varepsilon_r} \tag{7.48}$$

equation (7.48) is the same as that satisfied in vacuum by the same charge distribution with the only difference that there appears in addition ε_r. The presence of the dielectric, placed in the electric field between the armatures, does not alter the shape of the electric field as orientation and dependence of r, but only reduces the magnitude of the field intensity by ε_r times, in each dielectric medium. The statement is valid when placing the dielectric in an external field because in a simple polarized dielectric, the field inside is different both in size and orientation than that generated outside.

Relative permittivity and electric susceptibility. $\vec{D}, \vec{E}, \vec{P}$ relationship

But it is important that we now review these results. Consider the pair of parallel conducting plates shown in figure 7.10(a), which are very large in extent and connected by a battery. The medium between the two plates is free space. When the battery is connected, free charge will be transferred to these plates from the battery having the surface charge densities σ_f and $-\sigma_f$, respectively. As a result, an electric field will be present between the plates in free space $\vec{E_0} = \frac{\sigma_f}{\varepsilon_0}\vec{n}$, and because $\vec{P} = 0$, $\vec{D} = \varepsilon_0 \vec{E_0} = \sigma_f \vec{n}$.

Now consider the insertion of a block of dielectric between the plates, as shown in figure 7.10(b). The electric field polarizes the dielectric, causing net positive and negative bound charges (shown by small circles) to appear on opposite surfaces of the dielectric. Polarization charges induced on the surface of a dielectric material make a contribution to the macroscopic electric field inside the material. Their contribution is called the 'depolarizing field' because the sign of the induced charge always ensures that the field just inside the dielectric surface is less than the field just outside. This is illustrated in figure 7.10(b) which shows a block of dielectric material in the uniform electric field $\vec{E_0}$. The depolarizing fields point away from positive polarization charges towards the negative charges, i.e., from the left to the right, thus partly canceling the external field. The net macroscopic field is therefore less inside the dielectric than outside. This is represented in the figure by the lower density of field lines inside the dielectric. As a result of the presence of the depolarizing field, the capacitance of a capacitor is increased by filling it with dielectric material. For example, suppose that the plates of the parallel-plate capacitor in figure 7.11 have an area S and are separated by a distance d. When the capacitor is in vacuum (figure 7.11(a)), and there are free surface charge densities $\pm\sigma_f$ on the plates, the field between them is of magnitude formula, just as in the previous case. We have already shown that the capacity in vacuum is:

$$C_0 = \frac{Q_0}{U_0} = \frac{\sigma_f S}{E_0 d} = \frac{\sigma_f S}{\frac{\sigma_f}{\varepsilon_0} d} = \frac{\varepsilon_0 S}{d} \tag{7.49}$$

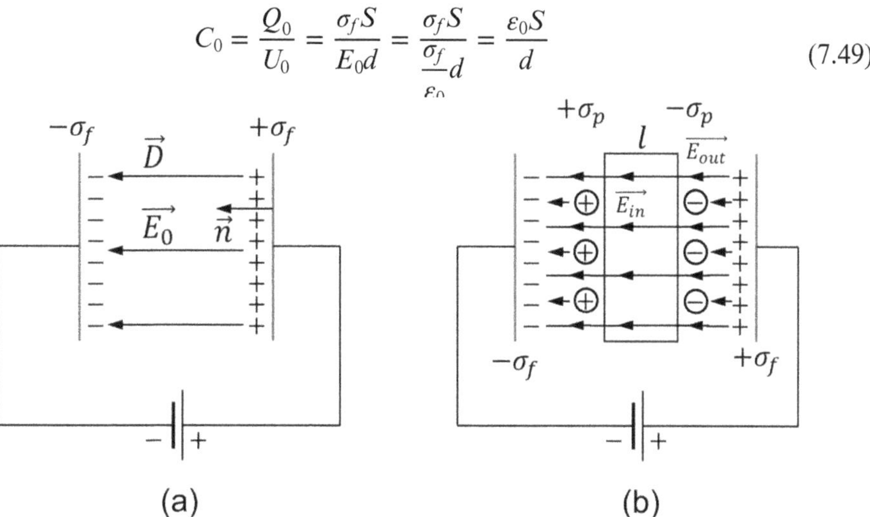

(a)   (b)

Figure 7.10. (a) A parallel-plate capacitor. (b) Increase of free charge by the polarization of a dielectric.

1. Now we keep **the same charge** densities $\pm\sigma_f$ on the conducting plates and fill the space between them with a slab of dielectric. Induced charges appear on the dielectric, causing a depolarizing field, and the magnitude of the macroscopic field is reduced by a factor ε_r, becoming:

$$E = \frac{E_0}{\varepsilon_r} = \frac{\sigma_f}{\varepsilon_0 \varepsilon_r} \tag{7.50}$$

It is found experimentally that for field below the field at which breakdown occurs, the factor ε_r is a constant depending only on the nature of the dielectric and not on size and shape of the capacitor. Charge and voltage on the capacitor remain proportional to one another, and the capacitance has a new value:

$$C = \frac{\varepsilon_0 \varepsilon_r S}{d} = \varepsilon_r C_0 \tag{7.51}$$

in other words, it increases by ε_r times. The dimensionless constant ε_r is called, as we already know, the **relative permittivity** of the dielectric material (or dielectric constant). The relative permittivity ε_r of a dielectric material is usually found by measuring the capacitance C_{air} of a capacitor with air between its plates and the capacitance C of the same capacitor when the space between the plates is filled with the dielectric. The ratio of capacitances $C/C_{\text{air}} = \frac{\varepsilon_0 \varepsilon_r}{\varepsilon_0 \varepsilon_{r_{\text{air}}}} = \simeq \varepsilon_r / \varepsilon_{r_{\text{air}}}$ and hence:

$$\varepsilon_r = \varepsilon_{r_{\text{air}}} \frac{C}{C_{\text{air}}} \tag{7.52}$$

Values of relative permittivity for a number of substances are given in table 7.1. One sees from the table that the capacitance of a capacitor is several times larger when it is filled with a solid or liquid dielectric than in vacuum. Many of the capacitors used as components in electronic circuits are in fact made with solid dielectrics such as polystyrene in the form of thin sheets coated with conducting films. The use of solid dielectrics gives a bonus to the value of the capacitance because the relative permittivity is larger than one, and also enables much smaller spacing to be

Table 7.1. The relative permittivity of various materials at room temperature.

Substance	Relative permittivity
Air at atmospheric pressure	1
Carbon tetrachloride (CCl_4)	2.24
Transformer oil	2.2
Paraffin wax	2–2.5
Polyethylene	2.3
Nylon	3.5
Porcelain	6
Mica	7

maintained between the conductors than is possible with air-filled capacitors. Typically, a 1 μF capacitor capable of sustaining 50 V across its conductors is about 1 cm in diameter and a few cm long. In contrast, the capacitance of air-filled capacitors is not usually more than a few hundred pF.

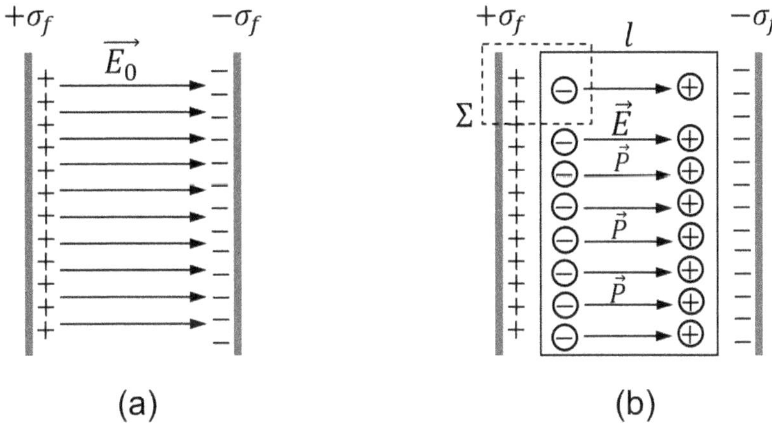

Figure 7.11. (a) The parallel-plate capacitor with air between the plates. (b) When a slab of dielectric is inserted into a parallel-plate capacitor, the electric field between the plates is reduced.

2. Now we keep the same potential difference U_0 between the plates of the capacitor by connecting it to a battery. The increase in capacitance must be accompanied by an increase in the conduction charge by a factor ε_r. Since $U = U_0$ in $Q = CU_0 = \varepsilon_0\varepsilon_r C_0 U_0 = \varepsilon_r Q_0$ is constant ($U = U_0$). The electric field strength in dielectric, like $U = U_0$ remains constant, $E = \dfrac{U}{d} = \dfrac{U_0}{d} = E_0$, because the battery supplies an extra charge density $(\varepsilon_r - 1)\sigma_f$, which cancels the opposite charge density σ_p on the dielectric surface, generating the depolarizing field. Now:

$$E = \frac{\sigma_f + (\varepsilon_r - 1)\sigma_f}{\varepsilon_0\varepsilon_r} = \frac{\varepsilon_r\sigma_f}{\varepsilon_0\varepsilon_r} = \frac{\sigma_f}{\varepsilon_0} = E_0 = \frac{U_0}{d} \tag{7.53}$$

So, the effective surface charge density on the dielectric σ_p, is $(\varepsilon_r - 1)\sigma_f$, and because the electric field $\overrightarrow{E}$ in dielectric is now $E = \dfrac{\sigma_f}{\varepsilon_0}$, hence σ_p is given by:

$$\sigma_p = \varepsilon_0(\varepsilon_r - 1)E \tag{7.54}$$

But the fact that the relative permittivity is found to be constant means that the polarization $\overrightarrow{P}$ which is directly related by σ_p ($\sigma_p = \overrightarrow{P}\overrightarrow{n}$) is proportional to the electric field $\overrightarrow{E}$. By applying again Gauss' law, we can see that this is so far the parallel-plate capacitor after it has been filled with a dielectric.

The surface Σ in figure 7.11(b) now encloses a negative polarization charge $-\sigma_p dS$, as well as the positive charge $\sigma_f dS$ on the conducting plate. The magnitude of the polarization charge within Σ is given by $-\sigma_p dS = -\overrightarrow{P}\,\overrightarrow{dS}$, and the total charge within Σ is therefore $\left[\sigma_f + (-\sigma_p)\right]dS = -(\sigma_f{-}P_n)dS$. By Gauss' law we have: $EdS = \frac{1}{\varepsilon_0}(\sigma_f - P_n)dS$ and substituting $\sigma_f = \varepsilon_0\varepsilon_r E$ from equation (7.50), we have $\overrightarrow{E}\,\overrightarrow{dS} = \frac{1}{\varepsilon_0}\left(\varepsilon_0\varepsilon_r\overrightarrow{E} - \overrightarrow{P}\right)\overrightarrow{dS}$, leading to:

$$\overrightarrow{P} = \varepsilon_0(\varepsilon_r - 1)\overrightarrow{E} \tag{7.55}$$

Since the vectors $\overrightarrow{P}$ and $\overrightarrow{E}$ point out in the same direction, this equation can be rewritten as:

$$\overrightarrow{P} = \varepsilon_0(\varepsilon_r - 1)\overrightarrow{E} = \varepsilon_0\chi_E\overrightarrow{E} \tag{7.56}$$

where the dimensionless constant of proportionality:

$$\chi_E = \varepsilon_r - 1 \tag{7.57}$$

is called the **electric susceptibility** of the material. Returning to the electric displacement $\overrightarrow{D}$, equation (7.47), since $\overrightarrow{P} = \varepsilon_0\chi_E\overrightarrow{E}$, the electric displacement can also be written as:

$$\overrightarrow{D} = \varepsilon_0\overrightarrow{E} + \overrightarrow{P} = \varepsilon_0\overrightarrow{E} + \varepsilon_0(\varepsilon_r - 1)\overrightarrow{E} = \varepsilon_0\varepsilon_r\overrightarrow{E} = \varepsilon\overrightarrow{E} \tag{7.58}$$

Because the flux of $\overrightarrow{D}$ out of a closed surface S is equal to the total free charge enclosed within S (as a result of Gauss' law), we can use this theorem to calculate the electric fields in the presence of dielectrics without first having known the distribution of polarization charges, as we already presented above when calculating the distributed capacitance of a coaxial cable with the dielectric formed by two cylindrical dielectric shells with different dielectric constants.

7.6 Boundary conditions for $\vec{D}$ and $\vec{E}$

Often we will have to solve field problems consisting of two (or more) regions with different material properties. To solve electromagnetic field problems involving a boundary between two different materials, we need to determine the transitional properties of the field in the two regions at this boundary. These are called **boundary conditions**.

When the space near a set of charges contains dielectric, but it is not completely filled by a single uniform dielectric material, then the electric field no longer has the same form as in vacuum. Suppose for example, that a slab of dielectric, which carries no free charges, is placed in a parallel-plate capacitor but that the thickness b of the slab is less than the distance a between the capacitor plates. When a potential difference is applied between the plates, polarization charges appear at the free surface of the dielectric, figure 7.12. According to Gauss' law, there is a discontinuity in the electric field because the fluxes of the field $\overrightarrow{E}$ through opposite faces of the closed box Σ shown in figure 7.12, containing the polarization charges, are not equal. But the flux law for the electric displacement $\overrightarrow{D}$, $\int_\Sigma \overrightarrow{D}\,\overrightarrow{dS} = \int_V \rho_{\text{free}}dV$ involved only the **free charge**. There is no free charge inside the box, and the fluxes of $\overrightarrow{D}$ across its opposite faces are equal. In other words, the vector $\overrightarrow{D}$ obeys the simple boundary

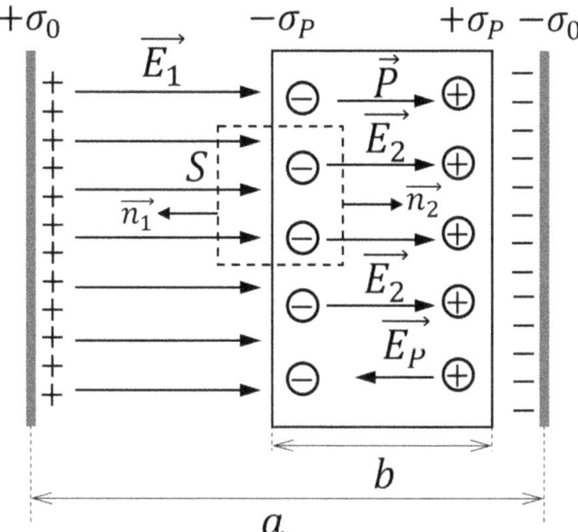

Figure 7.12. The discontinuity in the electric field at the free surface of the dielectric placed between the plates of a parallel-plate capacitor. There is no change in $\vec{D}$ across the dielectric surface enclosed by the box Σ.

conditions that $\vec{D}$ is continuous across a surface perpendicular to $\vec{D}$, if there is no free charges on the surface.

It is now easy to work out the electric field distribution in the capacitor and hence derive its capacitance. The electric displacement $\vec{D}$ is uniform through the capacitor. In the region where there is no dielectric material, the electric field is $\vec{E_1} = \vec{D}/\varepsilon_0$. The dielectric has a uniform polarization $\vec{P}$, and inside the electric field is $\vec{E_2} = \vec{D}/\varepsilon_0\varepsilon_r$. Potential difference between the plates is thus:

$$\Delta V = E_1(a - b) + E_2 b = E_1\left(a - b + \frac{b}{\varepsilon_r}\right)$$

The charge densities on the conducting plates are $\pm E_1\varepsilon_0$ and if their area is A, the plates carry the total charges $\pm Q = \pm E_1 A\varepsilon_0$. The capacitance is therefore $C = Q/\Delta V = \dfrac{\varepsilon_0 A}{a - b + \frac{b}{\varepsilon_r}}$.

The fields $\vec{D}$, $\vec{E}$ and $\vec{P}$ in the capacitor are illustrated in figure 7.13.

$\vec{D}$ is constant throughout the capacitor;

$\vec{E}$ is reduced inside the dielectric, where there are fewer lines;

$\vec{P}$ is zero except inside the dielectric.

$$\vec{D}_{\text{vac}} = \varepsilon_0\vec{E_0} = \sigma_{\text{free}}$$

$$\vec{D}_{\text{die}} = \varepsilon_0\vec{E_d} + \vec{P} = \varepsilon_0\vec{E_d} + \varepsilon_0(\varepsilon_r - 1)\vec{E_d} =$$

$$= \varepsilon_0\varepsilon_r\vec{E_d} = \varepsilon_0\varepsilon_r\frac{\vec{E_0}}{\varepsilon_r} = \sigma_{\text{free}}$$

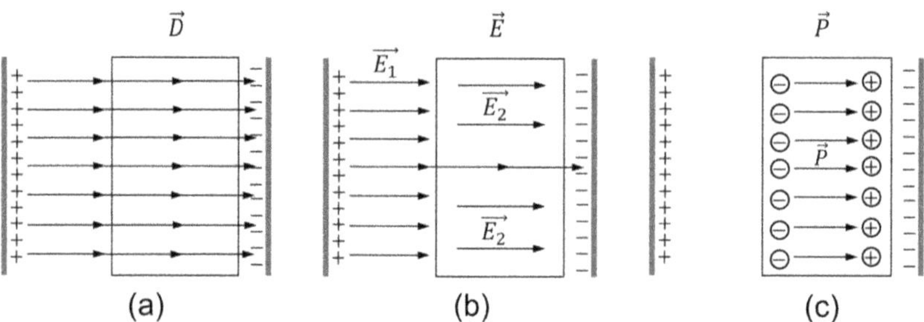

Figure 7.13. The lines of $\vec{D}$ (a), $\vec{E}$ (b) and $\vec{P}$ (c) in a parallel-plate capacitor, which is partially filled with a dielectric material.

The electric displacement is constant in vacuum and in dielectric. The electric field in vacuum is $\vec{E_0} = \frac{\sigma_f}{\varepsilon_0}$ and in dielectric it is smaller than vacuum being $\vec{E_d} = \frac{\sigma_f}{\varepsilon_0 \varepsilon_r}$. The lines of $\vec{E}$ are rare in dielectric because on the faces of dielectric appears the polarization charge $\pm\sigma_p$ which by its 'depolarization field' decreases the total electric field. The dielectric is uniformly polarized having the polarization $\vec{P}$, which is non-zero only inside the dielectric. We observe that by superposition of figures 7.13(b) and (c), figure 7.13(a) is obtained.

The parallel-plate capacitor is a specially simple case, but we can apply similar arguments to find the boundary conditions at any dielectric boundary. Imagine a disc enclosing part of the boundary surface between two different dielectrics having the relative permittivity ε_{r_1} and ε_{r_2} as indicated in figure 7.14. The thickness of the disc h is allowed to tend to zero, so that the only contribution to the outward flux of $\vec{D}$ from the disc comes from its flat faces. These faces have areas represented by the outward normal $\vec{n_{12}}$ and $\vec{n_{21}} = -\vec{n_{12}}$. If there is no free charge inside the disc, the total outward flux of $\vec{D}$ is zero:

$$\int_{\Sigma} \vec{D}\,\vec{dS} = 0 \Rightarrow \vec{D_2}\,\vec{n_{12}}\Delta S + \vec{D_1}\,\vec{n_{21}}\Delta S = 0 \tag{7.59}$$

Writing D_{2n} and D_{1n} for the component of $\vec{D_2}$ and $\vec{D_1}$ perpendicular to the surface boundary, and bearing in mind that $\vec{n_{12}}$ and $\vec{n_{21}}$ are pointing out in opposite directions ($\vec{n_{12}} = -\vec{n_{21}}$), it follows that $(D_{2n} - D_{1n})\Delta S = 0 \Rightarrow D_{2n} = D_{1n}$, i.e., the normal component of $\vec{D}$ is continuous across the boundary:

$$D_{2n} = D_{1n} \tag{7.60}$$

So, if there is no free charge on the interface between two different dielectrics, normal or perpendicular component of the electric displacement $\vec{D}$ is continuous.

At a boundary carrying a free surface charge density σ_f, there is a discontinuity in D_n, of magnitude σ_f:

$$D_{2n} = D_{1n} + \sigma_f \tag{7.61}$$

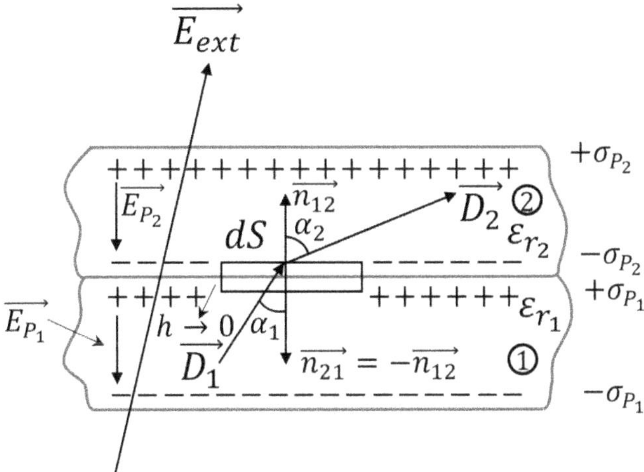

Figure 7.14. The change of $\vec{D}$ at a dielectric boundary.

If the perpendicular component of $\vec{D}$ is continuous, then the perpendicular component of electric field will be discontinuous, because:

$$\varepsilon_2 E_{2n} = \varepsilon_1 E_{1n} \Rightarrow \frac{E_{2n}}{E_{1n}} = \frac{\varepsilon_1}{\varepsilon_2} = \frac{\varepsilon_{r1}}{\varepsilon_{r2}} \tag{7.62}$$

or:

$$E_{2n}\varepsilon_2 = \varepsilon_1 E_{1n} + \sigma_f \tag{7.63}$$

if the boundary carries a free surface charge density σ_f.

A different boundary condition applies to the components of the electric fields parallel to the boundary: in this case *it is the component of the electric field which is continuous*. Consider the rectangular loop Γ shown in figure 7.15 with sides dl_1 and dl_2 lying close to the boundary between the two dielectrics with relative permittivities ε_{r_1} and ε_{r_2}. The side

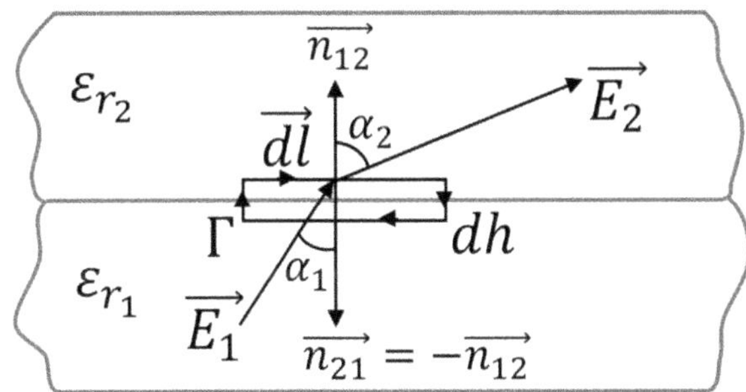

Figure 7.15. The change of $\vec{E}$ at a dielectric boundary.

dh perpendicular to the surface of this small rectangular path in figure 7.15, is very short ($dh \to 0$). According to the electrostatic potential theorem we have:

$$\int_\Gamma \vec{E}\,\vec{dl} = 0$$

$$\vec{E_2}\,\vec{dl_2} + \vec{E_1}\,\vec{dl_1} = 0$$

But:

$$\vec{dl_2} = -\vec{dl_1} = dl \Rightarrow (E_{2t} - E_{1t})dl = 0 \Rightarrow E_{2t} = E_{1t} \tag{7.64}$$

It follows that the components of the electric field parallel (tangential components) to the surface are continuous across the boundary. The parallel component of electric displacement, of course will be discontinuous, and we will write:

$$\frac{D_{2t}}{\varepsilon_2} = \frac{D_{1t}}{\varepsilon_1} \Rightarrow \frac{D_{2t}}{D_{1t}} = \frac{\varepsilon_2}{\varepsilon_1} = \frac{\varepsilon_{r_2}}{\varepsilon_{r_1}} \tag{7.65}$$

Summarizing, at a boundary between two dielectric media which carry no free surface charges, we have:

$$\begin{aligned}
D_{2n} &= D_{1n} \quad \text{or} \quad D_n \;\; \text{is continous} \\
E_{2t} &= E_{1t} \quad \text{or} \quad E_t \;\; \text{is continous}
\end{aligned} \tag{7.66}$$

When a boundary is neither parallel nor perpendicular to the electric field, the direction of the field changes across the boundary. Labeling the angles to the normal as in figure 7.14, the boundary becomes:

$$\begin{aligned}
\varepsilon_2 E_2 \cos \alpha_2 &= \varepsilon_2 E_1 \cos \alpha_1; \quad D_n \;\; \text{is continuous} \\
E_2 \sin \alpha_2 &= E_1 \sin \alpha_1; \quad E_t \;\; \text{is continuous}
\end{aligned} \tag{7.67}$$

$$\frac{\tan \alpha_2}{\varepsilon_2} = \frac{\tan \alpha_1}{\varepsilon_1} \Rightarrow \frac{\tan \alpha_2}{\tan \alpha_1} = \frac{\varepsilon_2}{\varepsilon_1} = \frac{\varepsilon_{r_2}}{\varepsilon_{r_1}} \tag{7.68}$$

This equation is reminiscent of Snell's law in optics. The resemblance is not a coincidence! Optical refraction can be explained in terms of the boundary conditions applying to changing electric and magnetic fields. In the chapter referring to electromagnetic waves, we shall derive Snell's law, and relate refractive index to the relative permittivity.

Exercise 1: dielectric between the plates of a parallel capacitor

Consider the pair of parallel conducting plates of area A and very large extent shown in figure 7.16. Suppose a voltage U_0 is applied between the plates, which are separated a distance d with air as the intervening medium. The battery is then removed and a dielectric slab ($\varepsilon = \varepsilon_0 \varepsilon_r$) of thickness d is inserted between the plates. The air gap between each plate and dielectric surface is assumed to be of negligible thickness but is shown in exaggerated proportions in the sketches. Determine $\vec{E}$, $\vec{D}$ and $\vec{P}$ before and after the insertion of the dielectric. Also determine the free surface charge density σ_f on the plates and polarization surface charge density σ_p on the surface of dielectric. We will assume that the normal is positive when it is directed from positive plate to negative plate and all the

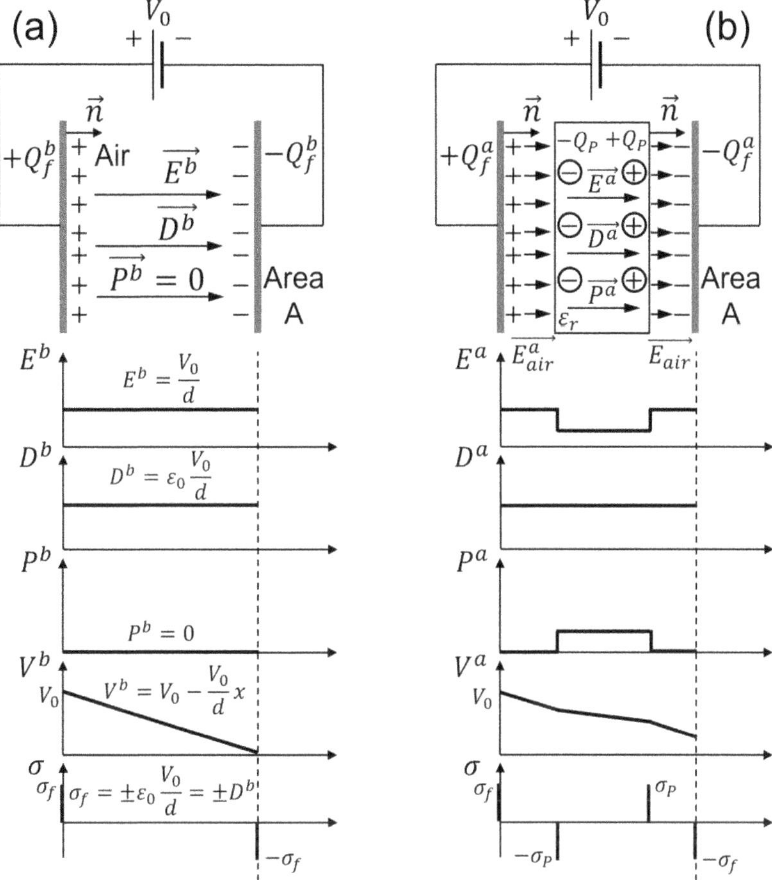

Figure 7.16. *The effect of a dielectric on a parallel-plate capacitor.*

vectors $\overrightarrow{E}$, $\overrightarrow{D}$ and $\overrightarrow{P}$ have the same direction, and we will therefore need only to deal with their magnitudes. Superscripts b and a will be used on all calculated quantities to symbolize before and after the insertion of dielectric, respectively.

Solution I: *before the insertion of the dielectric, the battery establishes an electric field between the plates (which is essentially normal to the plate surfaces, neglecting the fringing of the field at the plate edges) of:*

$$\overrightarrow{E^b} = \frac{V_0}{d}\overrightarrow{n} \left(\frac{V}{m}\right) \tag{7.69}$$

The electric displacement vector is:

$$\overrightarrow{D^b} = \varepsilon_0 \overrightarrow{E^b} = \varepsilon_0 \frac{V_0}{d}\overrightarrow{n} \left(\frac{C}{m^2}\right) \tag{7.70}$$

According to the boundary condition, the free surface charge density σ_f^b on each plate is numerically equal to the component of $\overrightarrow{D}$ that is normal to these (perfectly conducting) plates. On the positive plate this becomes:

$$\sigma_f^b = D^b = \varepsilon_0 \frac{V_0}{d} \left(\frac{C}{m^2} \right) \tag{7.71}$$

The total free charge on the positive plate having area A is:

$$Q_f^b = \sigma_f^b A = \frac{\varepsilon_0 A V_0}{d} \; (C) \tag{7.72}$$

Since air has $\varepsilon_r = 1$, there results:

$$\overrightarrow{P^b} = \varepsilon_0(\varepsilon_r - 1)\overrightarrow{E^b} = 0 \tag{7.73}$$

and then $\sigma_p = 0$.

Solution II: *when the battery is removed and the dielectric inserted, the total free charge on each plate remains uncharged and so, $Q_f^b = Q_f^a$. Thus the free surface charge density is uncharged. Therefore, the $\overrightarrow{D}$ field between the plates remains uncharged $\overrightarrow{D^a} = \overrightarrow{D^b}$, since no free charge has been added by the dielectric insertion. In the air gaps and in the dielectric there is:*

$$\overrightarrow{D^a} = \overrightarrow{D^b} = \frac{\varepsilon_0 V_0}{d}\overrightarrow{n} \left(\frac{C}{m^2} \right) \tag{7.74}$$

But in the air gaps the electric field is now:

$$\overrightarrow{E_{air}^a} = \frac{\overrightarrow{D^a}}{\varepsilon_0} = \frac{V_0}{d}\overrightarrow{n} \left(\frac{V}{m} \right) \tag{7.75}$$

and in the dielectric it is:

$$\overrightarrow{E_{die}^a} = \frac{\overrightarrow{D^a}}{\varepsilon_0 \varepsilon_r} = \frac{\varepsilon_0 V_0}{\varepsilon_0 d \varepsilon_r}\overrightarrow{n} = \frac{V_0}{\varepsilon_r d}\overrightarrow{n} = \frac{\overrightarrow{E_{air}^a}}{\varepsilon_r} = \frac{\overrightarrow{E^b}}{\varepsilon_r} \tag{7.76}$$

Thus the electric field in the dielectric is reduced from its value prior to the insertion of the dielectric ($\varepsilon_r > 1$). In the air gaps we have:

$$\overrightarrow{P_{air}^a} = 0 \tag{7.77}$$

but in the dielectric we have:

$$\overrightarrow{P^{a}} = \overrightarrow{D^{d}_{\text{die}}} - \varepsilon_0 \overrightarrow{E^{d}_{\text{die}}} = \left(\varepsilon_0 \frac{V_0}{d} - \varepsilon_0 \frac{V_0}{\varepsilon_r d} \right) \overrightarrow{n} =$$

$$= \varepsilon_0 \left(\frac{\varepsilon_r - 1}{\varepsilon_r} \right) \frac{V_0}{d} \overrightarrow{n} \Rightarrow$$

$$\Rightarrow \overrightarrow{P^{a}} = \varepsilon_0 (\varepsilon_r - 1) \overrightarrow{E^{d}_{\text{die}}} = \varepsilon_0 \chi_E \overrightarrow{E^{d}_{\text{die}}} \tag{7.78}$$

The polarization surface charge density can be found from the component of $\overrightarrow{P}$ normal to the dielectric surface. On the surface adjacent to the positive plate, $\overrightarrow{P}$ is directed into the dielectric $\sigma_p^a = -\overrightarrow{P}\,\overrightarrow{n_{da}} = -P_n = -\frac{\varepsilon_r - 1}{\varepsilon_r}\frac{V_0}{d}$ and on the surface adjacent to negative plate the polarization surface charge density is positive having the value:

$$\sigma_p = \frac{\varepsilon_r - 1}{\varepsilon_r} \frac{V_0}{d} \tag{7.79}$$

Exercise 2: potential and electric field of a polarized sphere

The solid sphere in figure 7.17 is supposed to be uniformly polarized. Compute the potential and the electric field, both inside and outside the sphere.

Solution: $\overrightarrow{P}$ as usual will denote the density of polarization constant in magnitude and direction through the volume of the sphere. The external field of the polarized sphere may be found if we will calculate the electric potential outside the sphere and then take the gradient of potential $(\overrightarrow{E} = -\nabla V)$. As regarding the potential outside of the sphere we can use the equivalence theorem between a small polarized body and an electric

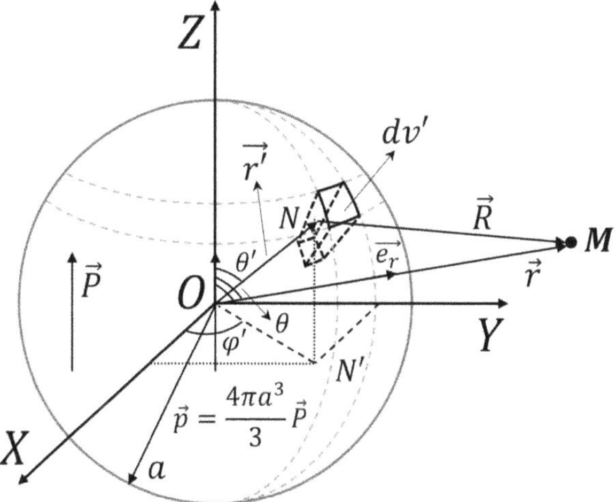

Figure 7.17. *The potential and the electric field of a polarized sphere.*

dipole. Indeed the volume element dv′ centered by $\overrightarrow{r'}$ position vector, has the dipole moment $\overrightarrow{dp} = \overrightarrow{P}\,dv'$ and creates at point M the elementary potential:

$$dV_P(M) = \frac{\overrightarrow{dp}\,\overrightarrow{R}}{4\pi\varepsilon_0 R^3} \tag{7.80}$$

But at a great distance from the sphere, $r \gg a > r'$, $\overrightarrow{R} \sim \overrightarrow{r}$, resulting that:

$$dV_P(M) \simeq \frac{\overrightarrow{dp}\,\overrightarrow{r}}{4\pi\varepsilon_0 r^3} = \frac{\overrightarrow{P}\,dv'}{4\pi\varepsilon_0 r^2}\,\overrightarrow{e_r} \tag{7.81}$$

Then, using the superposition principle, the potential at point M will be:

$$V_P(M) = \int_V \frac{\overrightarrow{P}\,dv'}{4\pi\varepsilon_0 r^2}\,\overrightarrow{e_r} = \overrightarrow{P}\int_V \frac{dv'}{4\pi\varepsilon_0 r^2}\,\overrightarrow{e_r}$$

$$= \overrightarrow{P}\int_V \frac{r'^2 \sin\theta'\,dr'\,d\theta'\,d\varphi'}{4\pi\varepsilon_0 r^2}\,\overrightarrow{e_r}$$

$$= \overrightarrow{P}\int_0^a \int_0^\pi \int_0^{2\pi} \frac{r'^2 \sin\theta'\,dr'\,d\theta'\,d\varphi'}{4\pi\varepsilon_0 r^2}\,\overrightarrow{e_r} = \frac{\overrightarrow{P}\,\dfrac{4\pi a^3}{3}}{4\pi\varepsilon_0 r^2}\,\overrightarrow{e_r} \tag{7.82}$$

$$= \frac{\overrightarrow{P}\,a^3}{3\varepsilon_0 r^2}\,\overrightarrow{e_r}$$

or:

$$V_{\overrightarrow{P}}(M) = \frac{\overrightarrow{p}\,\overrightarrow{e_r}}{4\pi\varepsilon_0 r^2} = \frac{p\cos\theta}{4\pi\varepsilon_0 r^2} \tag{7.83}$$

where $\overrightarrow{p} = \dfrac{\overrightarrow{P}\,4\pi a^3}{3}$ is the total dipole moment of the uniformly polarized sphere. Now, the electric field outside the sphere will be:

$$\overrightarrow{E_M} = -\nabla V_P(M) = -\frac{\partial}{\partial r}\left(\frac{p\cos\theta}{4\pi\varepsilon_0 r^2}\right)\overrightarrow{e_r} - \frac{1}{r}\frac{\partial}{\partial \theta}\left(\frac{p\cos\theta}{4\pi\varepsilon_0 r^2}\right)\overrightarrow{e_\theta}$$

$$= \frac{2p\cos\theta}{4\pi\varepsilon_0 r^3}\,\overrightarrow{e_r} + \frac{p\sin\theta}{4\pi\varepsilon_0 r^3}\,\overrightarrow{e_\theta} = E_r\,\overrightarrow{e_r} + E_\theta\,\overrightarrow{e_\theta} \tag{7.84}$$

where the radial component of the electric field is:

$$E_r = \frac{2p\cos\theta}{4\pi\varepsilon_0 r^3} = \frac{2P\dfrac{4\pi a^3}{3}}{4\pi\varepsilon_0 r^3}\cos\theta = \frac{2Pa^3\cos\theta}{3\varepsilon_0 r^3} \tag{7.85}$$

and the azimuthal component is:

$$E_\theta = \frac{p\sin\theta}{4\pi\varepsilon_0 r^3} = \frac{Pa^3}{3\varepsilon_0 r^3}\sin\theta \tag{7.86}$$

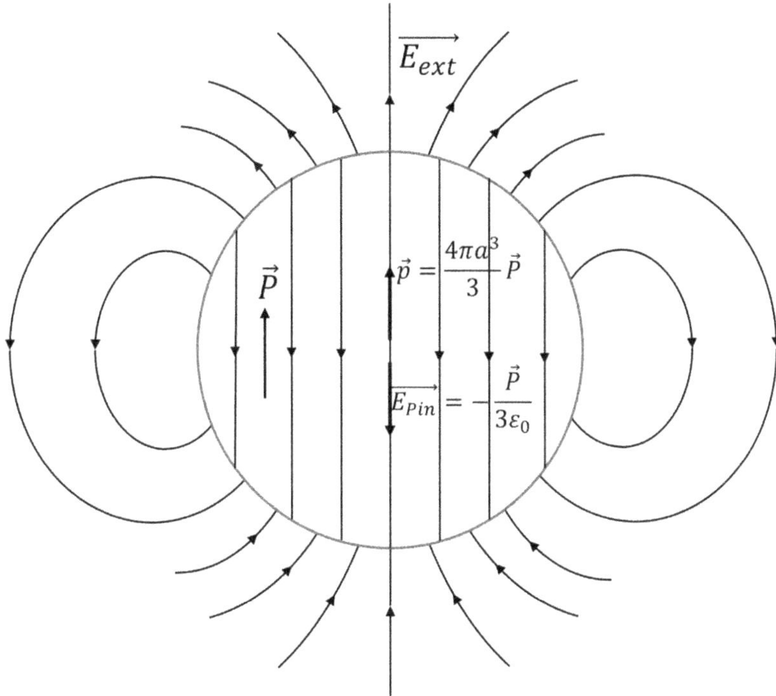

Figure 7.18. *The electric field outside and inside of an uniformly polarized sphere with the polarization vector $\vec{P}$.*

A representation of the external field lines is shown in figure 7.18. We can see that outside the sphere the electric field has the same configuration with the field lines of an electric dipole having the moment $\vec{p} = \frac{4\pi a^3}{3}\vec{P}$ located at the center of the uniformly polarized sphere. The electric field outside of the uniformly polarized sphere with the polarization vector $\vec{P}$ is exactly the same as that of an electric dipole located at the center of the sphere. The field of the uniformly polarized sphere inside is uniform, smaller than the external field on the sphere surface and oriented in opposite direction to that of the polarization vector $\vec{P}$. The internal field is a different matter. Let's look at the electric potential $V(x, y, z)$. We know the potential at all points on the spherical boundary because we know the external field. It is just the dipole potential $\frac{p\cos\theta}{4\pi\varepsilon_0 r^2}$ which on the spherical boundary of radius a becomes:

$$V_P(a) = \frac{p\cos\theta}{4\pi\varepsilon_0 a^2} = \frac{P\dfrac{4\pi a^3}{3}\cos\theta}{4\pi\varepsilon_0 a^2} = \frac{Pa\cos\theta}{3\varepsilon_0} = \frac{Pz}{3\varepsilon_0} \qquad (7.87)$$

Since $a\cos\theta = z$, we see that the potential of a point on the sphere depends only on its z coordinate, see figure 7.19. Equation (7.87) gives the potential at every point on the boundary of the region, inside which V_P must satisfy Laplace's equation. According to

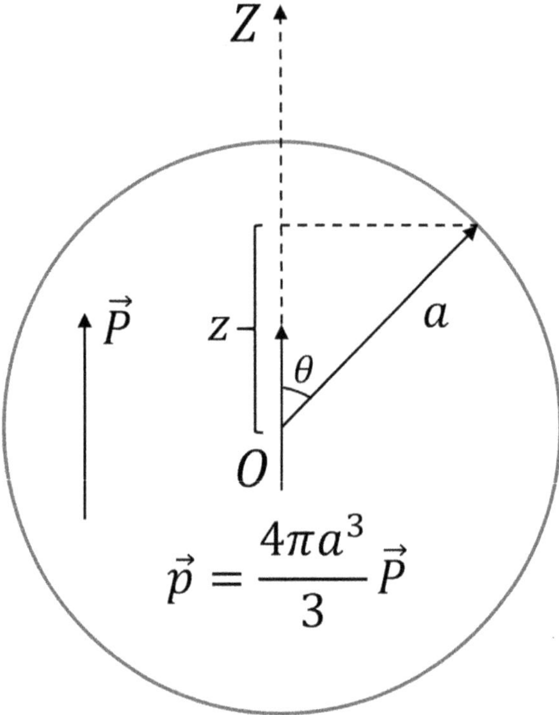

Figure 7.19. *The electric potential on the surface of a uniformly polarized sphere with the polarization vector $\vec{P}$, depends only on its z coordinate.*

*the Uniqueness theorem (see section 3.4), we proved previously that it is sufficient to determine V_P through interiors. If we can find a solution for inside, it must be **the solution of Laplace equation.** Now we can write:*

$$\frac{d^2 V_P}{dz^2} = 0 \Rightarrow \frac{dV_P}{dz} = C \Rightarrow V_P = Cz + C' \tag{7.88}$$

The constants C and C' can be determined knowing that:

$$\begin{cases} V_P = 0 \ \ at \ \ z = 0 \quad (electric \ potential \ at \ the \ midle \ of \ a \ dipole) \\ V_P(a) = \dfrac{Pa}{3\varepsilon_0} \ \ at \ \ z = a \quad (on \ the \ top \ of \ the \ polarized \ sphere) \end{cases}$$

*Then $C' = 0$ and $C = \frac{P}{3\varepsilon_0}$. Now the function Cz, where $C = \frac{P}{3\varepsilon_0}$, satisfies Laplace's equation. $V_P = \frac{P}{3\varepsilon_0}z$ **has actually handed us the solution to the potential in the interior of the sphere.** It is the potential of a uniform electric field oriented in the z direction.*

Indeed, $\overrightarrow{E_{P_{\text{in}}}} = -\frac{dV_P}{dz} = -\frac{P}{3\varepsilon_0}\overrightarrow{e_z}$. *As the direction of* $\overrightarrow{P}$ *was the only thing that distinguished the z axis, we can write our result in a more general form:*

$$\overrightarrow{E_{P_{\text{in}}}} = -\frac{\overrightarrow{P}}{3\varepsilon_0} \tag{7.89}$$

This is the macroscopic field $\overrightarrow{E_{\text{in}}}$ *in the polarized material. Figure 7.18 shows both the internal and external field. At the upper pole of the sphere, the strength of the upward-pointing external field is, from equation (7.85):*

$$\overrightarrow{E_{\text{ext}}}\Big|_{r=a}^{\theta-0} = \frac{2\overrightarrow{P}}{3\varepsilon_0} \quad (outside) \tag{7.90}$$

which is just twice the magnitude of the downward-pointing internal field:

$$\overrightarrow{E_{\text{int}}}\Big|_{r=a}^{\theta=0} = -\frac{\overrightarrow{P}}{3\varepsilon_0} \quad (inside) \tag{7.91}$$

This example illustrates the general rules for the behavior of the field components at the surface of a polarized medium. $\overrightarrow{E}$ *is discontinuous at the boundary of a polarized medium exactly as it would be at a surface in vacuum which carried a surface charge density* $\sigma_P = P_n$. *It follows that the normal component,* E_n, *must change abruptly by an amount* $\frac{P}{\varepsilon_0}$ *while the tangential component,* E_t, *parallel to the boundary remains continuous, that is, has the same value on both sides of the boundary. Indeed, at the north pole of our sphere the net change in* E_r *is:*

$$\overrightarrow{\Delta E_r} = \frac{2\overrightarrow{P}}{3\varepsilon_0} - \left(-\frac{\overrightarrow{P}}{3\varepsilon_0}\right) = \frac{3\overrightarrow{P}}{3\varepsilon_0} = \frac{\overrightarrow{P}}{\varepsilon_0} \tag{7.92}$$

Referring to the parallel component of $\overrightarrow{E}$, *equation (7.86), we can see that at the north pole* $E_\theta = \frac{P}{3\varepsilon_0}\sin 0 = 0$, *and zero is the parallel component of* $\overrightarrow{E_{\text{in}}}$, *inside the polarized sphere. So, we could check that the component of* $\overrightarrow{E}$ *parallel to the surface is continuous from inside to outside everywhere on the sphere. As regarding the electric displacement* $\overrightarrow{D}$, *we can show that the component of* $\overrightarrow{D}$ *normal to the surface is continuous from inside to outside everywhere on the sphere. Indeed,* **outside** $\overrightarrow{D_r} = \varepsilon_0\overrightarrow{E_r}$ *and at the upper pole of the sphere it is* $\overrightarrow{D_r}\Big|_{r=a}^{\theta=0} = \frac{2\overrightarrow{P}}{3}$, *directed along z axis.* **Inside** *the sphere:*

$$\overrightarrow{D_{r_{\text{in}}}} = \varepsilon_0\overrightarrow{E_{\text{in}}} + \overrightarrow{P} = \varepsilon_0\left(-\frac{\overrightarrow{P}}{3\varepsilon_0}\right) + \overrightarrow{P} = \overrightarrow{P} - \frac{\overrightarrow{P}}{3} = \frac{2\overrightarrow{P}}{3} = \overrightarrow{D_{r_{\text{out}}}} \tag{7.93}$$

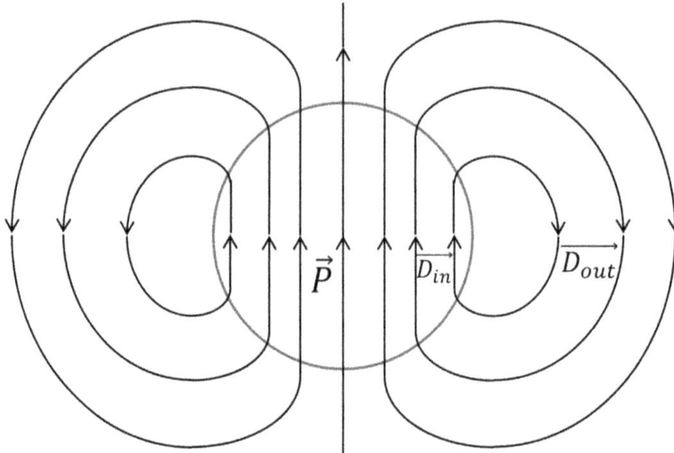

Figure 7.20. *The total displacement field $\vec{D}$, both inside and outside the uniformly polarized sphere.*

Assuming any sphere is uniformly polarized, figure 7.20 shows its displacement vector field lines. So, if a sphere is uniformly polarized with the polarization vector $\vec{P}$, the electric field behaves as in figure 7.18 having the normal component discontinuous at the sphere surface while the displacement vector $\vec{D}$ shown in figure 7.20 has the continuous normal component at the surface of the sphere. Onto this can be superposed any field from other sources, this representing many possible systems, but this will not affect the discontinuity in $\vec{E}$ at the boundary of the polarized medium. The rules just stated, therefore apply in any system, the discontinuity in $\vec{E}$ being determined solely by the existing polarization such as you can see further.

Exercise 3: a dielectric sphere in a uniform electric field

As an example, let us put a sphere of dielectric material with dielectric constant ε_r, into a homogeneous electric field $\vec{E_0}$, like the field between the parallel plates of a vacuum capacitor, see figure 7.21(a). Find the electric field and displacement vectors, verifying the boundary conditions.

__Solution:__ the sources of the field $\vec{E_0}$ remain fixed. The dielectric sphere develops some polarization $\vec{P}$. The total field $\vec{E}$ is the superposition of $\vec{E_0}$ and the field of this polarized sphere, see figure 7.21(b). Let the sources of this field (the charges on the plates), be far from the sphere so that they do not shift as the sphere is introduced. Then whatever the field may be in the vicinity of the sphere, it will remain practically $\vec{E_0}$ at a great distance. The total electric field $\vec{E}$ is no longer uniform in the

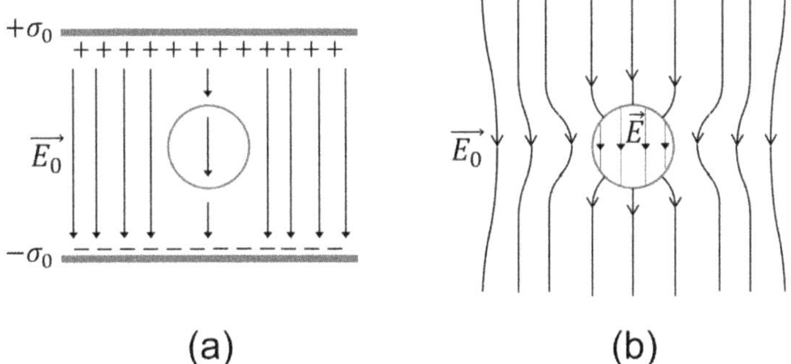

(a)   (b)

Figure 7.21. *(a) A dielectric sphere into electric field $\overrightarrow{E_0}$. (b) The total field $\vec{E}$ both inside and outside the dielectric sphere.*

neighborhood of the sphere. It is the sum of the uniform field $\overrightarrow{E_0}$ of the distant sources and a field $\overrightarrow{E_p}$ generated by the spherical polarized matter itself:

$$\vec{E} = \overrightarrow{E_0} + \overrightarrow{E_{P_{\text{in}}}} \tag{7.94}$$

The field $\overrightarrow{E_p}$ depends, such as we already saw, on the polarization $\vec{P}$ of the dielectric, which in turn depends on the value of $\vec{E}$ inside the sphere:

$$\vec{P} = \varepsilon_0 \chi_E \vec{E} = \varepsilon_0 (\varepsilon_r - 1)\vec{E} \tag{7.95}$$

We do not know yet what the total field $\vec{E}$ is. We only know that equation (7.95) has to hold at any point inside the sphere. If the sphere becomes uniformly polarized, an assumption that will need to be justified by our results, the relation between the polarization of the sphere and its field $\overrightarrow{E_{P_{\text{in}}}}$, at points inside, is given already by equation (7.89):

$$\overrightarrow{E_{P_{\text{in}}}} = -\frac{\vec{P}}{3\varepsilon_0} \tag{7.96}$$

Now we have enough equations to eliminate $\vec{P}$ and $\overrightarrow{E_{P_{\text{in}}}}$, which should give us a relation connecting $\vec{E}$ and $\overrightarrow{E_0}$. Using equations (7.94)–(7.96), we find:

$$\vec{E} = \overrightarrow{E_0} + \overrightarrow{E_{P_{\text{in}}}} = \overrightarrow{E_0} - \frac{\vec{P}}{3\varepsilon_0} = \overrightarrow{E_0} - \frac{1}{3}(\varepsilon_r - 1)\vec{E}$$

$$\vec{E}(3 + \varepsilon_r - 1) = 3\overrightarrow{E_0} \Rightarrow$$
$$\Rightarrow \vec{E} = \frac{3}{\varepsilon_r + 2}\overrightarrow{E_0} \tag{7.97}$$

Because ε_r is greater than one ($\varepsilon_r > 1$), the factor $\frac{3}{\varepsilon_r + 2}$ will be less than one and the field inside the dielectric is smaller than $\overrightarrow{E_0}$. Now the polarization of dielectric is:

$$\overrightarrow{P} = \varepsilon_0(\varepsilon_r - 1)\overrightarrow{E} = \frac{3\varepsilon_0(\varepsilon_r - 1)}{\varepsilon_r + 2}\overrightarrow{E_0} \tag{7.98}$$

The assumption of uniform polarization is now seen to be self-consistent. To compute the total field $\overrightarrow{E}$ outside the sphere, we must add vectorially to $\overrightarrow{E_0}$ the field of a central dipole with dipole moment equal to $\overrightarrow{P}$ times the volume of the sphere:

$$\overrightarrow{p} = \frac{4\pi a^3}{3}\overrightarrow{P} = \frac{4\pi a^3}{3}\frac{3\varepsilon_0(\varepsilon_r - 1)}{\varepsilon_r + 2}\overrightarrow{E_0} = \frac{4\pi a^3 \varepsilon_0(\varepsilon_r - 1)}{\varepsilon_r + 2}\overrightarrow{E_0} \tag{7.99}$$

So:

$$\overrightarrow{E_{\text{ext}}} = \overrightarrow{E_0} + \frac{2p\cos\theta}{4\pi\varepsilon_0 r^3}\overrightarrow{e_r} + \frac{p\sin\theta}{4\pi\varepsilon_0 r^3}\overrightarrow{e_\theta} \tag{7.100}$$

At the upper pole of the sphere, the strength of the total upward-pointing external field is:

$$\left.\overrightarrow{E_{\text{ext}}}\right|_{r=a}^{\theta=0} = \overrightarrow{E} + \frac{2p}{4\pi\varepsilon_0 a^3}\overrightarrow{e_r} = \overrightarrow{E_0} + \frac{2(\varepsilon_r - 1)}{(\varepsilon_r + 2)}\overrightarrow{E_0}$$

$$= \overrightarrow{E_0}\left(1 + \frac{2\varepsilon_r - 2}{\varepsilon_r + 2}\right) = \frac{3\varepsilon_r}{\varepsilon_r + 2}\overrightarrow{E_0} > \overrightarrow{E_0} \tag{7.101}$$

We can see that the electric field outside is greater than the uniform field $\overrightarrow{E_0}$ and also than the electric field inside the dielectric. Again, the normal component of $\overrightarrow{E}$ must change abruptly by the amount:

$$\Delta\overrightarrow{E} = \overrightarrow{E_{\text{ext}}} - \overrightarrow{E_{\text{in}}} = \left(\frac{3\varepsilon_r}{\varepsilon_r + 2} - \frac{3}{\varepsilon_r + 2}\right)\overrightarrow{E_0} = \frac{3(\varepsilon_r - 1)}{\varepsilon_r + 2}\overrightarrow{E_0} = \frac{\overrightarrow{P}}{\varepsilon_0} \tag{7.102}$$

*such as in the case of a uniformly polarized sphere, but here the electric field inside of the sphere is directed like $\overrightarrow{P}$ and $\overrightarrow{E_0}$. Some field lines of $\overrightarrow{E}$, both inside and outside the dielectric sphere, are shown in figure 7.22(a). As regarding the electric displacement, we can verify that it is continuous at the surface of the dielectric sphere. Indeed, **outside** at the upper pole of the sphere the electric displacement is:*

$$\left.\overrightarrow{D_{\text{ext}}}\right|_{r=a}^{\theta=0} = \varepsilon_0\left.\overrightarrow{E_{\text{ext}}}\right|_{r=a}^{\theta=0} = \frac{3\varepsilon_0\varepsilon_r}{\varepsilon_r + 2}\overrightarrow{E_0} \tag{7.103}$$

*and **inside** at the same point, the electric displacement is:*

$$\left.\overrightarrow{D_{\text{in}}}\right|_{r=a}^{\theta=0} = \varepsilon_0\overrightarrow{E} + \overrightarrow{P} = \varepsilon_0\frac{3}{\varepsilon_r + 2}\overrightarrow{E_0} + \frac{3\varepsilon_0(\varepsilon_r - 1)}{\varepsilon_r + 2}\overrightarrow{E_0}$$

$$= \frac{3\varepsilon_0}{\varepsilon_r + 2}(1 + \varepsilon_r - 1)\overrightarrow{E_0} = \frac{3\varepsilon_0\varepsilon_r}{\varepsilon_r + 2}\overrightarrow{E_0} \tag{7.104}$$

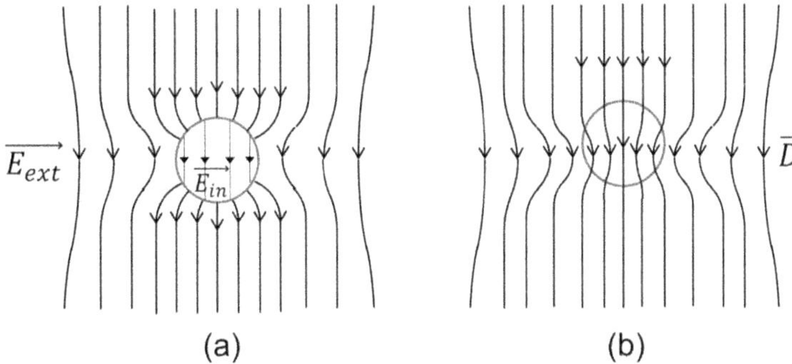

Figure 7.22. *The total electric field $\vec{E}$ (a) and total $\vec{D}$ (b), both inside and outside the dielectric sphere.*

The same field lines of $\overrightarrow{D}$, both inside and outside the dielectric sphere, are shown in figure 7.22(b).

Further reading

[1] Böttcher C J F 1973 *Theory of Electric Polarization: Dielectrics in Static Fields* 2nd edn (Amsterdam: Elsevier)

[2] Coelho R 1979 *Physics of Dielectrics for the Engineer* 1st edn (Amsterdam: Elsevier)

[3] Landsberg G 1987 *Cours élémentaire de physique—Tome 2, Electricité et magnétisme* (Moscow: Mir) edn

[4] Scaife B K P 1998 *Principles of Dielectrics (Monographs on the Physics and Chemistry of Materials)* revised edn (Oxford: Oxford University Press)

[5] Panofsky W K H and Phillips M 2005 *Classical Electricity and Magnetism Dover Books on Physics* 2nd edn (New York: Dover)

[6] Moliton A 2007 Electrostatics of dielectric materials *Basic Electromagnetism and Materials* (New York: Springer) pp 39–87

[7] Martinez-Vega J (ed) 2010 *Dielectric Materials for Electrical Engineering* (New York: Wiley)

[8] Feynman R P, Leighton R B and Sands M 2011 *The Feynman Lectures on Physics, Vol II: The New Millennium Edition: Mainly Electromagnetism and Matter* (New York: Basic Books)

[9] Zangwill A 2012 *Modern Electrodynamics* 1st edn (Cambridge: Cambridge University Press)

Electrostatics
Formalism of the electrostatic field in vacuum and matter
Ştefan Antohe and Vlad-Andrei Antohe

Chapter 8

Microscopic approach on the polarization of dielectrics

We learnt so far (in chapter 7) that the polarization of dielectrics occurs when an external electric field is applied to a dielectric substance. In particular, when an electric field is applied, it causes charges (both positive and negative) to be displaced one in respect to the other. Actually, the main action of the dielectric polarization is to connect macroscopic and microscopic characteristics of the dielectrics. This chapter will exclusively focus on the physical processes which take place at microscopic scale during the polarization of the dielectric materials.

8.1 Atomic and molecular dipoles. Induced dipole moments

The isolated neutral atom shown in figure 8.1 has Z electrons moving around to a nucleus carrying a charge $+Ze$. In figure 8.1(a), the size of the nucleus is enormously exaggerated. Its diameter in a real atom is only about one ten thousandth of the atomic diameter. The electrons are represented as being smeared out in a cloud around the nucleus, since according to quantum far mechanical theory this is the most useful way of thinking of the electrons when we are discussing their electrostatic properties. In the absence of an external electric field, the nucleus of an atom is at the center of the electronic cloud where it experiences no net electrostatic force, being in a stable equilibrium, so that if the nucleus was displaced from the center of the electron cloud a restoring force would arise from the mutual attraction of nucleus and electrons. What happens if the atom is placed in an external electric field pointing upward, as in figure 8.1(b)? The field pushes up the positively charged nucleus (in the sense of the applied electric field) and the negatively charged electronic cloud downward (in the opposite direction of the applied electric field). As a result, the center of the electronic cloud moves away from the nucleus until the external force is balanced by the internal restoring force. The atom as a whole is

doi:10.1088/978-0-7503-5859-0ch8   8-1   © IOP Publishing Ltd 2023

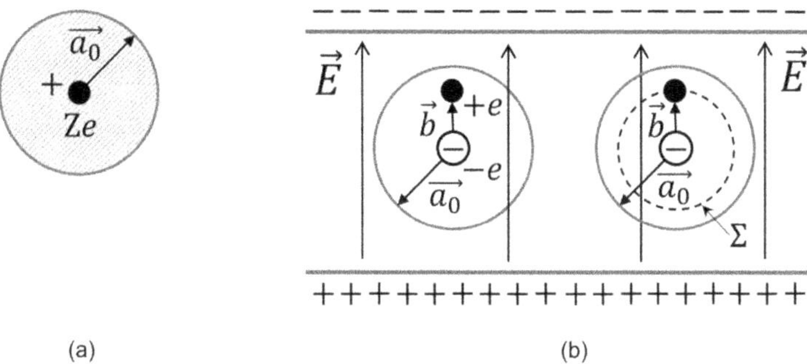

Figure 8.1. Polarization of an atom by an electric field: (a) an isolated atom of the element with the atomic number Z. (b) The electric field displaces the electronic cloud.

electrically neutral, so its center of mass must remain at rest. Most of the mass of the atom is in the nucleus, which hardly moves at all, when the electron cloud is displaced by the external field. In its new equilibrium position, in which the center of the electron cloud no longer coincides with the nucleus, **the atom is said to be polarized**.

We can use a makeshift model of the Z atom to estimate, in order of magnitude, the amount of distortion to be expected. Suppose that in the absence of an electric field the negative electric charge $-Ze$ is distributed with constant density ρ_e through a sphere of radius a, outside which it is zero. Then:

$$\rho_e = \frac{-Ze}{\frac{4\pi a^3}{3}} = \frac{-3Ze}{4\pi a^3} \tag{8.1}$$

Figure 8.1(a) shows this crude substitute for the real distribution of an electronic cloud. Assume that when the field $\vec{E}$ is applied, this ball of negative charge keeps its shape and density and is merely displaced relative to the nucleus, hence the nucleus ends up some distance $\vec{b}$ from the sphere' center, figure 8.1(b).

In equilibrium, the force on the nucleus due to the electric field $\vec{E}$, a force of $Ze\vec{E}$ acting in the upper side, must be balanced by the downward attraction force exerted on the nucleus by the negative charge cloud, which pulls the nucleus toward its center. To find the magnitude of the latter force, we recall that inside of a spherical charge distribution, at a point $\vec{b}$ cm from the center, the electric field is simply that due to the electric charge inside a sphere of radius $\vec{b}$ according to Gauss' law, figure 8.2.

$$\int_{\Sigma} \vec{E}\,\overrightarrow{dS} = \int_{V_{\Sigma}} \frac{\rho_e}{\varepsilon_0}dV \Rightarrow E4\pi b^2 = \frac{\rho_e}{\varepsilon_0}\frac{4\pi b^3}{3} \Rightarrow \vec{E_i} = -\frac{Ze}{4\pi\varepsilon_0 a^3}\vec{b} \tag{8.2}$$

Setting the force on the nucleus due to the external electric field $\vec{E}$, equal to that arising from the internal electric field $\vec{E_i}$, gives the equilibrium condition, $Ze\vec{E} = -\left[-\frac{(Ze)^2}{4\pi\varepsilon_0 a^3}\vec{b} \right]$, from which:

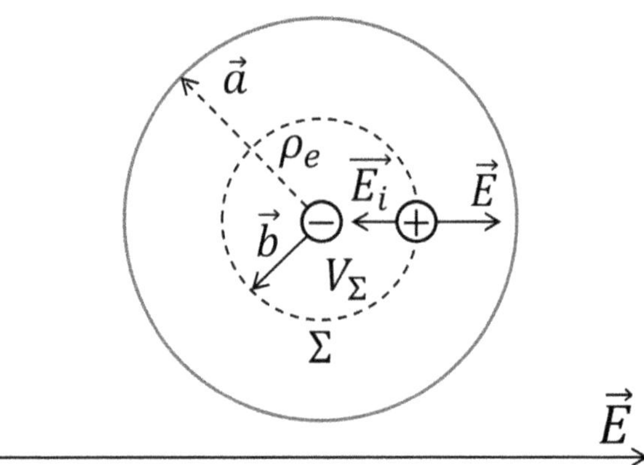

Figure 8.2. Equilibrium in the electric field.

$$\vec{b} = +\frac{4\pi\varepsilon_0 a^3}{Ze}\vec{E} \tag{8.3}$$

The resulting induced dipole moment is $Ze\vec{b}$, so that the relation between dipole moment and applied field, in this model is:

$$\vec{p} = Ze\vec{b} = 4\pi\varepsilon_0 a^3 \vec{E} = \varepsilon_0 \alpha \vec{E} \tag{8.4}$$

The direction of the dipole moment vector is from left to right in the case of figure 8.2, that is, in the same direction as the electric field. Notice that the dipole moment is simply proportional to the applied field. We can expect that this will be true in the real atom. The constant α is a property of the atom called, **atomic polarizability**. It is a property of the individual atom or molecule measuring the resistance of the atom or molecule to displacement of its electronic cloud. Notice that α has the dimensions of volume. For hydrogen atom it is: $\alpha = 4\pi a^3 = 4\pi(0.52 \cdot 10^{-8})^3 = 1.77 \times 10^{-24}$ cm^3, in agreement with this classical model, but an exactly quantum mechanical calculation of the polarizability of the hydrogen atom predicts the value: $\alpha = \frac{9}{2}a_0^3 = 0.66 \times 10^{-24}$ cm^3, where a_0 is the '*Bohr radius*' ($a_0 = 0.52 \times 10^{-8}$ cm), the characteristic distance in the H atom structure in its normal state. The electric polarizability values of several species of atoms, experimentally determined are given in table 8.1.

Table 8.1. Atomic polarizability values, in units of 10^{-24} cm^3.

Element	H	He	Li	Be	C	Ne	Na	Ar	K
α	0.66	0.21	12	9.3	1.5	0.4	27	1.6	34

The examples given are arranged in order of increasing number of electrons. Notice the wide variations in α. We observe that, hydrogen (H) and alkali metals lithium (Li), sodium (Na) and potassium (K), which occupy the first column of the periodic table, have large values of α and these increase steadily with increasing atomic number from H to K. The noble gases have much smaller atomic polarizability values but these also increase as we proceed, within the family, from helium (He) to neon (Ne) or krypton (Kr). Apparently, the alkali atoms, as a class, are easily deformed by an electric field, whereas the electronic structure of the noble gases atoms are much stiffer. It is the loosely bound outer, or 'valence' electron in the alkali atom structure that is responsible for the easy polarizability.

A molecule too, develops an induced dipole moment when an electric field is applied to it. The methane molecule (CH_4) depicted in figure 8.3 is made from four hydrogen atoms arranged in the corners of a tetrahedron around the central carbon atom, having an electrical polarizability of: $\alpha = 2.6 \times 10^{-24}$ cm^3. It is interesting to compare this with the sum of polarizability values of a carbon atom and four isolated hydrogen atoms. Taking the data from table 8.1, we find: $\alpha_{CH_4} = \alpha_C + 4\alpha_H = 4.14 \times 10^{-24}$ cm^3. Evidently, **binding of the atom into a molecule** has somewhat altered the electronic structure. Measurements of atomic and molecular polarizability values have long been used by chemists as dues to molecular structure. Molecules are necessarily less symmetrical than atoms. This raises the possibility of an induced dipole moment not parallel to the electric field that induced it. Then a molecule may be characterized by a **polarizability tensor**.

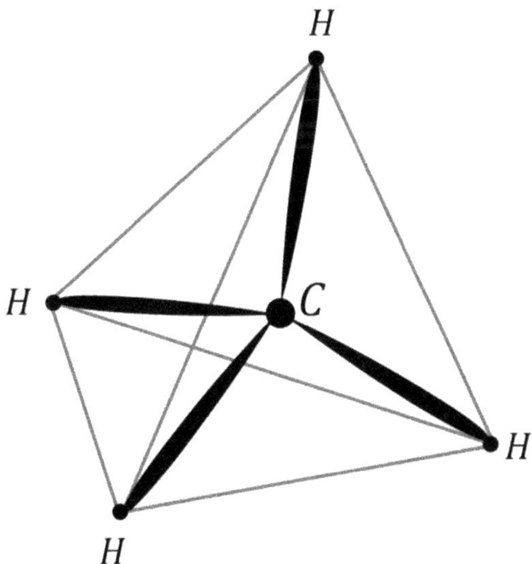

Figure 8.3. The methane molecule.

8.2 Polarization of a non-polar dielectric. Elastic polarization

The molecules in a solid or liquid dielectric are so close together that they are affected by another's short-range field and the average field $\overrightarrow{E_{loc}}$ acting on an individual molecule is not the same as the macroscopic field $\vec{E}$. However, if the molecules do not interact chemically, they have the same shape as in the gaseous state. The polarizability α of the molecules is unchanged, since it depends on their internal structure and each one acquires an average dipole moment $\vec{p} = \alpha\varepsilon_0 \overrightarrow{E_{\text{loc}}}$. If the molecular concentration is n, the dipole moment per unit volume, i.e., the polarization $\vec{P}$ is:

$$\vec{P} = n\varepsilon_0\alpha\,\overrightarrow{E_{\text{loc}}} \tag{8.5}$$

But, we know the relationship between the polarization $\vec{P}$ and the electric field $\vec{E}$ which produced the dielectric polarization, in terms of χ_E (dielectric susceptibility):

$$\vec{P} = \varepsilon_0\chi_E\vec{E} \tag{8.6}$$

And comparing equations (8.5) and (8.6), we see that the susceptibility χ_E (the macroscopic parameter) is related with the polarizability α (the microscopic parameter) and with the total electric field:

$$\chi_E = n\alpha\frac{\left|\overrightarrow{E_{\text{loc}}}\right|}{\left|\vec{E}\right|} \tag{8.7}$$

Thus, if we know $\left|\overrightarrow{E_{\text{loc}}}/\vec{E}\right|$, we calculate the susceptibility of a solid or liquid dielectric in terms of its molecular polarizability. In general, it is difficult to estimate $\overrightarrow{E_{\text{loc}}}$ for the fixed molecules in solids. Liquids and gases are more manageable, because the short-range fields may have a small average effect as neighboring molecules continually change their relative positions and orientations. Even in a liquid, dielectric behavior is complicated when polar molecules are present, since very large fields are generated by the permanent dipole moments (estimate the field at a distance of a few atomic diameters away from a water molecule—it is far, far bigger than any macroscopic field).

But in a non-polar liquid the short-range fields are not so strong, and it is not too bad an approximation to assume that an individual molecule is surrounded by a continuous medium in which atomic structure has been smeared out (a cavity). Like the real liquid, the continuous medium is polarized with a dipole moment per unit volume $\vec{P}$ so that polarization charges appear on any surface of the medium. Imagine that the molecule in question is removed, leaving a cavity embedded in the uniformly polarized medium. The field in the cavity is the sum of the macroscopic field $\vec{E}$ and the field generated by polarization charges on its surface, and this sum is the field $\overrightarrow{E_{\text{loc}}}$ which would act on a molecule filling the cavity. Let us assume that the cavity is a sphere of radius R, and as in figure 8.4 take the Z axis to be in the direction of $\vec{E}$. On the surface of this cavity there is a positive polarization charge σ_p on the down side of sphere and a negative polarization charge $-\sigma_p$ on the upper side

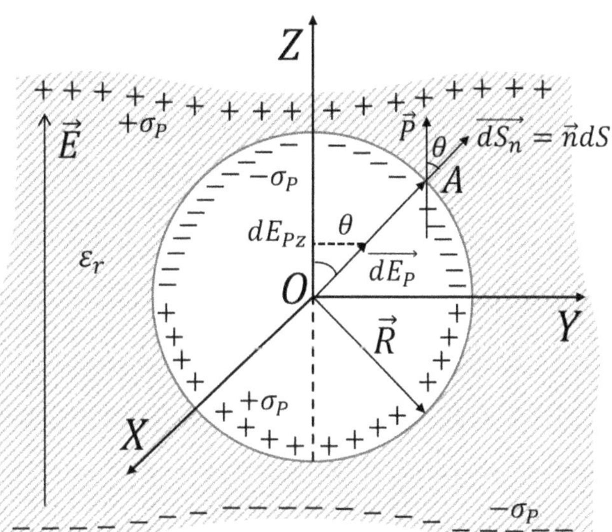

Figure 8.4. The total electric field inside a spherical cavity in a uniformly polarized dielectric material.

of the spherical cavity. The surface element dS located as point A carries the charge $\sigma_p dS = -\vec{P}\,\vec{n}\,dS = -PdS\cos\theta$, then the electric field, at the center O of the spherical cavity, due to the charge $\sigma_p(\theta)dS = -P\cos\theta dS$, is:

$$\overrightarrow{dE} = \frac{\sigma_p dS}{4\pi\varepsilon_0 R^3}\vec{R} = -\frac{P\cos\theta dS}{4\pi\varepsilon_0 R^3}\vec{R} \tag{8.8}$$

Then the total field created by the polarization charge will be:

$$E_z = -\int_S \frac{PR^2\sin\theta d\theta d\varphi \cos\theta}{4\pi\varepsilon_0 R^3} R_z \tag{8.9}$$

But we observe that $R_z = -R\cos\theta$ and just only the Z component of field is different to zero, due to the high spherical symmetry, so the field generated by all polarization charge is in the same direction as the macroscopic field $\vec{E}$:

$$E_z = \int_0^\pi \int_0^{2\pi} \frac{P\cos^2\theta \sin\theta d\theta d\varphi}{4\pi\varepsilon_0} = \frac{P}{2\varepsilon_0}\int_0^\pi \cos^2\theta[d(\cos\theta)]$$

$$= \frac{P}{2\varepsilon_0}\left[-\frac{\cos^3\theta}{3}\Big|_0^\pi\right] = \frac{P}{2\varepsilon_0}\left[\frac{1}{3}+\frac{1}{3}\right] = \frac{P}{3\varepsilon_0} \tag{8.10}$$

Now, the field in the cavity is uniform and is given by the sum of the externally applied field and the field created by the polarization charges on the surfaces of cavity:

$$\overrightarrow{E_{loc}} = \vec{E}+\frac{\vec{P}}{3\varepsilon_0} \tag{8.11}$$

Substituting this value for $\overrightarrow{E_{loc}}$ in equation (8.7), we find the susceptibility as:

$$\chi_E = n\alpha \frac{E_{loc}}{E} = n\alpha \frac{\left(E + \dfrac{P}{3\varepsilon_0}\right)}{E} = n\alpha \left(1 + \frac{P}{3\varepsilon_0 E}\right)$$

$$= n\alpha \left(1 + \frac{\varepsilon_0 \chi_E E}{3\varepsilon_0 E}\right) = n\alpha \left(1 + \frac{\chi_E}{3}\right) \tag{8.12}$$

Rearranging equation (8.12) we obtain:

$$n\alpha = \frac{\chi_E}{1 + \dfrac{\chi_E}{3}} = \frac{3(\varepsilon_r - 1)}{3 + \varepsilon_r - 1} \Rightarrow \frac{\varepsilon_r - 1}{\varepsilon_r + 2} = \frac{n\alpha}{3} \tag{8.13}$$

This approximation, which is known as the 'Clausius–Mossotti' formula, cannot be expected to be very accurate since it is based on such a crude representation of the neighboring molecules. However, the 'Clausius–Mossotti' formula does demonstrate that the polarization in the liquid dielectric causes the local field to be larger than the macroscopic field, and makes a reasonable estimate of the size of the increase. The molecular polarizability can be derived from the relative permittivity of a dielectric:

$$\alpha = \frac{3(\varepsilon_r - 1)}{n(\varepsilon_r + 2)} \tag{8.14}$$

Measuring the electric permittivity of a dielectric we can obtain the polarizability of its molecule. As regarding the measurement of ε_r, we know the method which uses the LC circuit in alternating current (see figure 8.5). When the capacitor C_{sample} has as dielectric the air, the maximum of voltage across the inductance L is obtained for resonance frequency:

$$\omega_0 = \frac{1}{\sqrt{L(C_{V_0} + C_{\text{air}})}}$$

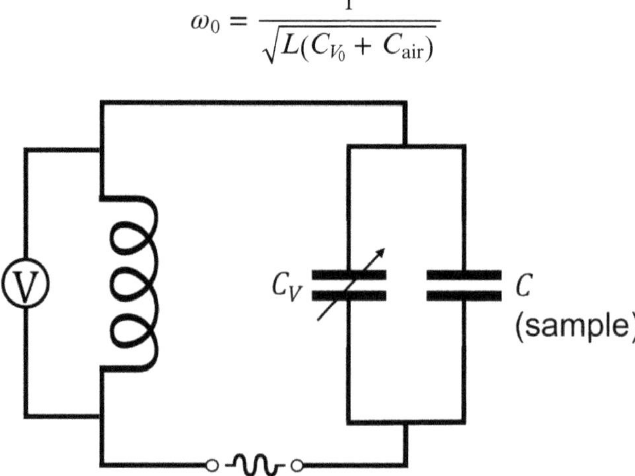

Figure 8.5. An LC circuit in alternating current used for the measurement of the electric permittivity of a dielectric.

Filling the space between the plates of the C_{sample} capacitor, the frequency ω_0 remains constant for the value C_V of the variable capacitor. Then:

$$C_{V_0} + C_{\text{air}} = C_V + C_{\text{sample}} \Rightarrow \varepsilon_r = \frac{C_{\text{sample}}}{C_{\text{air}}} = \frac{C_{V_0} - C_V}{C_{\text{air}}} + 1 \qquad (8.15)$$

8.3 Polarization of a polar dielectric. Orientational polarization

The local field acting on a molecule **in a gas** is almost the same as the external field $\vec{E}$. For a molecule with dipole moment $\vec{p}$, its potential energy in an external field $\vec{E}$ is:

$$W_p = -\vec{p} \cdot \vec{E} = -pE \cos \theta \qquad (8.16)$$

which shows how it varies with its orientation. The energy has a minimum value at $\cos \theta = 1$ ($\theta = 0$), when the dipole is aligned along the field. An amount of energy $2pE$ must be expended to reverse the dipole moment and point it in the direction opposite to the field. How much energy is this? If one valence electron in the molecule has been transferred from one atom to another, the resulting dipole moment is roughly: $e \times$ (atomic spacing) $\simeq 1.6 \times 10^{-19} \times 10^{-10} = 1.6 \times 10^{-29}$ C m. Typical polar molecules have dipole moments of this order of magnitude, and in an electric field $\vec{E}$ of 10^6 V m^{-1} (about as big as can be achieved in a gaseous dielectric):

$$2pE = 2 \times 1.6 \times 10^{-29} \text{ C m} \times 10^6 \text{ V m}^{-1} = 2 \times 10^{-4} \text{ eV} \qquad (8.17)$$

At ordinary temperatures, this is much less than the energy associated with thermal motion. The average kinetic energy $\frac{3}{2}kT$ of each molecule in a gas is about 0.04 eV at room temperature. The gas molecules are continually knocked about in collisions, changing their orientations and directions of motion, and transferring kinetic energy from one to another. Because the potential energy $-\vec{p} \cdot \vec{E}$ is so small compared to the kinetic energy, the random thermal motion is unaffected by the presence of the field, and the molecules are still to be found in all orientations. However, there is a slight preponderance of molecules with dipoles aligned up along the field, the direction in which their potential energy is minimum. To calculate the net polarization caused by this alignment, we must know the probability of finding a molecule at any particular orientation to the field, see figure 8.6. The element of solid angle is:

$$d\Omega = \int_0^{2\pi} \frac{dS}{r^2} = 2\pi \frac{r^2 \sin \theta d\theta}{r^2} = 2\pi \sin \theta d\theta \qquad (8.18)$$

Before the electric field is switched on, all orientations of the molecules in the gaseous polar dielectric are equally probable. The number of molecules whose dipole moments have directions which lie within a solid angle $d\Omega$ is simply proportional to $d\Omega = 2\pi \sin \theta d\theta$, therefore, the number of molecules from unit volume and unit solid angle which have the dipole moments pointing in directions in the range $d\theta$ is:

$$dn = \frac{n}{4\pi} d\Omega = \frac{n}{2} \sin \theta d\theta \qquad (8.19)$$

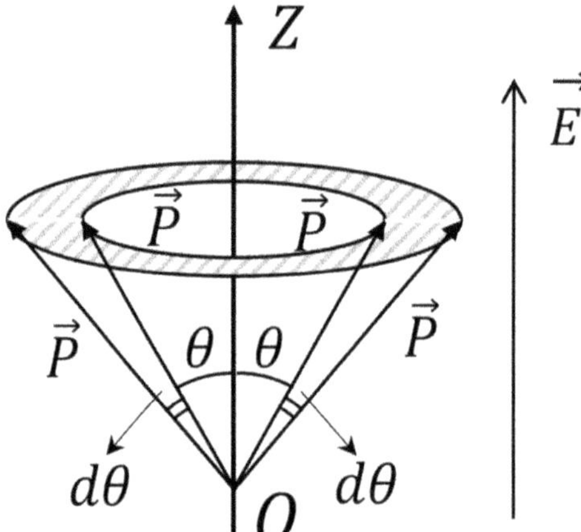

Figure 8.6. The shaded area represent the band between θ and $(\theta + d\theta)$ on the surface of a sphere of radius r.

After switching on an electric field $\vec{E}$ directed along Z axis, a molecule with dipole moment at θ, acquires an additional potential energy $W_p = -pE\cos\theta$. The probability distribution at thermal equilibrium is now modified by the presence of a Boltzmann factor $e^{-\frac{W_p}{kT}}$, so the probability to find in the presence of the field, a molecule with the dipole moment oriented in a solid angle $d\Omega$ is:

$$\mathcal{P} = Ae^{-\frac{W_p}{kT}}d\Omega \tag{8.20}$$

The constant of proportionality A is determined by the condition that, there is a probability equal to one to find at least one molecule having the moment dipole in the range $(\pi, 0)$, then:

$$\int_{\pi}^{0} \mathcal{P}d\Omega = 1 \Rightarrow A = \frac{1}{\int_{\pi}^{0} e^{-\frac{W_p}{kT}}d\Omega}$$

To calculate the polarization $\vec{P}$ of the polar dielectric introduced in the external field $\vec{E}$ we must take the average value of the components of the dipole moments from volume, along the electric field:

$$\vec{P} = n\langle p\cos\theta\rangle \tag{8.21}$$

Having the probability $\mathcal{P}$—equation (8.20), the average value of the dipole moment component is:

$$\langle p\cos\theta\rangle = \int_{\pi}^{0} p\cos\theta\,\mathcal{P}d\Omega = \frac{\int_{\pi}^{0} p\cos\theta\,e^{-\frac{W_p}{kT}}d\Omega}{\int_{\pi}^{0} e^{-\frac{W_p}{kT}}d\Omega} \tag{8.22}$$

But:

$$W_p = -\vec{p}\,\vec{E} = -pE\cos\theta$$

$$\langle p\cos\theta\rangle = \frac{p\displaystyle\int_{\pi}^{0}\cos\theta\, e^{\frac{pE\cos\theta}{kT}}2\pi\sin\theta d\theta}{\displaystyle\int_{\pi}^{0} e^{\frac{pE\cos\theta}{kT}}2\pi\sin\theta d\theta} = p\frac{\displaystyle\int_{\pi}^{0}\cos\theta\, e^{\frac{pE\cos\theta}{kT}}d(\cos\theta)}{\displaystyle\int_{\pi}^{0} e^{\frac{pE\cos\theta}{kT}}d(\cos\theta)} \tag{8.23}$$

Noting $\cos\theta = u$ and $\frac{pE}{kT} = x$, the equation (8.23) becomes:

$$\langle p\cos\theta\rangle = p\frac{\displaystyle\int_{-1}^{1} u e^{xu}du}{\displaystyle\int_{-1}^{1} e^{xu}du} \tag{8.24}$$

Using the logarithmic derivative we have:

$$\langle p\cos\theta\rangle = p\frac{d}{dx}\left[\lg\int_{-1}^{1} e^{xu}du\right] = p\frac{d}{dx}\left\{\lg\left[\frac{1}{x}(e^x - e^{-x})\right]\right\}$$

$$= p\frac{d}{dx}[\lg(e^x - e^{-x}) - \lg x] = p\left[\frac{e^x + e^{-x}}{e^x - e^{-x}} - \frac{1}{x}\right] \tag{8.25}$$

$$= p\left[\coth x - \frac{1}{x}\right] = p\mathcal{L}(x)$$

In equation (8.25), $\mathcal{L}(x)$ is so called '**Langevin's function**' and $x = \frac{pE}{kT}$. Such as we already saw, the dipole moment has a very small value, and for small applied electric field and high temperatures, $x \ll 1$ and then it could be made a Taylor's expansion of $\coth x$ in $\mathcal{L}(x)$:

$$\coth x = \frac{e^x + e^{-x}}{e^x - e^{-x}} = \frac{1 + x + \dfrac{x^2}{2} + \dfrac{x^3}{6} + 1 - x + \dfrac{x^2}{2} - \dfrac{x^3}{6} + \cdots}{1 + x + \dfrac{x^2}{2} + \dfrac{x^3}{6} - 1 + x - \dfrac{x^2}{2} + \dfrac{x^3}{6} + \cdots}$$

$$= \frac{2 + x^2}{2x + \dfrac{x^3}{3}} = \frac{\dfrac{1}{x} + \dfrac{x}{2}}{\left(1 + \dfrac{x^2}{6}\right)} = \left(\frac{1}{x} + \frac{x}{2}\right)\left(1 + \frac{x^2}{6}\right)^{-1}$$

$$= \left(\frac{1}{x} + \frac{x}{2}\right)\left(1 - \frac{x^2}{6}\right) = \frac{1}{x} + \frac{x}{2} - \frac{x}{6} - \frac{x^3}{12} \simeq \frac{1}{x} + \frac{x}{3}$$

So:

$$\begin{cases} \mathcal{L}(x) = \dfrac{1}{x} + \dfrac{x}{3} - \dfrac{1}{x} = \dfrac{x}{3}, & \text{for } x \ll 1 \\ \mathcal{L}(x) \to 1, & \text{for } x \to \infty \end{cases} \qquad (8.26)$$

A plot of $\mathcal{L}(x)$ is shown in figure 8.7. Returning to the average value of the component of the dipole moment along the electric field we obtain:

$$\langle p \cos \theta \rangle = p\mathcal{L}(x) \simeq p \cdot \frac{x}{3} = p\frac{pE}{3kT} = \frac{p^2 E}{3kT} \qquad (8.27)$$

and then the dipole moment per unit volume or the polarization $\vec{P}$ will be:

$$\vec{P} = n\langle p \cos \theta \rangle = \frac{np^2}{3kT}\vec{E} \qquad (8.28)$$

But all the molecules also acquire an induced dipole moment, which points along the field whatever may be the direction of the permanent dipole moment. If the molecular polarizability is α, there is an additional induced polarization $\vec{P_i} = \varepsilon_0 \alpha n \vec{E}$. Finally, we have the result that the net polarization of the gaseous polar dielectric is:

$$\vec{P} = \left(n\alpha\varepsilon_0 + \frac{np^2}{3kT} \right)\vec{E} \qquad (8.29)$$

and its susceptibility χ_E will be:

$$\chi_E = \frac{P}{\varepsilon_0 E} = n\alpha + \frac{np^2}{3k\varepsilon_0 T} \qquad (8.30)$$

Now, in a liquid or solid polar dielectric the '*Clausius–Mossotti*' equation will have another shape because there appears: $\vec{E}_{\text{loc}} = \vec{E} + \dfrac{\vec{P}}{3\varepsilon_0}$. Now, equation (8.30) becomes:

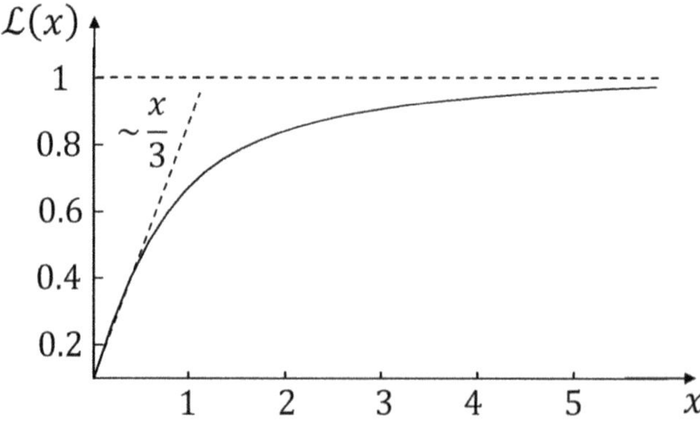

Figure 8.7. The plot of $\mathcal{L}(x)$.

$$\chi_E = \frac{\vec{P}}{\varepsilon_0 E_{\text{loc}}} = \frac{\varepsilon_0 \chi_E E}{\varepsilon_0 E \left(1 + \frac{\chi_E}{3}\right)} = n\alpha + \frac{np^2}{3\varepsilon_0 kT} \Rightarrow \frac{\varepsilon_r - 1}{\varepsilon_r + 2} = \frac{n\alpha}{3} + \frac{np^2}{9\varepsilon_0 kT} \quad (8.31)$$

the '*Clausius–Mossotti*' for a polar dielectric. By measuring the permittivity at different temperatures, it is possible to distinguish between permanent and induced dipole moments. In figure 8.8, $\chi_E = n\alpha + \frac{np^2}{3\varepsilon_0 kT} = a + \frac{b}{T}$, is plotted against $10^3/T$ for HCl gas at a fixed density of $1/22.4$ moles l^{-1} (i.e., the density found at NTP[1]). As expected from the equation (8.30), the plot is a straight line:

$$\chi_E = n\alpha + \frac{np^2}{3\varepsilon_0 k \cdot 10^3} \cdot \frac{10^3}{T} = a + b \cdot \frac{10^3}{T}$$

having the slope: $b = \frac{np^2}{3\varepsilon_0 k \cdot 10^3}$, which gives permanent dipole moment of its molecule being $p = 3.6 \times 10^{-30}$ C m. The intercept at $10^3/T = 0$ represents the susceptibility due to induced polarization at this density and it gives the polarizability α. To achieve the permanent dipole moment $p = 3.6 \times 10^{-30}$ C m charges $\pm e$ would be needed to be separated by a little more than 2×10^{-11} m $\simeq 0.2$ Å, or about one fifth of a typical atomic diameter. The susceptibilities and dipole moments of few gases are listed in table 8.2. It is interesting to observe that carbon disulfide (CS_2) has no dipole moment although, like water (see figure 8.9), it contains two identical atoms

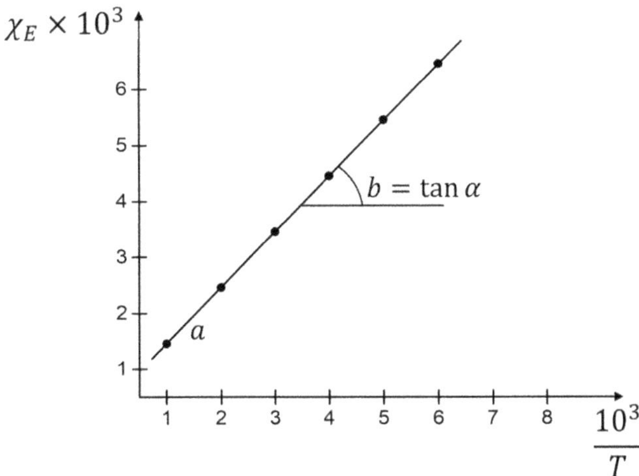

Figure 8.8. The temperature dependence of the susceptibility of HCl gas at a density of $1/22.4$ moles l^{-1}.

---

[1] Normal temperature and pressure (NTP) is commonly used as a standard condition for testing and documentation of fan capacities, i.e., air at 20 °C and 1 atm.

Table 8.2. Dipole moments of several polar molecules.

Substance	Susceptibility of gas at NTP	Dipole moments in C m
HCl	0.0046	3.6×10^{-30}
Methane (CH_4)	0.0015	0
Ethyl alcohol	0.0061	5.7×10^{-30}
H_2O	0.0126	6.2×10^{-30}
CS_2	0.0029	0

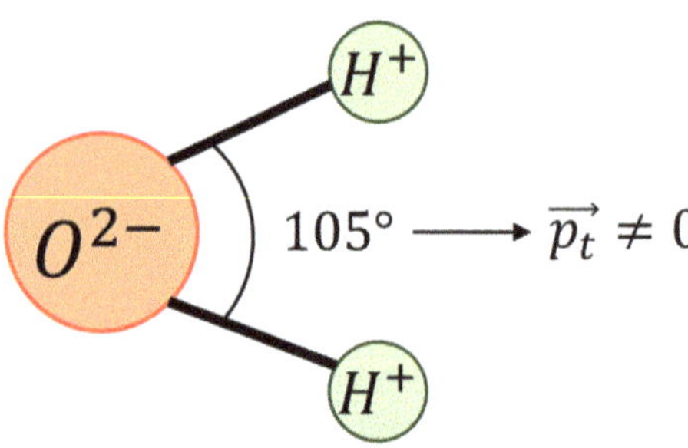

Figure 8.9. The water molecule has a permanent dipole moment, but the linear CS$_2$ molecule has none.

bound to a common partner. There is electron transfer across the C–S bonds in carbon disulfide, and dipole moment associated with each bond. But the atoms in the CS_2 molecule like on a straight line, and the dipole moment exactly cancel $\left(S \xrightarrow{P} C \xrightarrow{P} S \right)$ one another, whereas in water the two O–H bonds make an angle of 105°. A situation like this of CS_2 there is with methane (CH_4) where because of high symmetry of the four hybridized orbitals, even though there are four dipole moments along the C–H bonds, as a whole the molecule of CH_4 has none. By glancing at a table of susceptibilities one can learn something about the shape of the molecules.

Further reading

[1] Coelho R 1979 *Physics of Dielectrics for the Engineer* 1st edn (Amsterdam: Elsevier)
[2] Pinsky A A 1983 *Problems in Physics* 2nd edn (Moscow: Mir)
[3] Scaife B K P 1998 *Principles of Dielectrics (Monographs on the Physics and Chemistry of Materials)* revised edn (Oxford: Oxford University Press)
[4] Panofsky W K H and Phillips M 2005 *Classical Electricity and Magnetism Dover Books on Physics* 2nd edn (New York: Dover)
[5] Singh J 2005 Dielectric response: polarization effects *Smart Electronic Materials: Fundamentals and Applications* (Cambridge: Cambridge University Press) pp 264–95
[6] Feynman R P, Leighton R B and Sands M 2011 *The Feynman Lectures on Physics, Vol II: The New Millennium Edition: Mainly Electromagnetism and Matter* (New York: Basic Books)

[7] Zangwill A 2012 *Modern Electrodynamics* 1st edn (Cambridge: Cambridge University Press)
[8] Purcell E M and Morin D J 2013 *Electricity and Magnetism* 3rd edn (Cambridge: Cambridge University Press)
[9] Ilie C and Schrecengost Z 2016 Electric fields in matter *Electromagnetism* (San Rafael, CA: Morgan & Claypool Publishers) ch 5 pp 5-1-26

Chapter 9

Energy in the presence of dielectrics

The energy of a distribution of charge in free space was obtained so far by assembling the distribution little by little, bringing infinitesimal pieces of charge in from infinity. In the presence of dielectrics this formalism is not applicable, as additional work must be also done to induce polarization in the dielectric, that has to be accounted for as well. This chapter will try thus to answer the following question: '*What is the potential energy of charges in the presence of a dielectric material?*' To answer, we will be working out the energy stored in a single capacitor, before dealing with a general system of charges.

9.1 Energy of a capacitor

The capacitor is initially uncharged, and a potential difference U is built up between the plates by transferring a charge $Q = CU$ from one to the other. Energy supplied, for example by a battery, is needed to move the charge, and at the moment when the potential difference is U', the energy $U'dQ'$ is required to transfer a further charge dQ'. Now $dQ' = CdU'$ and the total energy stored is:

$$W = \int_0^U U'dQ' = \int_0^U U'CdU' = \frac{1}{2}CU^2 = \frac{1}{2}UQ = \frac{Q^2}{2C} \qquad (9.1)$$

This expression has the same form as equation $W_0 = \frac{1}{2}U_0Q = \frac{1}{2}C_0U_0^2$ for the energy of a vacuum capacitor. The actual value of the energy is of course affected by the presence of the dielectric material because $C = \varepsilon_r C_0$ depends on the relative permittivity, ε_r. Nevertheless, only Q, the free charge on the capacitor plates, appears in equation (9.1), which contains no explicit reference to polarization charges. This is because it is only the free charge which is directly moved across the potential difference by external forces.

     9-1   © IOP Publishing Ltd 2023

In a parallel-plate capacitor of area A and plate separation d, the energy stored when it is filled with dielectric is $\frac{1}{2}CU^2 = \frac{1}{2}\varepsilon_0\varepsilon_r\frac{A}{d}U^2$. The volume of the capacitor is Ad, and therefore the energy density is:

$$w = \frac{W}{Ad} = \frac{1}{2}\varepsilon_0\varepsilon_r\frac{A}{dAd}U^2 = \frac{1}{2}\varepsilon_0\varepsilon_r\left(\frac{U}{d}\right)^2 = \frac{1}{2}\varepsilon_0\varepsilon_r E^2 = \frac{1}{2}\vec{D}\vec{E}$$

(9.2)

$$\vec{D} = \varepsilon\vec{E} = \varepsilon_0\varepsilon_r\vec{E}$$

9.2 Electrostatic energy

The electrostatic energy is stored in whole space with the electric field. Just as with fields in vacuum, we can regard the energy density $\frac{1}{2}\vec{D}\vec{E}$ as residing in the field. As before, it is plausible that the energy density is $\frac{1}{2}\vec{D}\vec{E}$ in any electrostatic field, because we can imagine that the field is split up into a large number of parallel-plate capacitors by thin conductors placed along closely spaced equipotential. The total electrostatic energy stored in a volume v is:

$$W = \frac{1}{2}\int_v \vec{D}\vec{E}\,dv$$

(9.3)

This result is proved rigorously as described in the following pages. To find the energy stored by a general system of charges, we use the same technique as was applied to the capacitor in the previous section. We begin with no polarized dielectric with no free charges, and then assemble the free charge by bringing them up from infinity. External work is done in bringing up the free charge, which automatically accounts for the potential energy of any polarization charges which may appear. The energy required to assemble a set of charges is:

$$W = \frac{1}{2}\sum_{i=1}^{n} q_i V_i$$

(9.4)

where V_i is the potential at the position of q_i. In a parallel-plate capacitor, having as dielectric the air, the electrostatic potential energy is:

$$W_0 = \frac{1}{2}\frac{Q_0^2}{C_0} = \frac{1}{2}C_0U_0^2 = \frac{1}{2}U_0Q_0 = \frac{1}{2}\varepsilon_0E_0^2v = \frac{1}{2}\vec{D}\,\vec{E_0}v = \frac{1}{2}w_0v$$

where v is the volume of the space between the plates of condenser, and $\vec{E_0} = \frac{\sigma_0}{\varepsilon_0}\vec{n}$, whilst $\vec{D} = \varepsilon_0\vec{E_0} = \sigma_0\vec{n}$. If the free charge of the capacitor remains Q_0, then its energy when the space between the plates is filled with a dielectric becomes:

$$W = \frac{1}{2}\frac{Q_0^2}{C} = \frac{1}{2}CU^2 = \frac{1}{2}Q_0U = \frac{1}{2}\vec{D}\vec{E}v = \frac{1}{2}wv$$

where $C = \varepsilon_r C_0$, $D = \sigma_0$ and $E = \frac{E_0}{\varepsilon_r} = \frac{\sigma_0}{\varepsilon_0 \varepsilon_r}$. We observe that the energy decreases, when the dielectric was introduced between the plates, and the capacitor was disconnected from the battery. Indeed:

$$\Delta W = W - W_0 = \frac{1}{2}\vec{D}\vec{E}v - \frac{1}{2}\vec{D}\vec{E_0}v = \frac{1}{2}\left(\vec{E} - \vec{E_0}\right)\vec{D}v \tag{9.5}$$

But $\vec{P} = \vec{D} - \varepsilon_0\vec{E} \Rightarrow \vec{P}\vec{E_0} = \vec{D}\vec{E_0} - \varepsilon_0\vec{E}\vec{E_0} = \vec{D}\vec{E_0} - \vec{D}\vec{E} = -\left(\vec{D}\vec{E} - \vec{D}\vec{E_0}\right)$. With this we can see that effectively the energy of a capacitor decreases after the introduction of a dielectric between its plates with the quantity:

$$\Delta W = W - W_0 = -\frac{1}{2}\vec{P}\vec{E_0}v \tag{9.6}$$

The dielectric in the external electric field polarizes, and a part of initial energy is used to polarize the dielectric. That is why the energy of the capacitor decreases, when the dielectric is introduced between its plates. But generally, the energy may be calculated as a function of only the vector fields $\vec{D}$ and $\vec{E}$ using the density $w = \frac{1}{2}\vec{D}\vec{E}$, whatever the space where the electric field occurs:

$$W = \frac{1}{2}\int_v \vec{D}\vec{E}\,dv \tag{9.7}$$

Extending this result to the assembly of a point charge or, more generally, to the assembly of some volume distribution of charge, ρ_f, we obtain:

$$W = \frac{1}{2}\int_v \rho_f V\,dv \tag{9.8}$$

Note: equation (9.7) does not include the self-energy of charge, that is, the energy required to assemble each charge. Only the interaction energy between the charges is included in (9.7). Equation (9.8) can be written in an alternative form. Gauss' law in differential form, $\mathrm{div}\vec{D} = \rho_f$, can be substituted into (9.8) to yield:

$$W = \frac{1}{2}\int_v \mathrm{div}\vec{D}\,V\,dv \tag{9.9}$$

Using the identity: $\mathrm{div}\left(V\vec{D}\right) = V\,\mathrm{div}\vec{D} + \vec{D}\,\mathrm{grad}V$ (see appendix B.5), the relation (9.9) becomes:

$$W = \frac{1}{2}\int_v \mathrm{div}\left(V\vec{D}\right)dv - \frac{1}{2}\int_v \vec{D}\,\mathrm{grad}V\,dv$$

Applying the divergence theorem to the first of these volume integrals results in: $W = \frac{1}{2}\int_S V\vec{D}\,\vec{ds} - \frac{1}{2}\int_v \vec{D}\nabla V\,dv$. If we allow the volume v in equation (9.8) to include all space, the surface S in this equation goes to infinity. The absolute potential V

decays as $1/r$, and $\vec{D}$ decays as $1/r^2$. The surface area, however, increases as r^2, so that the first integral from the above equation goes to zero as S tends to infinity. Substituting $\vec{E} = -\nabla V$ in the remaining integral $W = -\frac{1}{2}\int_v \vec{D}\nabla V dv$, it becomes:

$$W = \frac{1}{2}\int_v \vec{D}\vec{E}\,dv \tag{9.10}$$

Note that volume v in equation (9.10) includes all space. The result in equation (9.10) indicates that the quantity $w_e = \frac{1}{2}\vec{D}\vec{E}$ is a volume energy density, or density of the energy stored in the field produced by the charge, since the units of $\vec{D}\vec{E}$ are joules per cubic meter (J m^{-3}). It must be emphasized that the location of the energy stored in the field cannot be determined with any more confidence than we could locate the potential energy stored in a gravitational field. Equation (9.10) simply states that if we integrate the quantity $w_e = \frac{1}{2}\vec{D}\vec{E}$ (which has dimensions of energy density) over all space we will obtain the total energy stored in the field. For a linear, homogenous and isotropic medium, $\vec{D} = \varepsilon\vec{E}$, and equation (9.10) can be rewritten as:

$$W_e = \frac{1}{2}\int_v \varepsilon E^2 dv \tag{9.11}$$

Exercise: energy stored by a sphere

In a sphere of radius a there is a uniformly distributed charge having the volume charge density ρ_f, see figure 9.1. Find the energy stored in the electric field of the sphere.
Solution: *the electric field inside of the sphere is:*

$$\int_{\Sigma_i} \vec{E_i}\,\vec{dS} = \int_{v_{\Sigma_i}} \frac{\rho_f dv}{\varepsilon_0\varepsilon_r} \Rightarrow \vec{E_i} = \frac{\rho_f}{3\varepsilon_0\varepsilon_r}\vec{r_i}$$

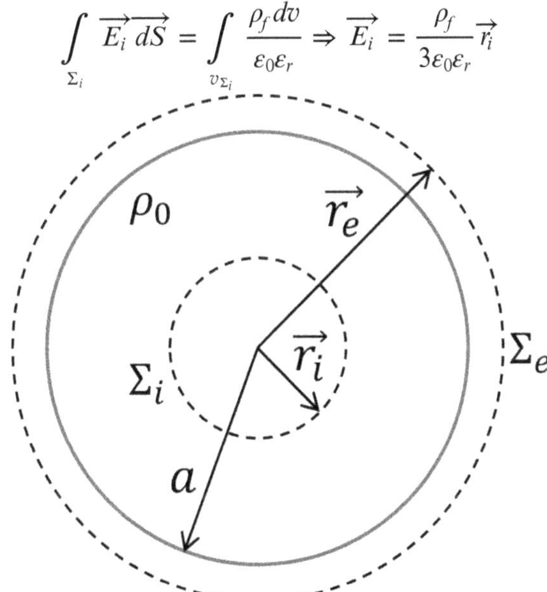

Figure 9.1. *Sphere of radius a, uniformly charged in its volume with volume charge density ρ_f.*

The electric field outside the sphere is:

$$\overrightarrow{E_e} = \frac{\rho_f a^3}{3\varepsilon_0 r^3}\overrightarrow{r}$$

Then the energy will be:

$$
\begin{aligned}
W &= \frac{1}{2}\int_v \varepsilon_0 E_e^2 dv + \frac{1}{2}\int_v \varepsilon_0\varepsilon_r E_i^2 dv \\[2mm]
&= \frac{1}{2}\int_a^\infty \int_0^\pi \int_0^{2\pi} \varepsilon_0 \frac{\rho_f^2 a^6}{9\varepsilon_0^2 r^4} r^2 \sin\theta\, dr\, d\theta\, d\phi \\[2mm]
&\quad + \frac{1}{2}\int_0^a \int_0^\pi \int_0^{2\pi} \varepsilon_0\varepsilon_r \frac{\rho_f^2}{9\varepsilon_0^2\varepsilon_r^2} r^2 r^2 \sin\theta\, dr\, d\theta\, d\phi = \frac{4\pi\rho_f^2}{18\varepsilon_0}a^5\left[1 + \frac{1}{5\varepsilon_r}\right]
\end{aligned}
\tag{9.12}
$$

$$\text{If}\quad \varepsilon_r = 1 \Rightarrow W = \frac{4\pi\rho_f^2 a^5}{18\varepsilon_0}\frac{6}{5} = \frac{4\pi\rho_f^2 a^5}{15\varepsilon_0} \tag{9.13}$$

9.3 Thompson's theorem

Now, we are interested in the potential nature of the electrostatic energy. If the electric charges are moving under the action of an electrostatic field, the electrostatic energy of the field decreases in proportion to the size of the mechanical work corresponding to the displacement of the electric charges, this resulting directly from the relation:

$$L_{12} = W_1 - W_2 \quad \text{or} \quad -L_{12} = \Delta W \tag{9.14}$$

If the charges can move freely, they will look for a distribution that corresponds to a minimum value of electrostatic energy in the electrostatic field. This property is stated by **W Thompson's – Lord Kelvin's Theorem**: *the charges on a system of conductors, located in a dielectric medium, are distributed on the surfaces of these conductors, so that the energy of the resulting electrostatic field is minimal.*

Let there be the charges q_i $(i = 1, 2, ...)$, carried by the conductors A_i $(i = 1, 2, ...)$ of a system of conductors to electrostatic equilibrium (see figure 9.2). Assuming the isotropic dielectric medium, for the given system we can write the general equations:

$$
\begin{aligned}
&\int_{\Sigma_i} \overrightarrow{D}\,\overrightarrow{dS} = q_i \quad (i = 1,\ 2,\ ...,\ n) \\[2mm]
&\overrightarrow{D} = \varepsilon\overrightarrow{E} \\[2mm]
&\overrightarrow{E} = -\operatorname{grad} V
\end{aligned}
\tag{9.15}
$$

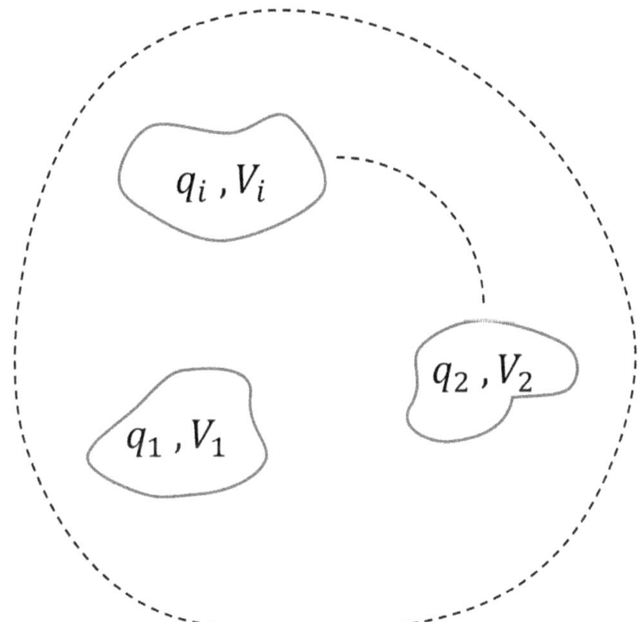

Figure 9.2. A system of charged conductors at electrostatic equilibrium carrying the charges $q_1, q_2, ...,q_i$.

and for the conductors' surfaces:

$$V_1 = \text{constant}; \; V_2 = \text{constant}; \; ...; \; V_i = \text{constant} \quad (i = 1, \; 2, \; ..., \; n) \qquad (9.16)$$

Consider another electrostatic field, characterized by vectors: $\overrightarrow{D'}$ and $\overrightarrow{E'}$, different from $\overrightarrow{D}$ and $\overrightarrow{E}$, which satisfy the same general equations (9.15), so they correspond to the same conductors system. To demonstrate Thompson's theorem, it is enough to show that $W' > W$, where:

$$W = \frac{1}{2} \int_v \overrightarrow{D}\,\overrightarrow{E}\, dv \; \text{ and } \; W' = \frac{1}{2} \int_v \overrightarrow{D'}\,\overrightarrow{E'}\, dv \qquad (9.17)$$

Let us suppose that these new fields are:

$$\overrightarrow{E'} = \overrightarrow{E} + \overrightarrow{E''} \; \text{ and } \; \overrightarrow{D'} = \overrightarrow{D} + \overrightarrow{D''}$$

Asking $\overrightarrow{D'}$ to verify the equations (9.15) results in:

$$\int_{\Sigma_i} \overrightarrow{D'}\,\overrightarrow{dS} = q_i \Rightarrow \int_{\Sigma_i} \left(\overrightarrow{D} + \overrightarrow{D''}\right)\overrightarrow{dS} = q_i$$

But taking into account the equations (9.15) results in:

$$\int_{\Sigma_i} \overrightarrow{D'}\,\overrightarrow{dS} = q_i \Rightarrow \int_{\Sigma_i} \overrightarrow{D''}\,\overrightarrow{dS} = 0 \Rightarrow \text{div}\,\overrightarrow{D''} = 0 \; \text{ and } \; \overrightarrow{D''} = \varepsilon\,\overrightarrow{E''} \qquad (9.18)$$

The electrostatic energy W' is:

$$W' = \frac{1}{2} \int_v \overrightarrow{D'}\,\overrightarrow{E'}\, dv = \frac{1}{2} \int_v \left(\overrightarrow{D} + \overrightarrow{D''}\right)\left(\overrightarrow{E} + \overrightarrow{E''}\right) dv$$

$$= \frac{1}{2} \int_v \overrightarrow{D}\,\overrightarrow{E}\, dv + \frac{1}{2} \int_v \overrightarrow{D''}\,\overrightarrow{E''}\, dv + \frac{1}{2} \int_v \left(\overrightarrow{D}\,\overrightarrow{E''} + \overrightarrow{D''}\,\overrightarrow{E}\right) dv$$

$$= W + \frac{1}{2} \int_v \varepsilon\, \overrightarrow{E''}^2\, dv + \frac{1}{2} \int_v \left(\varepsilon\overrightarrow{E}\,\overrightarrow{E''} + \varepsilon\overrightarrow{E''}\,\overrightarrow{E}\right) dv \qquad (9.19)$$

$$= W + \frac{1}{2} \int_v \varepsilon\, \overrightarrow{E''}^2\, dv + \frac{1}{2} \int_v \left(2\varepsilon\overrightarrow{E}\,\overrightarrow{E''}\right) dv$$

$$\Rightarrow W' = W + \frac{1}{2} \int_v \varepsilon\, \overrightarrow{E''}^2\, dv + \int_v \overrightarrow{E}\,\overrightarrow{D''}\, dv$$

But:

$$\overrightarrow{E} = -\mathrm{grad}\,V \Rightarrow \int_v \overrightarrow{E}\,\overrightarrow{D''}\, dv = \int_v -\nabla V \cdot \overrightarrow{D''}\, dv \qquad (9.20)$$

Taking into account the relation:

$$\mathrm{div}\left(V\overrightarrow{D''}\right) = \mathrm{grad}\,V \cdot \overrightarrow{D''} + V\,\mathrm{div}\,\overrightarrow{D''}$$

$$\Rightarrow -\mathrm{grad}\,V \cdot \overrightarrow{D''} = V\,\mathrm{div}\,\overrightarrow{D''} - \mathrm{div}\left(V\overrightarrow{D''}\right) \qquad (9.21)$$

and $\mathrm{div}\,\overrightarrow{D''} = 0$, according to equation (9.18), results in: $-\mathrm{grad}\,V \cdot \overrightarrow{D''} = -\mathrm{div}\left(V\overrightarrow{D''}\right)$ and:

$$\int_v \overrightarrow{E}\,D''\, dv = \int_v -\mathrm{grad}\,V \cdot \overrightarrow{D''}\, dv = \int_v -\mathrm{div}\left(V\overrightarrow{D''}\right) dv$$

$$= -\sum_{i=1}^{n} \int_{\Sigma_i} V\overrightarrow{D''}\,\overrightarrow{dS} = 0 \qquad (9.22)$$

Observation: The absolute potential V decays as $1/r$, and $\overrightarrow{D}$ decays as $1/r^2$. The surface area, however, increases as r^2, so that the last integral from the above equation goes to zero as Σ tends to infinity. Returning to equation (9.19), it results in:

$$W' = W + \frac{1}{2} \int_v \varepsilon\, \overrightarrow{E''}^2\, dv \qquad (9.23)$$

then $W' > W$ if in a region of the field $\overrightarrow{E'}$ and $\overrightarrow{E}$ are different.

According to Thompson's theorem, the electrostatic field corresponding to the equilibrium distribution of charges on conductors can be deduced from a minimum principle, just as in mechanics the stable equilibrium position of bodies can also be deduced from a minimum principle. *It follows that electrostatic energy has, in electrostatics, the same role that potential energy has in classical mechanics.* It also follows that, under conditions (9.15), the electrostatic field is unequivocally defined, because according to Thompson's theorem, when given the electric charges carried by conductors, there can be no two electrostatic fields corresponding to the same conductor system, at electrostatic equilibrium.

9.4 Ponderomotive actions in the electric field

We will now take a look on how ponderomotive actions (i.e., forces and torques) could be deduced from the electrostatic energy, with the aim of establishing the two theorems of the generalized forces in the electrostatic field.

9.4.1 Calculation of forces using Coulomb's formula

The interaction forces between charged bodies with electric charge can be calculated using Coulomb's Formula, given by the relation:

$$\vec{F} = \frac{q_1 q_2}{4\pi\varepsilon r^3}\vec{r} \tag{9.24}$$

It should be noted, however, that Coulomb's formula is valid only for point charges and can be used directly only when the linear dimensions of the bodies are very small in relation to the distances between these bodies. In the general case of continuous charge distributions, the use of Coulomb's Law is quite laborious, as it is necessary that the electric charge of a macroscopic body is divided into elementary charges, which could be considered as point charges, the resulting force being obtained on the basis of superposition principle.

Exercise: force between two plates

Calculate the interaction force between two parallel plates whose linear dimensions are very large in relation to the distance between them, for example the plates of a plane capacitor. **Solution:** *the distance between the plates being small and neglecting the edge effect, the electric field between the plates is uniform. If we assume a homogeneous and isotropic dielectric between the capacitor plates, then the displacement vector between the plates is uniform. A surface element dS, for example, on the upper plate, figure 9.3, carries the charge $dq_2 = -\sigma dS_2 = -\vec{D}\,\overrightarrow{dS_2}$. The interaction force between the elementary charge on the lower plate $dq_1 = \sigma dS_1 = \vec{D}\,\overrightarrow{dS_1}$ and the elementary charge on the upper plate is:*

$$d^2\vec{F} = \frac{dq_1 dq_2}{4\pi\varepsilon R^3}\vec{R} = -\frac{\vec{D}^2\,\overrightarrow{dS_1}\,\overrightarrow{dS_2}}{4\pi\varepsilon R^3}\vec{R} \tag{9.25}$$

Due to the symmetry, the resultant force has just the normal component to the plane of the armatures, being d^2F_z:

$$d^2\vec{F_z} = -\frac{D^2 dS_1 dS_2 R \cos\alpha}{4\pi\varepsilon R^3} = \frac{D^2 dS_1 dS_2 d}{4\pi\varepsilon(r^2 + d^2)^{3/2}} \tag{9.26}$$

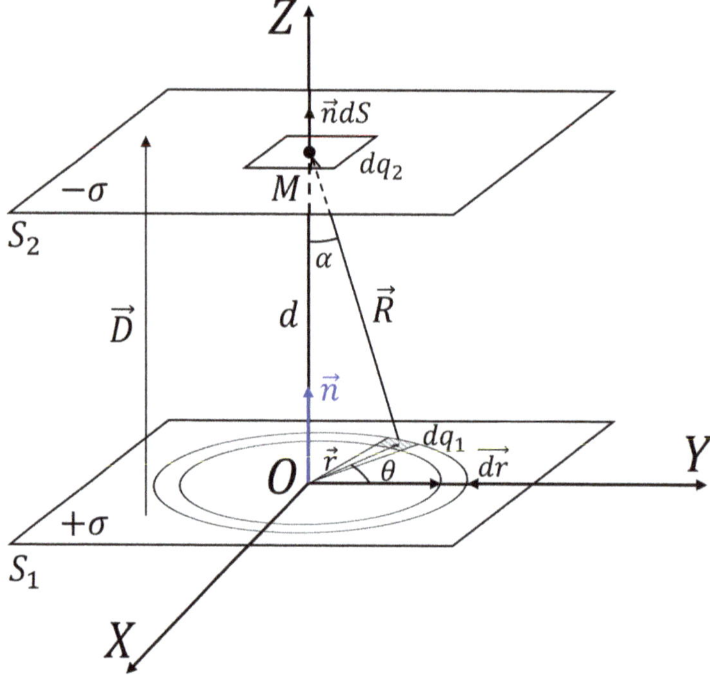

Figure 9.3. *The force of attraction between the plates of a plane capacitor.*

The total force that the lower armature will exert on the surface element dS_2 centered on M, on the upper armature, is:

$$\overrightarrow{dF_z} = -\frac{D^2 dS_2 d}{4\pi\varepsilon}\int_0^{r_0}\int_0^{2\pi}\frac{r\,dr\,d\theta}{(r^2+d^2)^{3/2}}\vec{n} = -\frac{D^2 dS_2 d}{2\varepsilon}\left[\frac{1}{d}-\frac{1}{\sqrt{r_0^2+d^2}}\right]\vec{n} \quad (9.27)$$

The plates were assumed to have the shape of circular plates of radius r_0, much greater than the distance, d, between them. In these conditions:

$$\overrightarrow{dF_z} = -\frac{D^2 dS_2}{2\varepsilon_0}\vec{n} \quad (9.28)$$

The force exerted by the lower plate on the upper one will be:

$$\overrightarrow{F_z} = -\int_S \frac{D^2}{2\varepsilon}dS_2\vec{n} = -\frac{D^2 S_2}{2\varepsilon}\vec{n} = -\frac{Q^2}{2\varepsilon S_2}\vec{n} \quad (9.29)$$

which is a force of attraction, as it results from the relation (9.29) taking into account the orientation of the considered normal. The force of attraction per unit area between the plates is:

$$P = \frac{F_z}{S_2} = \frac{1}{2}\frac{D^2}{\varepsilon} = \frac{1}{2}\vec{D}\vec{E} \quad (9.30)$$

an identical relation with that of the density of energy stored in the electric field between the plates of the capacitor. To have an idea on the order of magnitude of the electrostatic pressure experienced on the dielectric between the plates of capacitor will consider that the dielectric is the paper with $\varepsilon_r = 2.5$ and the dielectric rigidity of $E_r = 100 \text{ kV cm}^{-1}$. In this case we obtain:

$$P = \frac{1}{2}\varepsilon_0\varepsilon_r E_r^2 = \frac{1}{2}\frac{4\pi\varepsilon_0}{4\pi}\varepsilon_r E_{r_2}^2 = \frac{2.5}{18 \times 12.56} \times 10^5 \cong 1.1 \times 10^3 \text{ N m}^{-2}$$

9.4.2 Calculus of the ponderomotive actions on the base of the electrostatic energy

The application of Coulomb's Formula for calculating the interaction forces between bodies charged with electric charge, or which are placed in an electric field, can be done only in very particular cases, namely, when the dielectrics are homogeneous and the distribution of charges is known. In order to calculate these forces in any case, methods have been developed that are based on the mechanical work that would be performed at a certain movement of the bodies, on which these forces are exerted, so methods that will be based on the use of electrostatic energy. These methods use the concepts of generalized coordinates and generalized forces known in mechanics. When the geometric configuration of a body system is fully characterized by a number of independent scalar variables, these variables are called *the generalized coordinates of the system (or so-called position parameters)*. The minimum number of generalized coordinates required to determine the state of the system is equal to the number of degrees of freedom of the system. In the case of translational motion, the generalized coordinates are *the coordinates (x, y, z)*, with respect to the Cartesian coordinate system; in the case of a body performing a rotational movement, the generalized coordinate is the angle α; but a generalized coordinate can also be the area of a variable surface, or the volume in the case of a gas, etc. Generalized coordinates are often denoted by a_i precisely because they are not believed to represent only the spatial coordinates, but any of the sizes indicated above. When the configuration of the body system undergoes an elementary (infinitesimal) variation, then these generalized coordinates also have elementary variations, and the forces exerted on the bodies, denoted by A_i, called generalized forces (or force parameters), perform an elementary mechanical work given by the relationship:

$$dL = \sum_{i=1}^{n} A_i da_i \tag{9.31}$$

The scalar quantities A_i, which represent the coefficients of the elementary variations of the generalized coordinates da_i $(i = 1, 2, \ldots, n)$, from the expression (9.31) of the mechanical work, are called *generalized forces (Lagrangian forces)*, but they do not have to be understood as always proper forces. If for example a_i is a displacement, then A_i is indeed a force itself, but if a_i is an angle then A_i is a moment of force or a torque. Continuing the association, if a_i is an area, A_i is a surface tension, and if a_i is a volume, then A_i is a pressure.

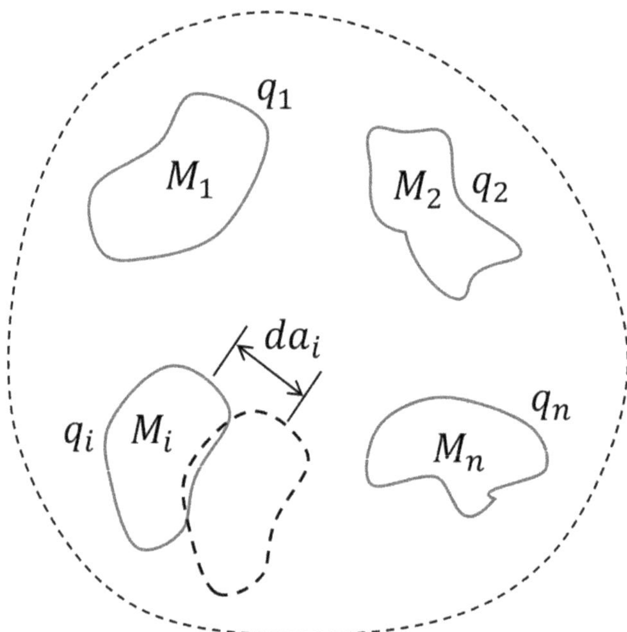

Figure 9.4. A system of charged conductors.

Consider a system consisting of n conductors charged with electric charge (see figure 9.4), and assume that one of the bodies of the system moves so that only a_i vary, all other bodies remaining fixed. The variation da_i, of the coordinate a_i is made under the action of the generalized force A_i, due to the interaction of the electric charge of the moving body, with the charges of all the other bodies of the system, considered fixed. Obviously, the possible variations of the body charges as well as the body movement are very small, so that the state of electrostatic equilibrium is not disturbed. Under these conditions, the mechanical work performed by the external energy sources for the variation with dq_i of the conductors' charges is:

$$-dL_{12} = \sum_{i=1}^{n} V_i dq_i \tag{9.32}$$

and it must cover not only the variation of the energy of the electric field but also the mechanical work done by the generalized forces A_i which change the position of the respective body. The law for the electrical energy conservation will be written as:

$$\sum_{i=1}^{n} V_i dq_i = dW + \sum_{i=1}^{n} A_i da_i \tag{9.33}$$

representing the most general equation from which the generalized forces can be determined. In the problems that appear in electricity, there are two cases that greatly simplify equation (9.33), namely:

1. The system of conductors is isolated and complete, i.e., the q_i charges of the bodies remain constant (q_i = constant), the conductors are disconnected from external sources.
2. The system is not isolated and the conductors are maintained at constant potentials V_i = constant (obviously by connecting them to the terminals of external sources of constant voltage).

By customizing the relation (9.33), in these two cases the two generalized force theorems are obtained in the electrostatic field.

1. First theorem of the generalized forces in the electrostatic field

The system of conductors being isolated (q_i = constant), results that $dq_i = 0$ and then equation (9.33) becomes:

$$(dW)_{q=\text{constant}} + \sum_{i=1}^{n} A_i da_i = 0 \tag{9.34}$$

i.e., the generalized force A_i in relation to the generalized coordinate is given by the partial derivative of the energy in relation to the generalized coordinate that varied, taken with changed sign:

$$A_i = -\left(\frac{\partial W}{\partial a_i}\right)_{q=\text{constant}} \tag{9.35}$$

If the energy of the system of conducting bodies is expressed as a function of elastances (and it is normal to be so because we work at q = constant), then the generalized force A in the direction of the generalized coordinate a is:

$$A = -\frac{1}{2}\sum_{i}\sum_{j} q_i q_j \frac{\partial S_{ij}}{\partial a_i} \tag{9.36}$$

The minus sign is interpreted in the sense that if da is the displacement produced under the action of electric force A and the work done by this force is positive $Ada > 0$, then the energy of the electric field decreases ($dW_e < 0$). Indeed, when external sources are disconnected, mechanical work can only occur due to internal energy reserves of the system, i.e., due to energy of the electric field. If da is a displacement made on the system by an external force A, then the mechanical work is negative ($Ada < 0$) and the energy of the electric field increases ($\frac{dW}{da} > 0$). So, the energy in the final state is greater than the energy in the initial state. The first theorem of generalized forces can be stated as:

Theorem 9.1. *Generalized force A corresponding to the generalized coordinate a is equal to the derivative with changed sign of the energy (expressed as a function of the charges and the generalized coordinate) in relation to the generalized coordinate, to constant charges of the conductors:*

$$A = -\left(\frac{\partial W}{\partial a}\right)_{q=\text{constant}} \tag{9.37}$$

For example, consider a charged plane capacitor, having the charge Q, the absolute electrical permittivity ε, the surface area of the armatures Σ and the distance between the plates x. The capacity of the capacitor being $C = \frac{\varepsilon\Sigma}{x}$, and its elastance $S = \frac{1}{C} = \frac{x}{\varepsilon\Sigma}$, results that the energy stored in the electric field between the capacitor plates is:

$$W = \frac{1}{2}SQ^2 = \frac{1}{2}\frac{x}{\varepsilon\Sigma}Q^2 \tag{9.38}$$

According to equations (9.37) and (9.38), the interaction force between the capacitor plates is:

$$\vec{F} = -\left(\frac{\partial W}{\partial x}\right)_{q=\text{constant}} = -\frac{1}{2}\frac{Q^2}{\varepsilon\Sigma}\vec{n} \tag{9.39}$$

where normal $\vec{n}$ to the capacitor plates is oriented from the positive to the negative plate. So between the capacitor plates there is an attractive force causing a pressure to be exerted on the dielectric, a pressure equal to:

$$p = \frac{F}{\Sigma} = \frac{Q^2}{2\varepsilon\Sigma^2} = \frac{\sigma^2\Sigma^2}{2\varepsilon\Sigma^2} = \frac{\sigma^2}{2\varepsilon} = \frac{D^2}{2\varepsilon} = W = \frac{1}{2}\vec{D}\vec{E} \tag{9.40}$$

Relations (9.39) and (9.40) are identical with relations (9.29) and (9.30) but the first ones show how easy it was to solve the proposed problem using the theorem of generalized forces, as comparing with Coulomb's Law. If the capacitor plates were free to move, the force of attraction between them determines the decrease of the distance between them, increasing the capacitance and then decreasing the capacitor energy $W = \frac{Q^2}{2C}$, in the conditions in which $Q = \text{constant}$.

2. Second theorem of the generalized forces in the electrostatic field

Another case of interest is that in which during the variation of the configuration of a system of conductors, their potentials are kept constant ($V_i = \text{constant}$). Such a regime is obtained when all the conductors in the system are connected to the terminals of constant voltage sources. When the system configuration changes, the capacitance and influence coefficients change, thus determining the variation of the charges on the conductors from system, in the conditions in which their potentials remain constant. The additional charges are transferred to the system by external voltage sources, so that the energy balances and equation (9.33) becomes:

$$\sum_{i=1}^{n} V_i dq_i = dW\bigg|_{V=\text{constant}} + \sum_{i=1}^{n} A_i da_i \tag{9.41}$$

However, if the potentials of the conductors are kept constant and the dielectric environment in which they are located is homogeneous and isotropic, then the energy stored in the electric field of the system of conductors is:

$$dW\bigg|_{V=\text{constant}} = \frac{1}{2}\sum_{i=1}^{n} V_i dq_i$$

With this, the equation (9.41) becomes:

$$2dW\bigg|_{V=\text{constant}} = dW\bigg|_{V=\text{constant}} + \sum_{i=1}^{n} A_i da_i \tag{9.42}$$

$$\Rightarrow dW\bigg|_{V=\text{constant}} = \sum_{i=1}^{n} A_i da_i$$

So the generalized force A_i that determines the variation of the generalized coordinate da_i is:

$$A_i = \left(\frac{\partial W}{\partial a_i}\right)_{V=\text{constant}} \tag{9.43}$$

In this case, as it is normal, the energy will be expressed as a function of the conductor potentials and the capacitance coefficients, then the generalized force A_i corresponding to the variation of the generalized coordinate a_i will be given by the equation:

$$A_i = \left(\frac{\partial W}{\partial a_i}\right)_{V=\text{constant}} = \frac{1}{2}\sum_{i=1}^{n}\sum_{j=1}^{n} V_i V_j \frac{\partial C_{ij}}{\partial a_i} \tag{9.44}$$

If in the system there is a displacement under the action of force A (in the conditions of maintaining constant potentials on the conductors), the mechanical work is positive and also the variation of the internal energy of the system is positive. In this case the energy transferred from the external sources to the system contributes equally to the variation of the internal energy of the system and to perform of the mechanical work.

Theorem 9.2. *The generalized force A corresponding to the generalized coordinate a is equal to the energy derivative (expressed as a function of the potentials and the generalized coordinate) in relation to the generalized coordinate, to constant potentials of the conductors:*

$$A = \left(\frac{\partial W}{\partial a}\right)_{V=\text{constant}} \tag{9.45}$$

Returning to the example presented above, the force of attraction between the plates of a planar capacitor is:

$$\vec{F} = \left(\frac{\partial W}{\partial x}\right)_{V=\text{constant}} = \frac{\partial}{\partial x}\left[\frac{1}{2}C(V_1 - V_2)^2\right]$$

$$= \frac{\partial}{\partial x}\left[\frac{1}{2}\frac{\varepsilon\Sigma}{x}(V_1 - V_2)^2\right]$$

$$= -\frac{1}{2}\frac{\varepsilon\Sigma}{x^2}(V_1 - V_2)^2\vec{n}$$

$$= -\frac{1}{2}\frac{Q^2}{Cx}\vec{n} = -\frac{1}{2}\frac{Q^2}{\varepsilon\Sigma}\vec{n}$$

$$(9.46)$$

where normal $\vec{n}$ to the capacitor plates is oriented from the positive to the negative plate. As it is normal, using the second theorem of generalized forces, the same value was obtained for the force of attraction between the capacitor plates, see above equation (9.39).

9.5 Theorems of generalized forces. Applications

In this section the theorems of the generalized forces presented above (see section 9.4) have been used to compute ponderomotive actions in two common applications, i.e., the electrometer and the analog voltmeter.

9.5.1 Thompson's electrometer

The electrometer is an electrostatic measuring instrument used to measure electrical charges and voltages. The Thompson electrometer consists of a plane capacitor with one movable plate, see figure 9.5. A guard ring is connected by a flexible wire to the top plate to eliminate the edge effects, the electric field being uniform at the center plate. The force between the capacitor plates is measured with a dynamometric device or a balance. The down plate and the guard ring are fixed, just the upper plate, of smaller diameter, being mobile. Its area is denoted by S, and the distance

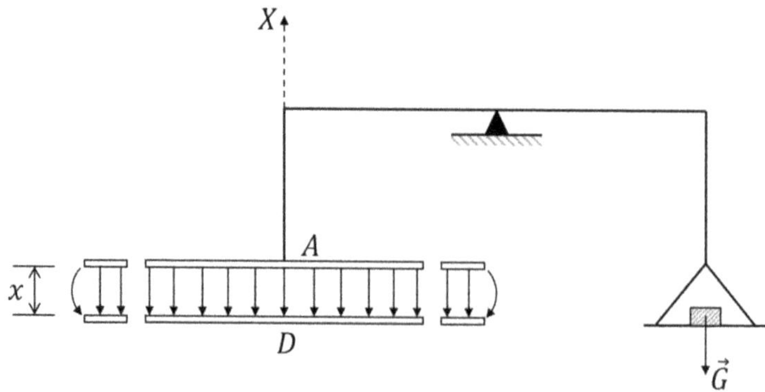

Figure 9.5. Thompson electrometer.

between the armatures is denoted by x and represents the generalized coordinate. The energy of the charged capacitor is given by one of the relations:

$$W = \frac{1}{2} S_e Q^2 = \frac{1}{2} \frac{x}{\varepsilon S} Q^2 \quad \text{or} \quad W = \frac{1}{2} C U^2 = \frac{1}{2} \frac{\varepsilon S}{x} U^2 \tag{9.47}$$

Consequently, taking into account the generalized forces theorems, the force of attraction between the plates will be given by:

$$\vec{F} = -\left(\frac{\partial W}{\partial x}\right)_{Q=\text{constant}} = -\frac{\partial}{\partial x}\left(\frac{1}{2} S_e Q^2\right) = -\frac{1}{2} \frac{Q^2}{\varepsilon S} \vec{n} \tag{9.48}$$

and:

$$\vec{F} = \left(\frac{\partial W}{\partial x}\right)_{V=\text{constant}} = \frac{\partial}{\partial x}\left(\frac{1}{2} C U^2\right) = -\frac{1}{2} \frac{\varepsilon S}{x^2} U^2 \vec{n}$$
$$= -\frac{1}{2} \frac{Q^2}{\varepsilon S} \vec{n} \tag{9.49}$$

respectively. The force $\vec{F}$, given by one of the ratios (9.48) or (9.49), is balanced by the weight force $\vec{G}$ of the marked masses placed on the plate of a balance. The generalized force is negative, i.e., it is opposite to the direction of variation of the generalized coordinate x, i.e., it is an attractive force between the capacitor plates, which tends as shown above to increase its capacity. According to equation (9.49), the force of attraction between the capacitor plates is a function of the square of the voltage applied between the capacitor plates, therefore it does not depend on its polarity. This force has an average value other than zero in the case of alternating voltages, therefore the device can also be used to measure the voltage in alternating current. Taking into account the relation $\vec{F} + \vec{G} = 0$, valid at equilibrium, with this instrument it is possible to measure both the voltage and the electric charge. If the device is calibrated directly in volts, it is called the electrostatic voltmeter, the principle of operation of which is described below.

9.5.2 Analog electrostatic voltmeter

A flat needle that can rotate about an axis OO', see figure 9.6, forms with a plate, flat of a special shape, a flat capacitor, whose capacity is proportional to the length l of the needle located between the plates, $C = k_1 l$, where k_1 is a proportionality constant. When the needle rotates in the direction of increasing the angle α, the capacitance of the capacitor determined by the needle and the plate will increase due to the special shape of the plate. If a voltage U is applied between the plate and the needle, the system will seek to increase its capacity, thus tending to the state of minimum energy, the needle moving clockwise, under the action of a motor torque, which is balanced with the help of a spiral spring. By applying the generalized forces theorems, it is deduced that the generalized force M corresponding to the generalized coordinate α, is the active moment M defined by the relation:

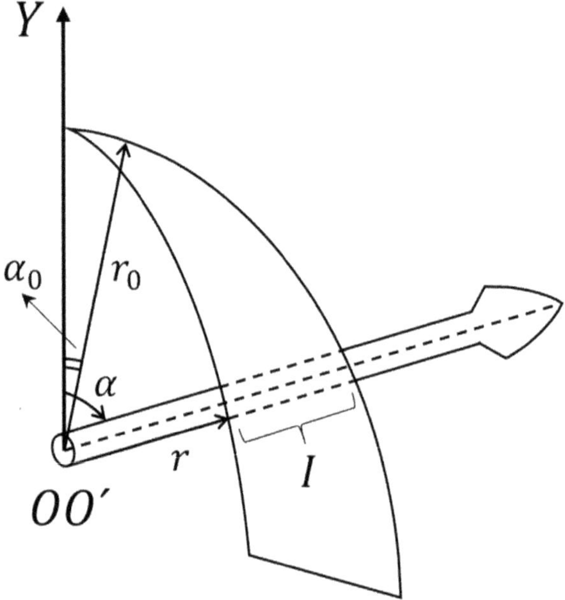

Figure 9.6. Scheme of the electrostatic voltmeter.

$$M = \left(\frac{\partial W}{\partial \alpha}\right)_{U=\text{constant}} = \frac{\partial}{\partial \alpha}\left(\frac{1}{2}CU^2\right) = \frac{1}{2}U^2\frac{\partial C}{\partial \alpha} \tag{9.50}$$

At equilibrium, the active torque M is equal to the resistive torque $M_r = k\alpha$, therefore the angle α will depend on the voltage U according to the relation:

$$\alpha = \frac{1}{2k}U^2\frac{\partial C}{\partial \alpha} \tag{9.51}$$

If it is desired that the indication of the electrometer is proportional with the applied voltage U ($\alpha = k_2 U$), then equation (9.51) allows one to determine the shape of the curve that delimits the fixed plate of the capacitor as follows:

$$\alpha = \frac{1}{2k}\frac{\alpha^2}{k_2^2}\frac{\partial C}{\partial \alpha} \Rightarrow \frac{\partial C}{\partial \alpha} = \frac{2kk_2^2}{\alpha} \tag{9.52}$$

Taking into account that $C = k_1 l = k_1(r_0 - r)$ results:

$$\frac{\partial C}{\partial \alpha} = -k_1\frac{dr}{d\alpha} \tag{9.53}$$

and considering the equation (9.52) the below differential equation is obtained:

$$\frac{dr}{d\alpha} = -\frac{2kk_2^2}{k_1}\frac{1}{\alpha} \tag{9.54}$$

having the solution:

$$r = r_0 + \frac{2kk_2^2}{k_1} \ln\left(\frac{\alpha_0}{\alpha}\right) \tag{9.55}$$

where α_0 is a very small angle made by the needle with the axis OY, in the initial position, where $r = r_0$. Since α increases clockwise, the ratio $\frac{\alpha_0}{\alpha}$ decreases, as a result $\ln\left(\frac{\alpha_0}{\alpha}\right)$ is negative and therefore r decreases. In practice, the curve described by equation (9.55), for the shape of the fixed flat reinforcement, can be easily achieved.

9.6 Mechanical stresses in the electric field

Faraday and Maxwell did not accept the theory of action at a distance, but considered that all physical interactions between bodies are transmitted continuously from one body to another through the electromagnetic field. Faraday interpreted this localization of the transmission of ponderomotive actions, in the field as a state of stress (tension), in the space occupied by the electric or magnetic field, thus giving a mechanical interpretation to the electric and magnetic phenomena. A mathematical expression of this view was given by Maxwell. Suppose that inside an enclosed surface Σ, of exterior normal $\vec{n}$, there are several charged bodies. The resulting electric force, exerted on these bodies, can be expressed in stationary regime in the form:

$$\vec{F_\Sigma} = \int_\Sigma \left[\left(\vec{D}\vec{n}\right)\vec{E} - \frac{\vec{D}\vec{E}}{2}\vec{n} \right] dS \tag{9.56}$$

The resulting force exerted by a charged body on the bodies in the surface Σ, is therefore equal to the surface integral of a vector quantity, called the Maxwellian tension, having the expression:

$$\vec{T_n} = \left(\vec{D}\vec{n}\right)\vec{E} - \frac{\vec{D}\vec{E}}{2}\vec{n} \tag{9.57}$$

From a physical point of view, everything happens like in the case when the environment outside the surface acts on the surface as an additional mechanical stress (unit effort or force per unit area, or pressure), of value $\vec{T_n}$.

    a) If the surface Σ is local, normal to the electric field lines, see figure 9.7(a), then $\vec{E} = E\vec{n}$ and $\vec{D} = D\vec{n}$. Thus:

$$\vec{T_n} = \left(\vec{D}\vec{n}\right)\vec{E} - \frac{\vec{D}\vec{E}}{2}\vec{n} = DE\vec{n} - \frac{1}{2}DE\vec{n} = \frac{\vec{D}\vec{E}}{2}\vec{n} = w_e\vec{n} \tag{9.58}$$

So, in the region where the surface is normal to the field lines, i.e., where the surface is confused practically with an equipotential surface, the stress $\vec{T_n}$ *is a tensile force*, oriented along the normal to the surface and equal with the energy density.

b) If the surface Σ is local, parallel with the electric field lines, see figure 9.7(b), then $E\vec{n} = D\vec{n} = 0$. Thus:

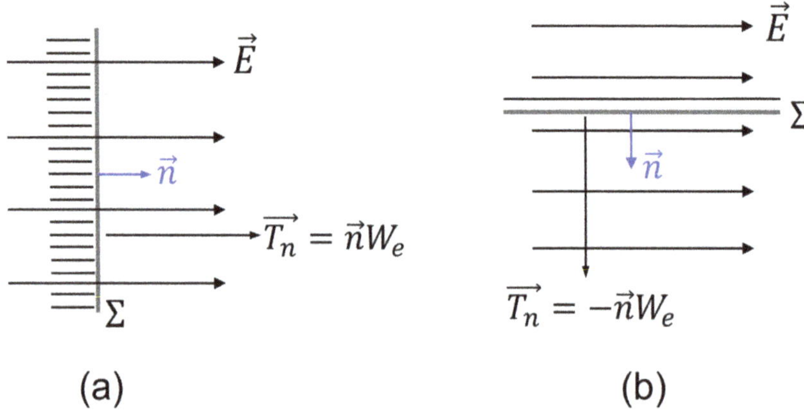

Figure 9.7. (a) Normal surface to the electric field lines. (b) Parallel surface to the electric field lines.

$$\overrightarrow{T_n} = \left(\vec{D}\vec{n}\right)\vec{E} - \frac{\vec{D}\vec{E}}{2}\vec{n} = -\frac{1}{2}DE\vec{n} = -w_e\vec{n} \tag{9.59}$$

So, when the surface Σ is parallel to the field lines, $\overrightarrow{T_n}$ is a *compression* perpendicular to the field lines, in the opposite direction to the outer normal and having the module equal with the energy density.

If a portion of the flow tube is considered, the mechanical image of the stresses exerted on it has the appearance of figure 9.8. Everything occurs as the field lines

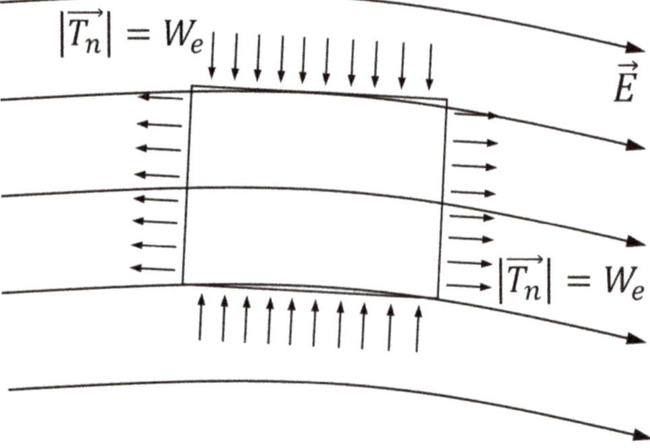

Figure 9.8. Tube of the field lines.

exert a stretch along them and a lateral compression of values equal to the energy density. The expression of Maxwellian stresses postulated in mathematical form by equation (9.57) leads practically, as it was found in the analyzed examples, to the fact that along the electric field lines tensions expressed by means of density of energy are exerted. These tensions are analogous to those existing in a bunch of elastic rubbers. These mechanical stresses, which occur in the electric field, can be calculated with the help of two equipotential surfaces and field lines that delimit a trunk of cone centered on any point in the electric field, as shown in figure 9.9. The bases of the cone trunk are equipotential surfaces and its lateral surface is formed by the field lines, i.e., practically the cone trunk represents a portion of a field line tube similar to the one in figure 9.8. Let's imagine the cone that this cone trunk is part of, see figure 9.9. The intensity of the electric field inside the cone trunk is a function of the distance x from the tip of the cone to the point where we express it. Thus, it is assumed that in the small base of the cone trunk the field intensity is $\vec{E}(x)$, and in the large base the intensity is $\vec{E}(x + dx)$. The radius of the bases are:

$$r = x \tan \alpha$$
$$R = (x + dx)\tan \alpha \tag{9.60}$$

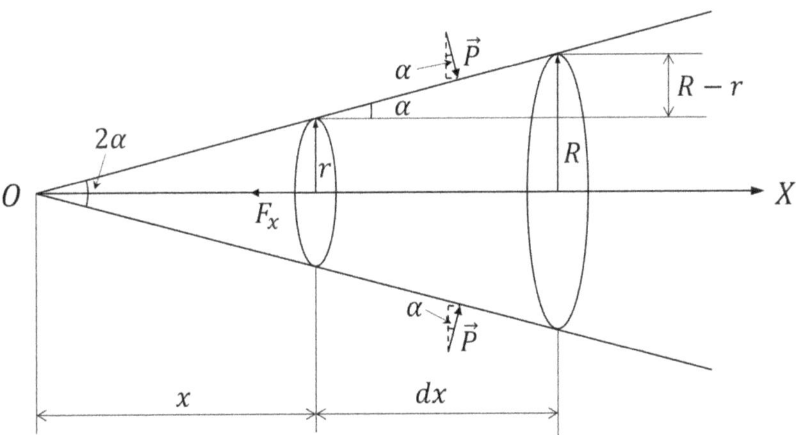

Figure 9.9. Tube of field lines in the shape of a truncated cone.

Given the conservation of flux along a tube of field lines (a consequence of Gauss' theorem), one can evaluate the field in the large base $\vec{E}(x + dx)$ as a function of the field in the small base $\vec{E}(x)$:

$$E(x)\pi r^2 = E(x + dx)\pi R^2 \Rightarrow E(x + dx) = E(x)\frac{r^2}{R^2} \tag{9.61}$$

Taking into account the equation (9.60), it results in:

$$E(x + dx) = E(x)\left(\frac{x}{x + dx}\right)^2 = E(x)\left(1 + \frac{dx}{x}\right)^{-2}$$
$$\cong E(x)\left(1 - 2\frac{dx}{x}\right) \tag{9.62}$$

On the small base will act the force:

$$F_{1(x)} = p_{1x} S_1 \Rightarrow F_{1(x)} = \frac{1}{2}\varepsilon E^2(x)\pi r^2 = \frac{1}{2}\varepsilon E^2(x)\pi x^2 \tan^2 \alpha \tag{9.63}$$

And on the large base the force:

$$F_{2(x)} = p_{2x} S_2 \Rightarrow F_{2(x)} = \frac{1}{2}\varepsilon E^2(x + dx)\pi R^2$$
$$= \frac{1}{2}\pi\varepsilon E^2(x)\left(1 - 2\frac{dx}{x}\right)^2 (x + dx)^2 \tan^2 \alpha$$
$$\cong \frac{1}{2}\pi\varepsilon E^2(x)x^2\left(1 - 4\frac{dx}{x}\right)\left(1 + 2\frac{dx}{x}\right)\tan^2 \alpha \tag{9.64}$$
$$\cong \frac{1}{2}\pi\varepsilon E^2(x)x^2\left(1 - 2\frac{dx}{x}\right)\tan^2 \alpha < F_{1(x)}$$

So, on the cone trunk a net force to the left will be exerted, equal to:

$$F(x) \simeq F_{1(x)} - F_{2(x)} \simeq \frac{1}{2}\varepsilon\pi E^2(x)x^2 \tan^2 \alpha\left(1 - 1 + 2\frac{dx}{x}\right)$$
$$= \varepsilon\pi E^2(x)x^2 \tan^2 \alpha\frac{dx}{x} \simeq \pi\varepsilon E^2 x dx \tan^2 \alpha \tag{9.65}$$

But the trunk of the cone, imaginarily delimited in the electric field, must be in static equilibrium. Therefore, the force $F(x)$ must be balanced by a pressure p to be exerted on the lateral surface of the truncated cone. This force $F'(x) = p \sin \alpha A_l$, according to figure 9.9, where A_l is the lateral area of the cone trunk, given by: $A_l = \pi G(R + r)$, where G is the generator of the cone trunk:

$$\cos \alpha = \frac{dx}{G} \Rightarrow G = \frac{dx}{\cos \alpha}, \quad R = r\left(\frac{x + dx}{x}\right) = r\left(1 + \frac{dx}{x}\right)$$
$$\Rightarrow r = x \tan \alpha$$
$$A_l = \pi\frac{dx}{\cos \alpha}r\left(2 + \frac{dx}{x}\right) = \pi\frac{dx}{\cos \alpha}x \tan \alpha\left(2 + \frac{dx}{x}\right) \tag{9.66}$$
$$\Rightarrow F'(x) = p\pi x\left(2 + \frac{dx}{x}\right)\frac{\sin \alpha}{\cos \alpha}\tan \alpha \simeq 2p\pi x dx \tan^2 \alpha$$

At equilibrium must satisfy the condition $F(x) = F'(x)$, resulting in:

$$p = \frac{1}{2}\varepsilon E^2 = \frac{1}{2}\vec{E}\,\vec{D} \qquad (9.67)$$

So, we find this hypothetical tension (pressure) exactly equal to the energy density (w_e) in the electric field, an elementary calculation leading us to the same expression as that of the Maxwellian tensions mathematically formulated by Faraday and Maxwell.

9.7 Mechanical forces between any two media

From the above considerations, it is noted that the surface of metal conductor (equipotential surface on which the field is normal) will exert forces (mechanical stresses), which in the electrostatic field are normal to the surfaces of conductors and have the same size as the density energy in the electric field. But this type of force can appear also at the interface between two dielectrics. It will be considered a surface element dS, on the interface between two dielectric media, having the absolute electrical permittivities, ε_1 and ε_2, see figure 9.10. It is assumed that the electric field vector enters the dielectric (1) at an angle α with respect to normal. Two forces are exerted on the surface element dS from the medium (1), namely a force $\overrightarrow{dF_i}$, stretching or traction along the field line and a pressure (compression) force perpendicular to the field lines $\overrightarrow{dF_p}$. These forces can be expressed as follows:

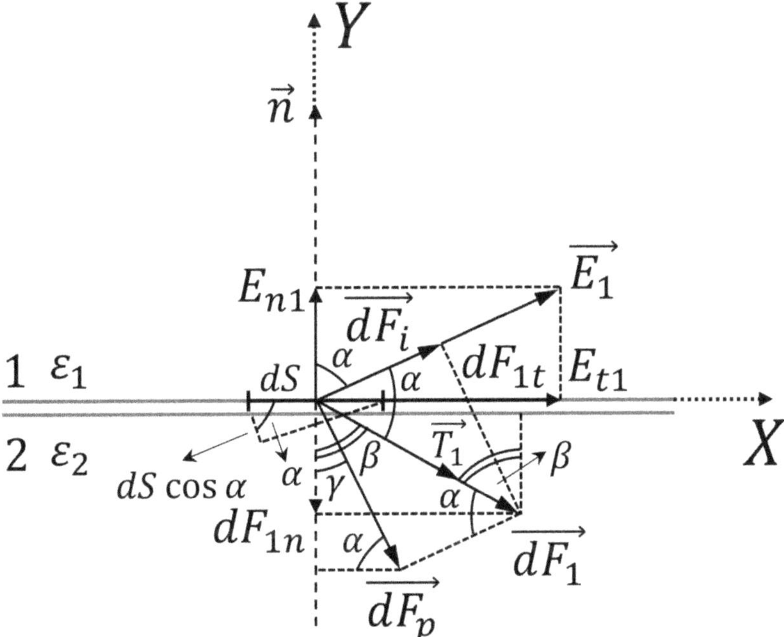

Figure 9.10. Mechanical forces at the separating surface between two dielectrics.

$$dF_i = \frac{1}{2}\vec{D}\vec{E}\,dS\cos\alpha \quad\text{and}\quad dF_p = \frac{1}{2}\vec{D}\vec{E}\,dS\sin\alpha \tag{9.68}$$

These forces dF_i and dF_p give the resultant:

$$dF_1 = \sqrt{dF_i^2 + dF_p^2} = p_1 dS \tag{9.69}$$

Having the components with respect to the considered axes:

$$dF_{1t} = dF_i\sin\alpha + dF_p\cos\alpha = 2p_1 dS\sin\alpha\cos\alpha = p_1 dS\sin 2\alpha$$
$$dF_{1n} = dF_i\cos\alpha - dF_p\sin\alpha = 2p_1 dS(\cos^2\alpha - \sin^2\alpha) \tag{9.70}$$
$$= p_1 dS\cos 2\alpha$$

From this relation it results that the force $\overrightarrow{dF_1}$ makes the angle 2α with the normal to the surface element dS and has the value $p_1 dS$. Indeed, its projection on the OX axis is $dF_1\sin\beta$, but this is $dF_1\sin(\pi - \beta)$ and according to (9.70):

$$dF_1\sin(\pi - \beta) = p_1 dS\sin 2\alpha \Rightarrow \begin{cases} dF_1 = p_1 dS \\ \pi - \beta = 2\alpha \end{cases}$$

This force can be imagined as a surface force and is nothing but the Maxwellian tension, equal to:

$$T_1 = \frac{dF_1}{dS} = p_1 = \frac{1}{2}\overrightarrow{E_1}\overrightarrow{D_1} = \frac{1}{2}\varepsilon_1 E_1^2 \tag{9.71}$$

This tension makes the angle 2α with normal at $\overrightarrow{dS}$ and as the angle between $\overrightarrow{E_1}$ and normal is α, it results that also between $\overrightarrow{E_1}$ and $\overrightarrow{T_1}$ is the angle α, too. But, relation (9.70) could be also written as:

$$dF_{1t} = \frac{1}{2}E_1 D_1 dS\cos\alpha\sin\alpha + \frac{1}{2}E_1 D_1 dS\sin\alpha\cos\alpha$$
$$= \frac{1}{2}(E_1\sin\alpha)(D_1\cos\alpha)dS + \frac{1}{2}(E_1\sin\alpha)(D_1\cos\alpha)dS \tag{9.72}$$
$$= D_{1n}E_{1t}dS = \varepsilon_1 E_{1n}E_{1t}dS$$

$$dF_{1n} = \frac{1}{2}E_1 D_1 dS\cos\alpha\cos\alpha - \frac{1}{2}E_1 D_1 dS\sin\alpha\sin\alpha$$
$$= \frac{1}{2}(E_{1n})(D_{1n})dS - \frac{1}{2}(E_{1t})(D_{1t})dS = \frac{1}{2}\varepsilon_1(E_{1n}^2 - E_{1t}^2)dS \tag{9.73}$$

Similarly, forces act from medium (2) to medium (1), forces which, by analogy, would have the components:

$$dF_{2t} = \varepsilon_2 E_{2n}E_{2t}dS$$
$$dF_{2n} = \frac{1}{2}\varepsilon_2(E_{2n}^2 - E_{2t}^2)dS \tag{9.74}$$

If we take into account the boundary equations for $\vec{E}$ and $\vec{D}$ at the separation surface between two media with different properties ($D_{1n} = D_{2n}$ and $E_{1t} = E_{2t}$), we can also find the boundary equations for the components of these forces:

$$dF_{2t} = \varepsilon_2 E_{2n} E_{2t} dS = \varepsilon_2 \frac{\varepsilon_1}{\varepsilon_2} E_{1n} E_{1t} dS = \varepsilon_1 E_{1n} E_{1t} dS = dF_{1t}$$

$$dF_{2n} = \frac{1}{2}\varepsilon_2\left(\frac{\varepsilon_1^2}{\varepsilon_2^2}E_{1n}^2 - E_{1t}^2\right)dS$$

$$(9.75)$$

The components of the force dF_2 produced by the field from medium (2) to medium (1) are opposite to those of the force dF_1 so that the tangential components of the two forces will cancel each other out as suggested by figure 9.10. It follows, however, that at the separation surface between the two media there is only a normal force on the surface element, a force equal to the resultant of the normal components of the two forces, $dF = dF_{1n} - dF_{2n}$. This force will cause a tension on the separation surface, equal to:

$$P = \frac{dF}{dS} = \frac{dF_{1n} - dF_{2n}}{dS} = \frac{\left[\frac{1}{2}\varepsilon_1(E_{1n}^2 - E_{1t}^2) - \frac{1}{2}\frac{\varepsilon_1^2}{\varepsilon_2}E_{1n}^2 + \frac{1}{2}\varepsilon_2 E_{1t}^2\right]dS}{dS}$$

$$= \frac{1}{2}\varepsilon_1 E_{1n}^2 - \frac{1}{2}\varepsilon_1 E_{1t}^2 - \frac{1}{2}\frac{\varepsilon_1^2}{\varepsilon_2}E_{1n}^2 + \frac{1}{2}\varepsilon_2 E_{1t}^2$$

$$= \frac{1}{2}(\varepsilon_2 - \varepsilon_1)E_{1t}^2 - \frac{\varepsilon_1}{2}\left(\frac{\varepsilon_1}{\varepsilon_2}E_{1n}^2 - E_{1n}^2\right)$$

$$= \frac{1}{2}(\varepsilon_2 - \varepsilon_1)E_{1t}^2 + \frac{\varepsilon_2 - \varepsilon_1}{2}\left(E_{1n}^2 \cdot \frac{\varepsilon_1}{\varepsilon_2}\right)$$

$$= \frac{1}{2}(\varepsilon_2 - \varepsilon_1)\left(E_{1t}^2 + \frac{\varepsilon_1}{\varepsilon_2}E_{1n}^2\right)$$

$$(9.76)$$

tension having the sense given by the values of the absolute permittivities of the two dielectrics. With this, we can analyze a few interesting cases:

    a) The lines of the electric fields are perpendicular to the separation surface between the dielectrics ($E_{1t} = 0$ and $E_{1n} = E_1$), in this case the relation (9.76) becomes:

$$P = \frac{1}{2}(\varepsilon_2 - \varepsilon_1)\left(\frac{\varepsilon_1}{\varepsilon_2}E_1^2\right)$$

$$(9.77)$$

    b) The lines of the electric fields are parallel with the separation surface between the dielectrics ($E_{1n} = 0$ and $E_{1t} = E_1$), in this case the relation (9.76) becomes:

$$P = \frac{1}{2}(\varepsilon_2 - \varepsilon_1)E_1^2$$

$$(9.78)$$

c) If $\varepsilon_2 \gg \varepsilon_1$, the relation (9.76) becomes:

$$P = \frac{1}{2}\varepsilon_2 E_1^2 = p_1 \tag{9.79}$$

A physical interpretation of these relations can be easily given by means of a plane capacitor between the plates of which a dielectric is inserted. If a glass plate is inserted entirely between the plates of the plane capacitor, then on the faces of the glass plate normal tensions given by relation (9.77) will be exerted. If the plate is only partially inserted, then the plate will be attracted inside with a force equal to $p_1 S$, given by relation (9.79). In this case a normal tension is exerted on the plate, which will cause the plate to be attracted between the plates. In the case of fluid dielectric media, the electric forces act as a hydrostatic pressure, tending to increase the volume of the dielectric brought into the field.

Further reading

[1] Maxwell J C 1881 *An Elementary Treatise on Electricity* ed W Garnett (Oxford: Clarendon)
[2] Jackson J D 1998 *Classical Electrodynamics* 3rd edn (New York: Wiley)
[3] Feynman R P, Leighton R B and Sands M 2011 *The Feynman Lectures on Physics, Vol II: The New Millennium Edition: Mainly Electromagnetism and Matter* (New York: Basic Books)
[4] Zangwill A 2012 *Modern Electrodynamics* 1st edn (Cambridge: Cambridge University Press)
[5] Purcell E M and Morin D J 2013 *Electricity and Magnetism* 3rd edn (Cambridge: Cambridge University Press)
[6] Peck E R 2013 *Electricity and MagnetismDover Books on Physics* (New York: Dover)
[7] Bettini A 2016 Electrostatic field in a vacuum *A Course in Classical Physics 3—Electromagnetism. Undergraduate Lecture Notes in Physics* (Berlin: Springer) pp 1–60
[8] Griffiths D J 2017 *Introduction to Electrodynamics* 4th edn (Cambridge: Cambridge University Press)
[9] Li Q (ed) 2020 *Advanced Dielectric Materials for Electrostatic CapacitorsSeries: Energy Engineering* (London: Institution of Engineering and Technology)

Electrostatics
Formalism of the electrostatic field in vacuum and matter
Ştefan Antohe and Vlad-Andrei Antohe

Appendix A

Three-orthogonal curvilinear coordinate systems

Most frequently, the rectangular (or Cartesian) coordinate systems are used to formulate physical problems. In particular, the location of a point in a three-dimensional (3D) space (with respect to an origin) is usually specified by giving its three Cartesian coordinates (x, y, z) or equivalently, by specifying the position vector $\vec{r}$ of that point. However, it is often more convenient to describe the position of that point in the 3D space, using another set of non-rectangular (or curvilinear) coordinates, more appropriate to the physical problem in question, common examples being cylindrical and spherical coordinate systems. The latter ones are especially useful when resolving complex problems of electrostatics based on superposition principle, and they are both particular cases of the three-orthogonal curvilinear coordinate systems, whose general properties are examined in detail within this appendix.

Definition A.1. *A three-orthogonal curvilinear coordinate system* is formed by three families of surfaces, two-by-two orthogonal, each surface from a family being characterized by the parameters q_1, q_2 or q_3, which are separately kept constant at any point on that surface. A family of surfaces described by the constants q_1, each corresponding to a different value of q_1, is defined by holding that variable constant while varying q_2 and q_3 over their entire range.

Generalizing, the intersections between the surfaces of constant q_j and constant q_k define a series of lines, called coordinate lines, along which only q_i changes (see figure A.1). In particular, the intersection of the surfaces $q_2 = $ constant and $q_3 = $ constant defines the line of variation of q_1; the intersection of the surfaces $q_1 = $ constant and $q_2 = $ constant defines the line of variation of q_3; the intersection of the surfaces $q_1 = $ constant and $q_3 = $ constant defines the line of variation of q_2; and ultimately the intersection of all three coordinated axes defines the origin of the system. More generally, a point P (in a 3D space) is specified by the three coordinate lines on

doi:10.1088/978-0-7503-5859-0ch10

© IOP Publishing Ltd 2023

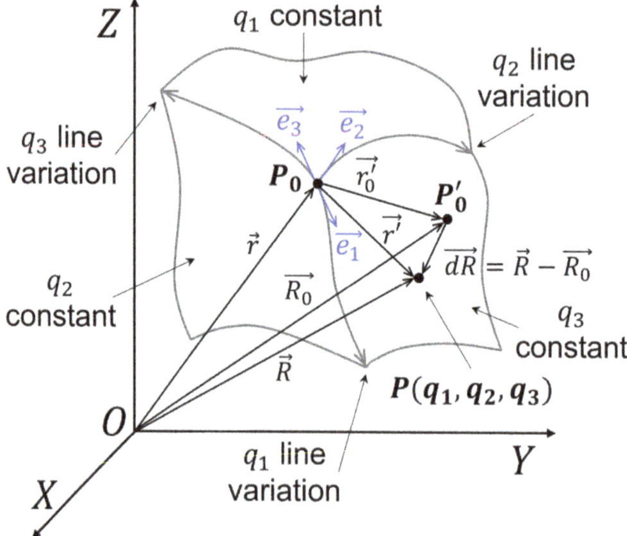

Figure A.1. Coordinate surfaces and coordinate lines of a general three-orthogonal curvilinear coordinate system.

which it lies. In order to approach the problem of expressing the vectors in generalized coordinates, we construct at point P_0 a set of basis (elementary) vectors: $\vec{e_1}$, $\vec{e_2}$ and $\vec{e_3}$, each tangent to one of the three coordinate lines. Then, if dl_1, dl_2 and dl_3 are the magnitudes of the infinitesimal displacements along the three coordinate lines, respectively, at point P, an arbitrary infinitesimal displacement vector $\vec{dR}$ can be expressed as:

$$\vec{dR} = dl_1\vec{e_1} + dl_2\vec{e_2} + dl_3\vec{e_3} \tag{A.1}$$

Note that, generally, dl_i is not equal to the increments of the coordinate dq_i. In fact, one of the first tasks is to express the infinitesimal displacements in terms of the generalized coordinates. In order to do this, it should be mentioned first that the new generalized coordinates q_1, q_2 and q_3 are related to the Cartesian coordinates by:

$$\begin{aligned} X &= x(q_1, q_2, q_3) \\ Y &= y(q_1, q_2, q_3) \\ Z &= z(q_1, q_2, q_3) \end{aligned} \tag{A.2}$$

and thus the position vector $\vec{R}$ of the point P will be given by:

$$\vec{R} = x(q_1, q_2, q_3)\vec{i} + y(q_1, q_2, q_3)\vec{j} + z(q_1, q_2, q_3)\vec{k} \tag{A.3}$$

Going further, it is exploited the fact that the magnitude of an infinitesimal scalar is an invariant independent of all three coordinate lines. For example, in Cartesian coordinates $\overrightarrow{dR}$ will be given by equation (A.4), while in a new coordinate system $\overrightarrow{dR}$ will be given by equation (A.1).

$$\overrightarrow{dR} = dx\,\vec{i} + dy\,\vec{j} + dz\,\vec{k} \tag{A.4}$$

In this context, it can be alternatively written that:

$$\begin{aligned}
\overrightarrow{dR} &= \frac{\partial \overrightarrow{R}}{\partial q_1}dq_1 + \frac{\partial \overrightarrow{R}}{\partial q_2}dq_2 + \frac{\partial \overrightarrow{R}}{\partial q_3}dq_3 \\[6pt]
&= \left|\frac{\partial \overrightarrow{R}}{\partial q_1}\right| dq_1\,\vec{e_1} + \left|\frac{\partial \overrightarrow{R}}{\partial q_2}\right| dq_2\,\vec{e_2} + \left|\frac{\partial \overrightarrow{R}}{\partial q_3}\right| dq_3\,\vec{e_3} \\[6pt]
&= h_1 dq_1\,\vec{e_1} + h_2 dq_2\,\vec{e_2} + h_3 dq_3\,\vec{e_3} = dl_1\,\vec{e_1} + dl_2\,\vec{e_2} + dl_3\,\vec{e_3}
\end{aligned} \tag{A.5}$$

Obviously, the following quantities can be expressed from (A.5):

$$\begin{aligned}
h_1 &= \left|\frac{\partial \overrightarrow{R}}{\partial q_1}\right| = \sqrt{\left(\frac{\partial x}{\partial q_1}\right)^2 + \left(\frac{\partial y}{\partial q_1}\right)^2 + \left(\frac{\partial z}{\partial q_1}\right)^2} \\[8pt]
h_2 &= \left|\frac{\partial \overrightarrow{R}}{\partial q_2}\right| = \sqrt{\left(\frac{\partial x}{\partial q_2}\right)^2 + \left(\frac{\partial y}{\partial q_2}\right)^2 + \left(\frac{\partial z}{\partial q_2}\right)^2} \\[8pt]
h_3 &= \left|\frac{\partial \overrightarrow{R}}{\partial q_3}\right| = \sqrt{\left(\frac{\partial x}{\partial q_3}\right)^2 + \left(\frac{\partial y}{\partial q_3}\right)^2 + \left(\frac{\partial z}{\partial q_3}\right)^2}
\end{aligned} \tag{A.6}$$

The quantities expressed in (A.6) are referred as: '*the metric coefficients of transformation*' or '*Lame's coefficients*'.

Furthermore, by comparing the last equation from (A.5) with (A.1) results that the length elements (infinitesimal displacements in terms of the generalized coordinates) can be written as:

$$\begin{aligned}
dl_1 &= h_1 dq_1 \\
dl_2 &= h_2 dq_2 \\
dl_3 &= h_3 dq_3
\end{aligned} \tag{A.7}$$

Having the infinitesimal line elements, the element of surface area bounded by the infinitesimal displacement vectors $\overrightarrow{dl_j}$ and $\overrightarrow{dl_k}$ can be derived taking their cross product: $\overrightarrow{dl_j} \times \overrightarrow{dl_k} = ds_i\,\vec{e_i}$. Consequently, the components of a surface element vector can be now written as:

$$\begin{aligned}
ds_1 &= h_2 h_3 dq_2 dq_3 \\
ds_2 &= h_1 h_3 dq_1 dq_3 \\
ds_3 &= h_1 h_2 dq_1 dq_2
\end{aligned} \tag{A.8}$$

Finally, an expression for the volume element whose sizes are the three infinitesimal displacements vectors $\overrightarrow{dl_1}$, $\overrightarrow{dl_2}$, $\overrightarrow{dl_3}$, can be obtained by extending the geometric interpretation of the scalar triple product in Cartesian coordinates to curvilinear coordinates $dv = \overrightarrow{dl_1}(\overrightarrow{dl_2} \times \overrightarrow{dl_3})$. For a three-orthogonal curvilinear coordinate system the volume element is thus given by:

$$dv = h_1 h_2 h_3 dq_1 dq_2 dq_3 \tag{A.9}$$

A.1 Rectangular coordinate system

Figure A.2 has been particularly drawn in order to calculate the displacement vector in Cartesian coordinates between two points P_1 and P_2. In this context, the two position vectors $\overrightarrow{R_1}$ and $\overrightarrow{R_2}$ of points P_1 and P_2, respectively, in respect to the origin of the reference coordinate system (CS$_0$), are:

$$\begin{aligned}
\overrightarrow{R_1} &= \overrightarrow{R_0} + \overrightarrow{r_1} = (x_0 + x_1)\overrightarrow{i} + (y_0 + y_1)\overrightarrow{j} + (z_0 + z_1)\overrightarrow{k} \\
\overrightarrow{R_2} &= \overrightarrow{R_0} + \overrightarrow{r_2} = (x_0 + x_2)\overrightarrow{i} + (y_0 + y_2)\overrightarrow{j} + (z_0 + z_2)\overrightarrow{k}
\end{aligned} \tag{A.10}$$

Therefore, in respect to CS$_0$, the difference between the two vectors within equations (A.10) derives to:

$$\overrightarrow{dR} = \overrightarrow{R_2} - \overrightarrow{R_1} = (x_2 - x_1)\overrightarrow{i} + (y_2 - y_1)\overrightarrow{j} + (z_2 - z_1)\overrightarrow{k} = dx\,\overrightarrow{i} + dy\overrightarrow{j} + dz\overrightarrow{k} \tag{A.11}$$

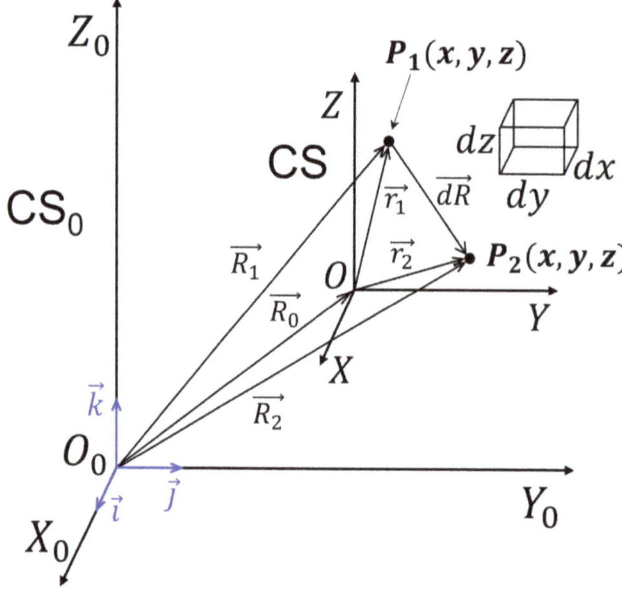

Figure A.2. Coordinate surfaces and coordinate lines of a Cartesian coordinate system (CS).

while in respect to CS, the displacement vector $\overrightarrow{dR}$ becomes:

$$\overrightarrow{dR} = \overrightarrow{r_2} - \overrightarrow{r_1} = (x_2 - x_1)\overrightarrow{i} + (y_2 - y_1)\overrightarrow{j} + (z_2 - z_1)\overrightarrow{k} = dx\overrightarrow{i} + dy\overrightarrow{j} + dz\overrightarrow{k} \quad (A.12)$$

It can be thus observed that the length elements are the same in respect with each coordinate system (CS_0 and CS): dx, dy and dz, and consequently:

$$h_1 = h_2 = h_3 = 1 \quad (A.13)$$

Summarizing, according to figure A.2 in rectangular (Cartesian) coordinate system, the three-orthogonal plane families can be defined with a set of basis (elementary) vectors: $\overrightarrow{i}$, $\overrightarrow{j}$, $\overrightarrow{k}$ and three coordinates: $q_1 = x$, $q_2 = y$, $q_3 = z$. Using the generalized Lame's coefficients (A.6) and particularizing equations (A.7), (A.8) and (A.9), length (A.14a), surface (A.14b) and volume (A.14c) elements, respectively, can be successively written as:

$$\text{Length elements: } dl_1 = dx; \quad dl_2 = dy; \quad dl_3 = dz \quad (A.14a)$$

$$\text{Surface elements: } dS_1 = dydz; \quad dS_2 = dxdz; \quad dS_3 = dxdy \quad (A.14b)$$

$$\text{Volume element: } dV = dxdydz \quad (A.14c)$$

A.2 Cylindrical coordinate system

A cylindrical coordinate system is defined according to figure A.3, by: (i) a cylindrical surface family of constant $q_1 = r$, representing the radius of a straight cylinder from the family; (ii) a vertical plane surface family of constant $q_2 = \theta$,

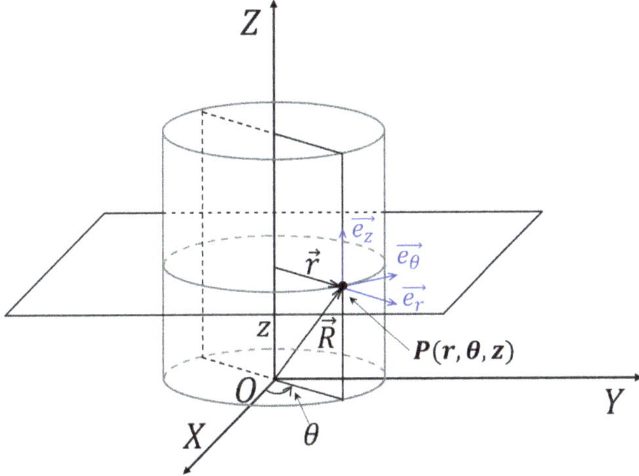

Figure A.3. Coordinate surfaces and coordinate lines of a cylindrical coordinate system.

representing the angle made with the (XOZ) plane; and (iii) a horizontal plane surface family of constant $q_3 = z$, representing its perpendicular distance from the (XOY) reference plane.

In particular, the r coordinate-line is the intersection between the vertical surface of constant θ and the horizontal plane surface of constant z. Correspondingly, the θ coordinate-line is a circle representing the intersection of the cylindrical surface of constant r and the horizontal planes of constant z. Ultimately, the z coordinate-line is the intersection of cylindrical surfaces of constant r and the vertical plane of constant θ. The set of basis (elementary) vectors in this case is: $\vec{e_r}$, $\vec{e_\theta}$, $\vec{e_z}$, each of them tangent to r-line, θ-line and z-line, respectively, and two-by-two orthogonal.

The three Cartesian coordinates of a point are related to its cylindrical coordinates by equations:

$$\begin{aligned} x &= r\cos\theta \\ y &= r\sin\theta \\ z &= z \end{aligned} \tag{A.15}$$

Then from equation (A.4) the position vector $\vec{R}$ with respect to the Cartesian coordinate system is given by:

$$\vec{R} = r\cos\theta\,\vec{i} + r\sin\theta\,\vec{j} + z\vec{k} \tag{A.16}$$

and from relations (A.6), the Lame' coefficients can be written as:

$$\begin{aligned} h_1 &= \sqrt{\cos^2\theta + \sin^2\theta + 0} = 1 \\ h_2 &= \sqrt{r^2\sin^2\theta + r^2\cos^2\theta + 0} = r \\ h_3 &= \sqrt{0 + 0 + 1} = 1 \end{aligned} \tag{A.17}$$

Concluding, in cylindrical coordinate system, the three-orthogonal plane families can be defined with a set of basis (elementary) vectors: $\vec{e_r}$, $\vec{e_\theta}$, $\vec{e_z}$ and three coordinates: $q_1 = r$, $q_2 = \theta$, $q_3 = z$. Using the generalized Lame's coefficients (A.6) and particularizing equations (A.7), (A.8) and (A.9), length (A.18a), surface (A.18b) and volume (A.18c) elements, respectively, can be successively written as:

Length elements: $dl_1 = dr$; $\quad dl_2 = rd\theta$; $\quad dl_3 = dz$ $\hfill$ (A.18a)

Surface elements: $dS_1 = rd\theta dz$; $\quad dS_2 = drdz$; $\quad dS_3 = rdrd\theta$ $\hfill$ (A.18b)

Volume element: $dV = rdrd\theta dz$ $\hfill$ (A.18c)

A.3 Spherical coordinate system

A spherical coordinate system is defined according to figure A.4, by: (i) a spherical surface family of constant $q_1 = r$, representing the radius of one sphere from the family; (ii) a conical surface family of constant $q_2 = \theta$, representing the angle between the generatrix of a cone surface from the family and the Z axis; and (iii) a

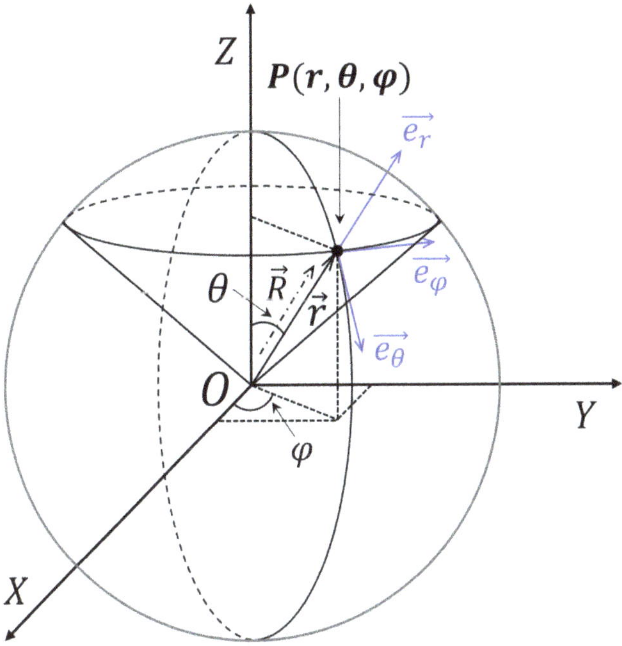

Figure A.4. Coordinate surfaces and coordinate lines of a spherical coordinate system. In this particular case $\vec{R}$ matches $\vec{r}$.

meridian plane surface family of constant $q_3 = \varphi$, representing the angle between a meridian plane from the family and (XOZ) plane.

In particular, the r coordinate-line is the intersection of a conical surface with a meridian plane, the θ coordinate-line is a half-circle and represents the intersection of a spherical surface with a meridian plane, while the φ coordinate-line is a circle representing the intersection of a spherical surface with a conical surface. In this case, the basis (elementary) vectors are: $\vec{e_r}$, $\vec{e_\theta}$, $\vec{e_\varphi}$, each of them tangent to r-line, θ-line and φ-line, respectively, and two-by-two orthogonal.

The three Cartesian coordinates of a point are related to its spherical coordinates by equations:

$$
\begin{aligned}
x &= r \sin \theta \cos \varphi \\
y &= r \sin \theta \sin \varphi \\
z &= r \cos \theta
\end{aligned}
\tag{A.19}
$$

Then from equation (A.4) the position vector $\vec{R}$ with respect to the Cartesian coordinate system is given by:

$$
\vec{R} = r \sin \theta \cos \varphi \, \vec{i} + r \sin \theta \sin \varphi \, \vec{j} + r \cos \theta \, \vec{k}
\tag{A.20}
$$

and from relations (A.6), the Lame' coefficients can be written as:

$$h_1 = \sqrt{\sin^2\theta\cos^2\varphi + \sin^2\theta\sin^2\varphi + \cos^2\theta} = 1$$
$$h_2 = \sqrt{r^2\cos^2\theta\cos^2\varphi + r^2\cos^2\theta\sin^2\varphi + r^2\sin^2\theta} = r \qquad \text{(A.21)}$$
$$h_3 = \sqrt{r^2\sin^2\theta\sin^2\varphi + r^2\sin^2\theta\cos^2\varphi} = r\sin\theta$$

Underlining, in spherical coordinate system, the three-orthogonal plane families can be defined with a set of basis (elementary) vectors: $\vec{e_r}$, $\vec{e_\theta}$, $\vec{e_\varphi}$ and three coordinates: $q_1 = r$, $q_2 = \theta$, $q_3 = \varphi$. Using the generalized Lame's coefficients (A.6) and particularizing the equations (A.7), (A.8) and (A.9), length (A.22a), surface (A.22b) and volume (A.22c) elements, respectively, can be successively written as:

$$\text{Length elements: } dl_1 = dr; \quad dl_2 = rd\theta; \quad dl_3 = r\sin\theta d\varphi \qquad \text{(A.22a)}$$

$$\text{Surface elements: } dS_1 = r^2\sin\theta d\theta d\varphi; \quad dS_2 = r\sin\theta dr d\varphi$$
$$dS_3 = rdrd\theta \qquad \text{(A.22b)}$$

$$\text{Volume element: } dV = r^2\sin\theta dr d\theta d\varphi \qquad \text{(A.22c)}$$

Further reading

[1] Blanpied W A 1971 *Modern Physics: An Introduction to Its Mathematical Language* (New York: Holt, Rinehart and Winston)

[2] Savelyev I V 1982 Fundamentals of theoretical physics *Mechanics Electrodynamics* 1 (Moscow: Mir)

[3] Clayton P R, Keith W W and Syed N A 1997 *Introduction to Electromagnetic Fields* (New York: McGraw-Hill)

[4] Jackson J D 1998 *Classical Electrodynamics* 3rd edn (New York: Wiley)

[5] Matthews P C 1998 Curvilinear coordinates *Vector Calculus. Springer Undergraduate Mathematics Series* (London: Springer) ch 6

[6] Griffiths D J 2017 *Introduction to Electrodynamics* 4th edn (Cambridge: Cambridge University Press)

[7] Antoni M 2019 *Calculus with Curvilinear Coordinates. Problems and Solutions* (Berlin: Springer)

IOP Publishing

Electrostatics
Formalism of the electrostatic field in vacuum and matter
Ştefan Antohe and Vlad-Andrei Antohe

Appendix B

Elements of vector analysis

Vector calculus or vector analysis is about differentiation and integration of scalar and vector fields primarily in the three-dimensional (3D) space, the three operators, namely, *gradient, divergence* and *curl* being generally used to study complicated partial differential equations. Vector analysis is extensively used in physics and engineering, including the description and physical interpretation of electromagnetic fields.

B.1 Scalar and vector fields

Definition B.1. A scalar function defined for any point in a 3D space is called a *scalar field* if it associates with any point of position vector $\vec{r}$ a scalar or real number $\varphi(\vec{r}) = \varphi(x, y, z)$.

We've already encountered examples without calling them 'scalar fields', i.e., the temperature $T(x, y)$ in a metal plate, mass, pressure, density, or the electrostatic potential $\phi = \phi(x, y, z)$. Scalar fields can be thus represented in space by the level surfaces $\varphi(x, y, z) = \text{constant}$, or in plane by the level curves (i.e., isotherms, isobars, etc). Formally, 'scalar' is a word used to distinguish the field from a vector field, as the scalar field is *invariant* under the rotation of the coordinate system.

Definition B.2. If the vector function $\vec{G}(\vec{r}) = \vec{G}(x, y, z)$ associates with any point $P(\vec{r}) = P(x, y, z)$ from a 3D space a vector $\vec{G}(\vec{r})$, then the function $\vec{G}(\vec{r})$ defines a *vector field.*

In practice, examples of 'vector fields' are: force field, speed field, acceleration field, electric field or magnetic field. In contrast to the scalar field, for a vector field the values of the components change in a new coordinate system. Essentially, the vector fields are represented by a set of marks showing the orientation and magnitude of the vector $\vec{G}(\vec{r})$. In a Cartesian coordinate system (see appendix A.1), the vector field is represented by its components according to equation:

$$\vec{G}(\vec{r}) = G_x(x, y, z, t)\,\vec{i} + G_y(x, y, z, t)\vec{j} + G_z(x, y, z, t)\vec{k} \tag{B.1}$$

in the case of a time-dependent vector field, or according to equation:

$$\vec{G}(\vec{r}) = G_x(x, y, z)\,\vec{i} + G_y(x, y, z)\vec{j} + G_z(x, y, z)\vec{k} \tag{B.2}$$

in the case of a time-independent vector field. Following equation (B.2), the derivative of the vector function can be written as:

$$d\vec{G}(\vec{r}) = dG_x(x, y, z)\,\vec{i} + dG_y(x, y, z)\vec{j} + dG_z(x, y, z)\vec{k} \tag{B.3}$$

or each component can be written with partial derivatives as:

$$dG_x = \frac{\partial G_x}{\partial x}dx + \frac{\partial G_x}{\partial y}dy + \frac{\partial G_x}{\partial z}dz$$

$$dG_y = \frac{\partial G_y}{\partial x}dx + \frac{\partial G_y}{\partial y}dy + \frac{\partial G_y}{\partial z}dz \tag{B.4}$$

$$dG_z = \frac{\partial G_z}{\partial x}dx + \frac{\partial G_z}{\partial y}dy + \frac{\partial G_z}{\partial z}dz$$

B.2 Gradient of a scalar field

Let's consider the scalar function $\varphi(\vec{r}) = \varphi(x, y, z)$. Its variation $d\varphi$ along the displacement element $\vec{dr} = dx\,\vec{i} + dx\vec{j} + dx\vec{k}$ is given by:

$$d\varphi = \frac{\partial \varphi}{\partial x}dx + \frac{\partial \varphi}{\partial y}dy + \frac{\partial \varphi}{\partial z}dz \tag{B.5}$$

Alternatively, the variation $d\varphi$ of the scalar function with respect to the displacement element $\vec{dr}$, can be written as the scalar product of two vectors, namely, grad φ (or $\nabla\varphi$ using the first order differential '*nabla*' operator, and referred as '*gradient of φ*)', which is given by the relation:

$$\text{grad } \varphi = \nabla\varphi = \frac{\partial \varphi}{\partial x}\,\vec{i} + \frac{\partial \varphi}{\partial y}\vec{j} + \frac{\partial \varphi}{\partial z}\vec{k} \tag{B.6}$$

while the displacement vector $\vec{dr}$ is given by:

$$\vec{dr} = dx\,\vec{i} + dy\vec{j} + dz\vec{k} \tag{B.7}$$

Then $d\varphi = \nabla\varphi \cdot \overrightarrow{dr}$, the variation of a scalar function with respect to the displacement vector $\overrightarrow{dr}$ will be always equal to the scalar product between the *gradient* of the scalar function and the displacement vector. If $\overrightarrow{dr}$ is chosen along on a level surface, $\varphi = $ constant, then: $d\varphi = 0 \Rightarrow \nabla\varphi \cdot \overrightarrow{dr} = 0 \Rightarrow \nabla\varphi \perp \overrightarrow{dr}$, the vector *gradient* is always perpendicular to that level surface. When the displacement $\overrightarrow{dr}$ makes angle θ with the vector *gradient* $\nabla\varphi$, it results in:

$$d\varphi = \nabla\varphi\,\overrightarrow{dr} = \nabla\varphi dr \cos\theta < \nabla\varphi \cdot dr \tag{B.8}$$

So at any point, $\nabla\varphi$ lies along the direction in which φ is most rapidly increasing (see figure B.1). In a three-orthogonal curvilinear system (see appendix A), the *gradient* operator can be defined as:

$$\nabla = \frac{1}{h_1}\frac{\partial}{\partial q_1}\overrightarrow{e_1} + \frac{1}{h_2}\frac{\partial}{\partial q_2}\overrightarrow{e_2} + \frac{1}{h_3}\frac{\partial}{\partial q_3}\overrightarrow{e_3} = \lim_{v_\Sigma \to 0}\frac{\displaystyle\int_{S_\Sigma}\varphi\,\overrightarrow{dS}}{v_\Sigma} \tag{B.9}$$

It can be thus noticed that the shape of the *gradient* changes according to the coordinate system used. Replacing the metric coefficients in relation (B.9), the *gradient* can be written as in equations (B.10), (B.11) and (B.12), respectively:

(a) **Cartesian coordinate system:**

$$\nabla_{car} = \frac{\partial}{\partial x}\overrightarrow{i} + \frac{\partial}{\partial y}\overrightarrow{j} + \frac{\partial}{\partial z}\overrightarrow{k} \tag{B.10}$$

(b) **Cylindrical coordinate system:**

$$\nabla_{cyl} = \frac{\partial}{\partial r}\overrightarrow{e_r} + \frac{1}{r}\frac{\partial}{\partial\theta}\overrightarrow{e_\theta} + \frac{\partial}{\partial z}\overrightarrow{e_z} \tag{B.11}$$

(c) **Spherical coordinate system:**

$$\nabla_{sph} = \frac{\partial}{\partial r}\overrightarrow{e_r} + \frac{1}{r}\frac{\partial}{\partial\theta}\overrightarrow{e_\theta} + \frac{1}{r\sin\theta}\frac{\partial}{\partial\varphi}\overrightarrow{e_\varphi} \tag{B.12}$$

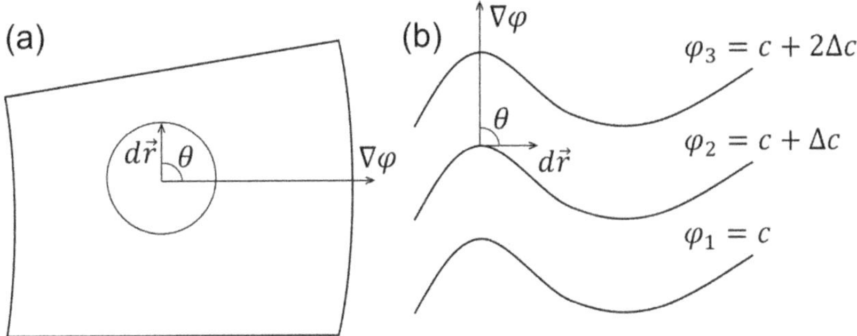

Figure B.1. Direction and sense of the *gradient* operator for a scalar function.

We will now review the *gradient* of a few usual scalar functions. Let's consider: $\vec{r} = x\vec{i} + y\vec{j} + z\vec{k}$. Then, the equations below are valid:

$$r = (x^2 + y^2 + z^2)^{1/2} \Rightarrow \quad \nabla r = \frac{x\vec{i} + y\vec{j} + z\vec{k}}{(x^2 + y^2 + z^2)^{1/2}} = \frac{\vec{r}}{r} \tag{B.13a}$$

$$r^n = (x^2 + y^2 + z^2)^{1/2} \Rightarrow \quad \nabla r^n = \frac{n}{2}(2x\vec{i} + 2y\vec{j} + 2z\vec{k})r^{\frac{n-2}{2}}$$

$$= nr^{n-2}\vec{r} \tag{B.13b}$$

$$r^{-n} = (x^2 + y^2 + z^2)^{-n/2} \Rightarrow \quad \nabla\frac{1}{r^n} = \frac{-n\vec{r}}{r^{n+2}} \tag{B.13c}$$

B.3 Potential vector field

As demonstrated in section B.2, when a *gradient* operator $\nabla\varphi$ is applied to any scalar function $\varphi(\vec{r})$, a vector field is always obtained. However, the reciprocal is not always valid in the sense that not any vector field results from a scalar field, when applying the *gradient* operator.

Definition B.3. The vector field $\vec{G}$ resulting from a scalar field φ by applying the *gradient* operator is called a '*potential vector field*', while the corresponding scalar field is called the '*potential of the field*':

$$\vec{G} = \text{grad } \varphi = \nabla\varphi \tag{B.14}$$

The potential vector field is a *conservative field*, being thus characterized by a few properties as follows:

(1) Line integral of a potential vector field is independent on the path of integration being equal with the variation of potential (difference between the potential at final point and potential at initial point). In other words, line integral of a potential vector field between two points P_1 and P_2 does not depend on the integration path, but only on its limits (see figure B.1). Indeed,

$$\int_{P_{1(\Gamma_1)}}^{P_2} \vec{G}\,\vec{dl} = \int_{P_{1(\Gamma_1)}}^{P_2} \nabla\varphi\,\vec{dl} = \int_{P_{1(\Gamma_1)}}^{P_2} d\varphi = \varphi(P_2) - \varphi(P_1)$$

$$\int_{P_{1(\Gamma_2)}}^{P_2} \vec{G}\,\vec{dl} = \int_{P_{1(\Gamma_2)}}^{P_2} \nabla\varphi\,\vec{dl} = \int_{P_{1(\Gamma_2)}}^{P_2} d\varphi = \varphi(P_2) - \varphi(P_1) \tag{B.15}$$

(2) The reciprocal can be applied too, saying that: if the line integral of a vector field, $\int_{P_1}^{P_2} \vec{G}\,\vec{dl}$ is independent on the path of integration, the vector field is a potential vector field resulting from the *gradient* of its potential φ.

(3) When the line integral drawn in the vector field on a close path Γ is zero, $\oint_\Gamma \vec{G}\,\overrightarrow{dl} = 0$, the vector field is a potential vector field, deriving from the *gradient* of its potential. If the components of $\vec{G}$ are:

$$G_x = \frac{\partial \varphi}{\partial x}; \quad G_y = \frac{\partial \varphi}{\partial y}; \quad G_z = \frac{\partial \varphi}{\partial z} \tag{B.16}$$

then, they satisfy the relations:

$$\frac{\partial G_x}{\partial y} = \frac{\partial G_y}{\partial x}; \quad \frac{\partial G_x}{\partial z} = \frac{\partial G_z}{\partial x}; \quad \frac{\partial G_y}{\partial z} = \frac{\partial G_z}{\partial y} \tag{B.17}$$

then, the potential vector field has always the *curl* (see section B.4) zero:

$$\nabla \times \vec{G} = 0 \tag{B.18}$$

B.4 Curl of a vector field

The integral form of the electrostatic potential theorem is:

$$\oint_\Gamma \vec{E}\,\overrightarrow{dl} = 0 \tag{B.19}$$

Trying to describe the properties of the electric field at a point, just using this line integral of the electric field along the closed path Γ, the following question arises: *'what happens with this circulation when the closed path Γ decreases to be closely drawn around that point?'*

Let there be the closed path Γ limiting the arbitrary surface S_Γ. Choosing a sense of integration along the closed path Γ ($\overrightarrow{dl}$), the orientation of the normal to the surface will be done according to the *'right hand screw rule'*[1]. The circulation of $\vec{E}$ along the closed path Γ will be given by the left side of equation (B.19).

If the surface S_Γ is divided in two surfaces bounded by the closed paths Γ_1 and Γ_2 containing the common path AB (see figure B.2), one can observe that circulation of $\vec{E}$ along the closed path Γ is equal to the sum of the circulations of $\vec{E}$ along the two closed paths Γ_1 and Γ_2:

$$\oint_{\Gamma_1} \vec{E}\,\overrightarrow{dl} + \oint_{\Gamma_2} \vec{E}\,\overrightarrow{dl} = \oint_\Gamma \vec{E}\,\overrightarrow{dl} \tag{B.20}$$

because during the two line integrals, the common path AB is followed in opposite directions. If the surface S_Γ is further divided into surface elements: $\Delta S_{\Gamma_1} \cdots \Delta S_{\Gamma_n}$, bounded by the closed paths: $\Gamma_1 \cdots \Gamma_n$ (see figure B.2(c)), the circulation of the vector

---

[1] In general terms for right-handed individuals, the *'right hand screw rule'* states that if the right thumb points out along the Z axis in the positive direction, then the curl of the fingers represents a motion from the first (or X) axis to the second (or Y) axis. When viewed from the top (or Z) axis, the movement is counter-clockwise.

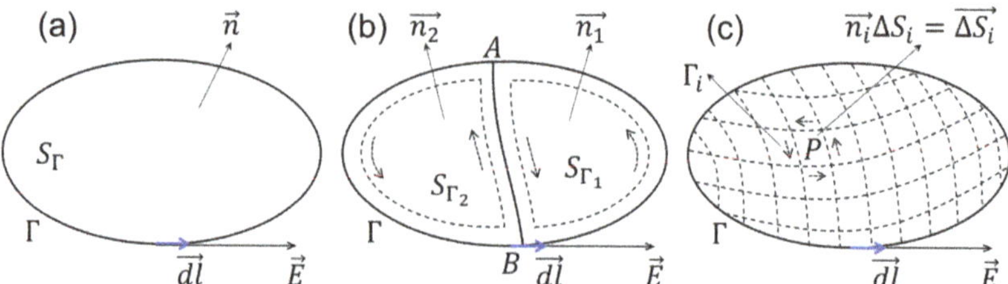

Figure B.2. Closed paths used to define the *curl* of a vector field.

field $\vec{E}$ along the initial closed path Γ can be written as the sum of the elementary circulations of $\vec{E}$ along the closed paths Γ_i:

$$\oint_{\Gamma} \vec{E}\,\vec{dl} = \sum_{i=1}^{N} \oint_{\Gamma_i} \vec{E}\,\vec{dl_i} \tag{B.21}$$

Note, by decreasing the path Γ_i, the circulation of the vector field $\vec{E}$ decreases, at limit becoming zero, then following this approach the information on the behavior of the electrostatic field at that point will be lost. One can observe that simultaneously with decreasing of the path Γ_i, the elementary surface bounded by it (ΔS_i) also decreases, and then the limit of the ratio between the circulation of $\vec{E}$ along the elementary path Γ_i and the bounded surface ΔS_i has a finite value which describes the local properties of the field:

$$\lim_{\Delta S_i \to 0} \frac{\oint_{\Gamma_i} \vec{E}\,\vec{dl_i}}{\Delta S_i} \tag{B.22}$$

The limit is valid only with one orientation of the surface element $\overrightarrow{\Delta S_i}$.

Choosing an orientation for $\overrightarrow{\Delta S_i}$ generally given by the unit vector $\vec{n_i}$ (normal to the surface element), the limit of the ratio between the circulation of $\vec{E}$ along the elementary path Γ_i and the bounded surface $\overrightarrow{\Delta S_i}$, defined by the equation (B.21) is a scalar quantity associated with P point and to the direction $\vec{n_i}$, in the vectorial field $\vec{E}$, represents the component with respect to the direction $\vec{n_i}$ of a vector called the *curl of the vector field* $\vec{E}$ ($\nabla \times \vec{E}$):

$$\vec{n_i} \cdot \nabla \times \vec{E} = \lim_{\Delta S_i \to 0} \frac{\oint_{\Gamma_i} \vec{E}\,\vec{dl_i}}{\Delta S_i} \tag{B.23}$$

Definition B.4. The *curl of a vector field* is another vector field having the magnitude equal to the limit value of circulation of the vector field per surface unit from the plane around the point, being always perpendicular to the plane for which the circulation is maximum.

Curl in a three-orthogonal curvilinear system

Starting from equation (B.23) as well as from figure B.3, we can calculate the three general components of the *curl* of a vector field in a three-orthogonal curvilinear coordinate system, as following:

$$(1) \quad \vec{e_1} \cdot \nabla \times \vec{E} = \lim_{dS_{\Gamma_1} \to 0} \frac{\oint_{\Gamma_1} \vec{E}\, \vec{dl}}{dS_{\Gamma_1}}$$

$$= \lim_{\substack{dq_2 \to 0 \\ dq_3 \to 0}} \frac{E_{2(1)}h_2 dq_2 + E_{3(2)}h_3 dq_3 - E_{2(3)}h_2 dq_2 - E_{3(4)}h_3 dq_3}{h_2 h_3 dq_2 dq_3}$$

$$= \frac{1}{h_2 h_3} \lim_{\substack{dq_2 \to 0 \\ dq_3 \to 0}} \frac{1}{dq_2 dq_3} \left[\begin{array}{l} E_{2(1)}h_2 dq_2 + E_{3(4)}h_3 dq_3 + \dfrac{\partial(h_3 E_3)}{\partial q_2} dq_2 dq_3 - \\[2mm] -E_{2(1)}h_2 dq_2 - \dfrac{\partial(h_2 E_2)}{\partial q_3} dq_2 dq_3 - E_{3(4)}h_3 dq_3 \end{array} \right]$$

$$= \frac{1}{h_2 h_3} \left[\frac{\partial(h_3 E_3)}{\partial q_2} - \frac{\partial(h_2 E_2)}{\partial q_3} \right]$$

$$\Rightarrow (\nabla \times \vec{E})_{\vec{e_1}} = \frac{1}{h_2 h_3} \left[\frac{\partial(h_3 E_3)}{\partial q_2} - \frac{\partial(h_2 E_2)}{\partial q_3} \right] \vec{e_1}$$

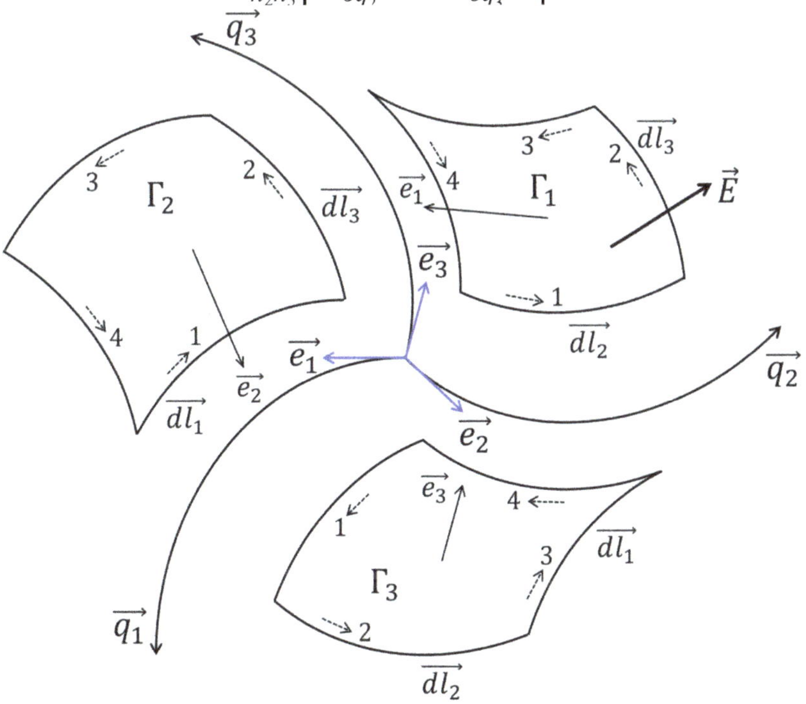

Figure B.3. Closed paths Γ_i used to calculate the components of the *curl* of a vector field in a three-orthogonal curvilinear coordinate system.

$$(2) \quad \overrightarrow{e_2}\cdot\nabla \times \overrightarrow{E} = \lim_{dS_{\Gamma_2}\to 0}\frac{\oint_{\Gamma_2}\overrightarrow{E}\,\overrightarrow{dl}}{dS_{\Gamma_2}}$$

$$= \lim_{\substack{dq_1\to 0\\ dq_3\to 0}}\frac{E_{1(1)}h_1dq_1 + E_{3(2)}h_3dq_3 - E_{1(3)}h_1dq_1 - E_{3(4)}h_3dq_3}{h_1h_3dq_1dq_3}$$

$$= \frac{1}{h_1h_3}\lim_{\substack{dq_1\to 0\\ dq_3\to 0}}\frac{1}{dq_1dq_3}\left[\begin{array}{l} -E_{1(1)}h_1dq_1 + E_{3(2)}h_3dq_3 + \dfrac{\partial(h_1E_1)}{\partial q_3}dq_3dq_1 \\[2ex] +E_{1(1)}h_1dq_1 - \dfrac{\partial(h_3E_3)}{\partial q_1}dq_1dq_3 - E_{3(2)}h_3dq_3 \end{array}\right]$$

$$= \frac{1}{h_1h_3}\left[\frac{\partial(h_1E_1)}{\partial q_3} - \frac{\partial(h_3E_3)}{\partial q_1}\right]$$

$$\Rightarrow (\nabla \times \overrightarrow{E})_{\overrightarrow{e_2}} = \frac{1}{h_1h_3}\left[\frac{\partial(h_1E_1)}{\partial q_3} - \frac{\partial(h_3E_3)}{\partial q_1}\right]\overrightarrow{e_2}$$

$$(3) \quad \overrightarrow{e_3}\cdot\nabla \times \overrightarrow{E} = \lim_{dS_{\Gamma_3}\to 0}\frac{\oint_{\Gamma_3}\overrightarrow{E}\,\overrightarrow{dl}}{dS_{\Gamma_3}}$$

$$= \lim_{\substack{dq_1\to 0\\ dq_2\to 0}}\frac{E_{1(1)}h_1dq_1 + E_{2(2)}h_2dq_2 - E_{1(3)}h_1dq_1 - E_{3(4)}h_2dq_2}{h_1h_2dq_1dq_2}$$

$$= \frac{1}{h_1h_2}\lim_{\substack{dq_1\to 0\\ dq_2\to 0}}\frac{1}{dq_1dq_2}\left[\begin{array}{l} E_{1(1)}h_1dq_1 + E_{2(2)}h_2dq_2 + \dfrac{\partial(h_2E_2)}{\partial q_1}dq_1dq_2- \\[2ex] -E_{1(3)}h_1dq_1 - \dfrac{\partial(h_1E_1)}{\partial q_2}dq_1dq_2 - E_{3(4)}h_2dq_2 \end{array}\right]$$

$$= \frac{1}{h_1h_2}\left[\frac{\partial(h_2E_2)}{\partial q_1} - \frac{\partial(h_1E_1)}{\partial q_2}\right]$$

$$\Rightarrow (\nabla \times \overrightarrow{E})_{\overrightarrow{e_3}} = \frac{1}{h_1h_2}\left[\frac{\partial(h_2E_2)}{\partial q_1} - \frac{\partial(h_1E_1)}{\partial q_2}\right]\overrightarrow{e_3}$$

Concluding, the *curl* of a vector field in a three-orthogonal curvilinear coordinate system is given by:

$$\begin{aligned} \nabla \times \overrightarrow{E} = {} & \frac{1}{h_2h_3}\left(\frac{\partial(h_3E_3)}{\partial q_2} - \frac{\partial(h_2E_2)}{\partial q_3}\right)\overrightarrow{e_1} \\[2ex] & + \frac{1}{h_1h_3}\left(\frac{\partial(h_1E_1)}{\partial q_3} - \frac{\partial(h_3E_3)}{\partial q_1}\right)\overrightarrow{e_2} \\[2ex] & + \frac{1}{h_1h_2}\left(\frac{\partial(h_2E_2)}{\partial q_1} - \frac{\partial(h_1E_1)}{\partial q_2}\right)\overrightarrow{e_3} \end{aligned} \tag{B.24}$$

Replacing the metric coefficients in relation (B.24), the *curl* of a vectorial field for each coordinate system can be written as in equations (B.25), (B.26) and (B.27), respectively:

(a) **Cartesian coordinate system:**

$$(\nabla \times \vec{E})_{\text{car}} = \left(\frac{\partial E_z}{\partial y} - \frac{\partial E_y}{\partial z}\right)\vec{i} + \left(\frac{\partial E_x}{\partial z} - \frac{\partial E_z}{\partial x}\right)\vec{j} + \left(\frac{\partial E_y}{\partial x} - \frac{\partial E_x}{\partial y}\right)\vec{k} \tag{B.25}$$

(b) **Cylindrical coordinate system:**

$$(\nabla \times \vec{E})_{\text{cyl}} = \frac{1}{r}\left(\frac{\partial E_z}{\partial \theta} - \frac{\partial(E_\theta \cdot r)}{\partial z}\right)\vec{e_r} + \left(\frac{\partial E_r}{\partial z} - \frac{\partial E_z}{\partial r}\right)\vec{e_\theta} + \frac{1}{r}\left(\frac{\partial(E_\theta \cdot r)}{\partial r} - \frac{\partial E_r}{\partial \theta}\right)\vec{e_z} \tag{B.26}$$

(c) **Spherical coordinate system:**

$$(\nabla \times \vec{E})_{\text{sph}} = \frac{1}{r^2 \sin\theta}\left(\frac{\partial(r \sin\theta E_\varphi)}{\partial \theta} - \frac{\partial(rE_\theta)}{\partial \varphi}\right)\vec{e_r} + \frac{1}{r \sin\theta}\left(\frac{\partial E_r}{\partial \varphi} - \frac{\partial(r \sin\theta E_\varphi)}{\partial r}\right)\vec{e_\theta} + \frac{1}{r}\left(\frac{\partial(rE_\theta)}{\partial r} - \frac{\partial E_r}{\partial \theta}\right)\vec{e_\varphi} \tag{B.27}$$

B.4.1 Stokes' theorem

Equation (B.20) established above can be rewritten as:

$$\oint_\Gamma \vec{E}\,\vec{dl} = \sum_{i=1}^{n}\oint_{\Gamma_i} \vec{E}\,\vec{dl} = \sum_{i=1}^{n}\Delta S_i \frac{\oint_{\Gamma_i} \vec{E}\,\vec{dl_i}}{\Delta S_i} \tag{B.28}$$

Taking into account the definition of the *curl* of the vector field $\vec{E}$ given by equation (B.22), the right side of equation (B.24) represents the sum of the products between the component of the *curl* of $\vec{E}$ on the direction of the normal $\vec{n_i}$ to the surface element ΔS_i, and its area ΔS_i. This sum extended on all the surface elements forming the surface S_Γ bounded by the closed path Γ, is at limit the surface integral of the *curl* of vector field $\vec{E}$ on the whole surface S_Γ:

$$\oint_\Gamma \vec{E}\,\vec{dl} = \lim_{n\to\infty}\sum_{i=1}^{n}\Delta S_i \cdot \vec{n_i}\cdot\nabla\times\vec{E} = \int_{S_\Gamma}(\nabla\times\vec{E})\vec{n}\,dS$$
$$= \int_{S_\Gamma}(\nabla\times\vec{E})\,\vec{dS} \tag{B.29}$$

Then, the *Stokes' theorem*, given by equation:

$$\oint_\Gamma \vec{E}\,\vec{dl} = \int_{S_\Gamma}(\nabla\times\vec{E})\,\vec{dS} \tag{B.30}$$

states that the line integral of a vector field $\vec{E}$ along the closed path Γ is equal to the flux of the *curl* of the vector field through the surface bounded by the path Γ.

B.5 Divergence of a vector field

The *divergence* of an arbitrary vector field $\vec{G}(\vec{r})$ written div $\vec{G}$ (or $\nabla\vec{G}$) represents the outward flux of $\vec{G}$ per unit volume at the point $\vec{r}$. The flux of $\vec{G}$ out of a closed surface S enclosing a small volume dV is $\int_S \vec{G}\,\vec{dS}$, taking $\vec{dS}$ to be the outward normal to the element of area dS. In the limit, as dV tends to zero, the outward flux of $\vec{G}$ per unit volume is independent of the shape of surface S, provided that $\vec{G}$ is smoothly varying:

$$\text{div}\,\vec{G} = \nabla\vec{G} = \lim_{dV\to 0}\frac{\int_S \vec{G}\,\vec{dS}}{dV} \tag{B.31}$$

Observation: We saw that the flux of an electric field is a quantity that characterizes the field at macroscopic level. Thinking that this flux may be used to characterize the field properties at each point $(\vec{r})$ we can divide a whole volume V in a big number of volume elements dV and taking the flux through the surface dS which limits the infinitesimal volume element, so we hope to find a physical quantity which is a local property of the field. But only the surface integral of a vector field over the closed infinitesimal surface element enclosing a point $P(\vec{r})$, is not a local quantity because this surface integral decreases when the surface element decreases. That is why taking the ratio between the flux through a surface element and the volume element dV enclosed by the surface, we obtain at limit a physical quantity independent of the shape of S, which could be a specific physical quantity, associated with point P from the field.

In order to derive the differential operator $\nabla\vec{G}$, we must start with the definition of the operator, in other words we must calculate the limit of the ratio between the outward flux of $\vec{G}$ through the closed surface S and the enclosed volume element dV, when it tends to zero. Let's thus consider an infinitesimal volume element dV in a three-orthogonal curvilinear coordinate system and a vector field $\vec{G}(\vec{r})$ which passes through the surface dS enclosing the volume element (see figure B.4). The volume element is: $dV = h_1h_2h_3dq_1dq_2dq_3$, while the surface elements are: $dS_1 = h_2h_3dq_2dq_3$, $dS_2 = h_1h_3dq_1dq_3$ and $dS_3 = h_1h_2dq_1dq_2$.

The total outward flux $\Phi_{\vec{G}}$ through the surface S enclosing the volume element dV is the sum of the fluxes Φ_1, Φ_2 and Φ_3 on the three directions:

$$\Phi_{\vec{G}} = \Phi_1 + \Phi_2 + \Phi_3 \tag{B.32}$$

The outward flux Φ_1 along the direction q_1 is the difference between the mean value of outward flux of component $G_{1m}(2)$ through the second face having the area $h_2h_3dq_2dq_3$ and the mean value of inward flux of the first component $G_{1m}(1)$ in the first face of the same area $h_2h_3dq_2dq_3$:

$$\Phi_1 = \Phi_{1m}(2) - \Phi_{1m}(1) = G_{1m}(2)h_2h_3dq_2dq_3 - G_{1m}(1)h_2h_3dq_2dq_3$$

$$= G_1h_2h_3dq_2dq_3 + \frac{\partial[G_1(1)h_2h_3]}{\partial q_1}dq_1dq_2dq_3 + \frac{\partial[G_1(1)h_2h_3]}{\partial q_2}\frac{dq_2}{2}$$

$$dq_2dq_3$$

$$+ \frac{\partial[G_1(1)h_2h_3]}{\partial q_3}\frac{dq_3}{2}dq_2dq_3 - G_1h_2h_3dq_2dq_3 - \frac{\partial[G_1(1)h_2h_3]}{\partial q_2}\frac{dq_2}{2} \tag{B.33}$$

$$dq_2dq_3-$$

$$- \frac{\partial[G_1(1)h_2h_3]}{\partial q_3}\frac{dq_3}{2}dq_2dq_3 = \frac{\partial(G_1h_2h_3)}{\partial q_1}dq_1dq_2dq_3$$

$\vec{q_3}$ — D′ (2) $G_1(2)$ — C′ — $\vec{G}$ — D (1) — C — $G_1(1)$ — $\vec{G}$ — $\vec{G}(q_1, q_2, q_3)$ — $\vec{G}$ — $\vec{dl_3}$ — A′ — B′ — $\vec{G}$ — A — $\vec{dl_2}$ — B — $\vec{dl_1}$ — (q_1, q_2, q_3) — $\vec{q_1}$ — $\vec{q_2}$

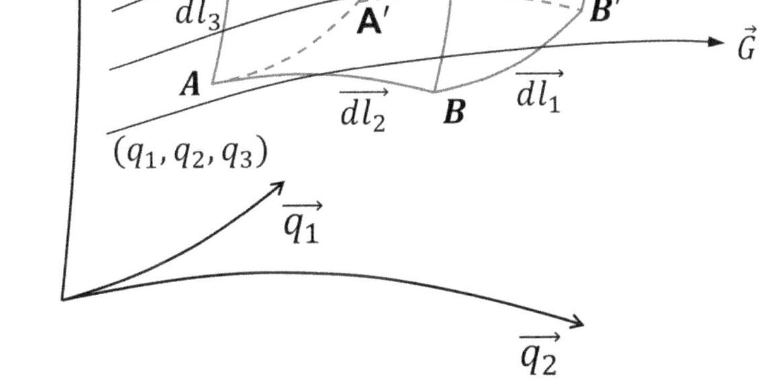

Figure B.4. Volume element in a three-orthogonal curvilinear coordinate system used to compute the *divergence* of a vector field $\nabla\vec{G}$.

The net outward flux along $\vec{e_1}$ direction (unit vector of line of variation of q_1) is thus given by equation (B.33a), while the outward fluxes for the other two components through the volume can be calculated in exactly the same manner, leading to equations (B.33b) and (B.33c):

$$\Phi_1 = \frac{\partial(G_1 h_2 h_3)}{\partial q_1} dq_1 dq_2 dq_3 \tag{B.33a}$$

$$\Phi_2 = \frac{\partial(G_2 h_1 h_3)}{\partial q_2} dq_1 dq_2 dq_3 \tag{B.33b}$$

$$\Phi_3 = \frac{\partial(G_3 h_1 h_2)}{\partial q_3} dq_1 dq_2 dq_3 \tag{B.33c}$$

Consequently, following the equations (B.33) the *divergence* of the vector field $\vec{G}$ will be:

$$\operatorname{div}\vec{G} = \nabla\vec{G} = \lim_{dV \to 0} \frac{\int \vec{G}\,\vec{dS}}{dV} = \lim_{\substack{dq_1 \to 0 \\ dq_2 \to 0 \\ dq_3 \to 0}} \frac{\Phi_1 + \Phi_2 + \Phi_3}{h_1 h_2 h_3 dq_1 dq_2 dq_3} \tag{B.34}$$

$$\Rightarrow \nabla\vec{G} = \frac{1}{h_1 h_2 h_3}\left[\frac{\partial(h_2 h_3 G_1)}{\partial q_1} + \frac{\partial(h_1 h_3 G_2)}{\partial q_2} + \frac{\partial(h_1 h_2 G_3)}{\partial q_3} \right]$$

Replacing the metric coefficients in relation (B.34), the *divergence* of a vectorial field for each coordinate system can be written as in equations (B.35), (B.36) and (B.37), respectively:

(a) Cartesian coordinate system:

$$(\nabla\vec{G})_{\text{car}} = \frac{\partial G_x}{\partial x} + \frac{\partial G_y}{\partial y} + \frac{\partial G_z}{\partial z} \tag{B.35}$$

(b) Cylindrical coordinate system:

$$(\nabla\vec{F})_{\text{cyl}} = \frac{1}{r}\left[\frac{\partial(rG_r)}{\partial r} + \frac{\partial(G_\theta)}{\partial \theta} + \frac{\partial(rG_z)}{\partial z} \right] \tag{B.36}$$

(c) Spherical coordinate system:

$$(\nabla\vec{G})_{\text{sph}} = \frac{1}{r^2 \sin\theta}\left[\frac{\partial(r^2 \sin\theta\ G_r)}{\partial r} + \frac{\partial(r \sin\theta\ G_\theta)}{\partial \theta} + \frac{\partial(rG_\varphi)}{\partial \varphi} \right] \tag{B.37}$$

B.5.1 Gauss–Ostrogradsky theorem

A mathematical identity following directly from the definition of the differential operator *divergence* is the '*divergence theorem*' (known also as '*Gauss–Ostrogradsky theorem*'), which is very useful in the theory of the fields and in particular for studying the electromagnetism, because it makes the correlation between the flux of a vector field through a closed surface and the *divergence* of the field in the volume enclosed.

The total flux of a vector field $\vec{G}$ out of a closed surface S limiting the finite volume V (with arbitrary shape) is given by the surface integral $\Phi = \int_S \vec{G}\,\overrightarrow{dS}$ extended on the whole surface S (see figure B.5(a)). Dividing the volume V into two smaller volumes V_1 and V_2, by diaphragm D (see figure B.5(b)), one can see that the total flux of a vector field $\vec{G}$ out of a closed surface S limiting the finite volume V is equal to the sum of the surface integrals of the vector field $\vec{G}$ out of the closed surfaces S_1 and S_2 limiting the finite volumes V_1 and V_2:

$$\int_{S_1} \vec{G}\,\overrightarrow{dS_1} + \int_{S_2} \vec{G}\,\overrightarrow{dS_2} = \int_S \vec{G}\,\overrightarrow{dS} \tag{B.38}$$

Indeed, the two volumes are bounded by the surfaces S_1 and S_2 including the diaphragm D and any flux leaving the volume V_1 through D is an entering flux into V_2 through D, then the net flux through diaphragm D is zero. It is very well known that the leaving flux is positive and the entering flux is negative, always the positive normal to a closed surface being oriented from inside to outside.

Dividing further the volume V in many elementary volume elements $V_1, \ldots, V_i, \ldots, V_N$, limited by the elementary closed surfaces $S_1, \ldots, S_i, \ldots, S_N$ (see figure B.5(c)), based on the above observation it can be concluded that:

$$\Phi = \sum_{i=1}^{N} \Phi_i = \sum_{i=1}^{N} \int_{S_i} \vec{G}\,\overrightarrow{dS_i} = \int_S \vec{G}\,\overrightarrow{dS} \tag{B.39}$$

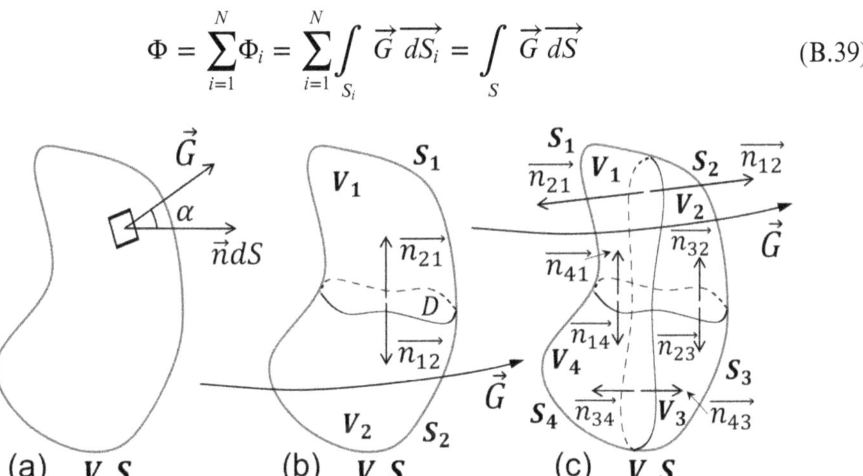

Figure B.5. The flux of a vector field through a closed surface S is equal to the sum of the elementary fluxes of the vector field through the closed surfaces bounding the volume elements resulting from the division of the initial volume V bounded by the closed surface S.

Introducing the definition relation of the *divergence* of the vector field $\vec{G}$ given within equation (B.31), into the relation (B.39), the latter can be rewritten as:

$$\Phi = \sum_{i=1}^{N} \int_{S_i} \vec{G}\,\overrightarrow{dS_i} = \sum_{i=1}^{N} V_i \left[\frac{\int_{S_i} \vec{G}\,\overrightarrow{dS_i}}{V_i} \right] \tag{B.40}$$

At limit, when $N \to \infty$ and $V_i \to 0$, then the right term of equation (B.40) becomes the *divergence* of the vector field $\vec{G}$, while the discrete sum is converted into a volume integral:

$$\Phi = \int_{S} \vec{G}\,\overrightarrow{dS} = \int_{V} \nabla\vec{G}\,dV \tag{B.41}$$

equation (B.41) represents the '**Gauss–Ostrogradsky theorem**', valid for any vector field satisfying the equation (B.31). The total flux of a vector field $\vec{G}$ out of a closed surface S equals the volume integral of $\nabla\vec{G}$ over the volume enclosed within S:

$$\int_{S} \vec{G}\,\overrightarrow{dS} = \int_{V} \nabla\vec{G}\,dV \tag{B.42}$$

B.6 Laplace operator

In mathematics, *Laplace operator*'[2] (or *Laplacian*) is a second-order differential operator given by the *divergence* of the *gradient* of a function on a three-orthogonal curvilinear system and it is usually denoted as $\nabla \cdot \nabla$, ∇^2 or Δ:

$$\begin{aligned} \text{div grad} &= \nabla \cdot \nabla = \nabla^2 = \Delta \\ &= \frac{1}{h_1 h_2 h_3} \\ &\cdot \left[\frac{\partial}{\partial q_1}\left(\frac{h_2 h_3}{h_1} \frac{\partial}{\partial q_1} \right) + \frac{\partial}{\partial q_2}\left(\frac{h_1 h_3}{h_2} \frac{\partial}{\partial q_2} \right) + \frac{\partial}{\partial q_3}\left(\frac{h_1 h_2}{h_3} \frac{\partial}{\partial q_3} \right) \right] \end{aligned} \tag{B.43}$$

Replacing the metric coefficients in relation (B.43), the *Laplacian* can be written for each coordinate system as in equations (B.44), (B.45) and (B.46), respectively:

(a) Cartesian coordinate system:

$$\Delta_{\text{car}} = \frac{\partial^2}{\partial x^2} + \frac{\partial^2}{\partial y^2} + \frac{\partial^2}{\partial z^2} \tag{B.44}$$

---

[2] *Laplace*' operator is named after the French mathematician *Pierre-Simon de Laplace* (1749–1827), who first applied the operator to the study of celestial mechanics, where the operator gives a constant multiple of the mass density when it is applied to the gravitational potential due to the mass distribution with that given density.

(b) Cylindrical coordinate system:

$$\Delta_{\mathrm{cyl}} = \frac{1}{r}\left[\frac{\partial}{\partial r}\left(r\frac{\partial}{\partial r}\right) + \frac{\partial}{\partial \theta}\left(\frac{1}{r}\frac{\partial}{\partial \theta}\right) + \frac{\partial}{\partial z}\left(r\frac{\partial}{\partial z}\right)\right] \tag{B.45}$$

(c) Spherical coordinate system:

$$\Delta_{\mathrm{sph}} = \frac{1}{r^2 \sin\theta} \cdot \left[\frac{\partial}{\partial r}\left(r^2 \sin\theta\frac{\partial}{\partial r}\right) + \frac{\partial}{\partial \theta}\left(\sin\theta\frac{\partial}{\partial \theta}\right) + \frac{\partial}{\partial \varphi}\left(\frac{1}{\sin\theta}\frac{\partial}{\partial \varphi}\right)\right] \tag{B.46}$$

References

[1] Blanpied W A 1971 *Modern Physics: An Introduction to Its Mathematical Language* (New York: Holt, Rinehart and Winston)

[2] Savelyev I V 1982 Fundamentals of theoretical physics *Mechanics Electrodynamics* 1 (Moscow: Mir)

[3] Clayton P R, Keith W W and Syed N A 1997 *Introduction to Electromagnetic Fields* (New York: McGraw-Hill)

[4] Jackson J D 1998 *Classical Electrodynamics* 3rd edn (New York: Wiley)

[5] Matthews P C 1998 *Curvilinear coordinates Vector Calculus. Springer Undergraduate Mathematics Series* (London: Springer) ch 6

[6] Joag P S 2016 *An Introduction to Vectors, Vector Operators and Vector Analysis* (Cambridge: Cambridge University Press)

[7] Griffiths D J 2017 *Introduction to Electrodynamics* 4th edn (Cambridge: Cambridge University Press)

IOP Publishing

Electrostatics
Formalism of the electrostatic field in vacuum and matter
Ştefan Antohe and Vlad-Andrei Antohe

Appendix C

Universal physical constants

Universal (fundamental) physical constants are physical quantities that are universal in Nature, have fixed numerical values in time and directly involve some kind of physical measurement. There are many physical constants in science, but table C.1 enumerates the few most commonly used in physics and chemistry. The exact values of the given universal constants are approximated with up to three decimal places and the corresponding measuring units are indicated in the 'International System of Units' (IS).

Table C.1. Universal physical constants mostly used in physics and chemistry.

Universal constant	Symbol	Value	IS unit
Speed of light in vacuum	c	3×10^8	m s^{-1}
Planck constant	h	6.63×10^{-34}	J Hz^{-1}
Reduced Planck constant	$\hbar = h/2\pi$	1.05×10^{-34}	J s
Atomic mass unit	u	1.66×10^{-27}	kg
Boltzmann constant	k_B	1.38×10^{-23}	J K^{-1}
Vacuum electric permittivity	ε_0	8.85×10^{-12}	F m^{-1}
Coulomb constant	$k_e = \frac{1}{4\pi\varepsilon_0}$	9×10^9	N m^2 C^{-2}
Standard atmospheric pressure	atm	$101\ 325$	Pa
Vacuum magnetic permeability	μ_0	$4\pi \times 10^{-7} = 1.25 \times 10^{-6}$	H m^{-1}
Josephson constant	$K_J = 2e/h$	$483\ 597.84 \times 10^9$	Hz V^{-1}
Molar gas constant	R	8.314	J mol^{-1} K^{-1}
Avogadro's number	N_A	6.023×10^{23}	mol^{-1}
Faraday constant	$F = N_A e$	$96\ 485.332$	C mol^{-1}
Charge of electron	e	-1.6×10^{-19}	C
Electron radius	r_e	2.817×10^{-15}	m
Mass of electron	m_e	9.1×10^{-31}	kg
Mass of proton	m_p	1.672×10^{-27}	kg
Mass of neutron	m_n	1.675×10^{-27}	kg

(Continued)

Table C.1. (*Continued*)

Universal constant	Symbol	Value	IS unit
Electron volt	eV	1.602×10^{-19}	J
Bohr radius	a_0	0.529×10^{-10}	m
Bohr magneton	$\mu_B = e\hbar/2m_e$	9.28×10^{-24}	J T^{-1}
Nuclear magneton	$\mu_N = e\hbar/2m_p$	5.05×10^{-27}	J T^{-1}
Quantum of circulation	$h/2m_e$	3.636×10^{-4}	m^2 s^{-1}
Conductance quantum	$G_0 = 2e^2/h$	7.75×10^{-5}	S
Magnetic flux quantum	$\Phi_0 = h/2e$	2.07×10^{-15}	Wb
Impedance of vacuum	$Z_0 = \mu_0 c$	376.73	Ω
Newtonian gravitational constant	G	6.67×10^{-11}	m^3 kg^{-1} s^{-2}
Standard acceleration of gravity	g	9.806	m s^{-2}

Further reading

[1] Spiridonov O P 1986 Universal Physical Constants *Translated in English by Yevgeni Strelchenko* (Moscow: Mir)

[2] Quigg C 2013 Appendix C: physical constants Gauge Theories Of Strong *Weak, and Electromagnetic Interactions* 2nd edn (Princeton, NJ: Princeton University Press) pp 457–8

www.ingramcontent.com/pod-product-compliance
Ingram Content Group UK Ltd.
Pitfield, Milton Keynes, MK11 3LW, UK
UKHW051939150726
7214IPUK00005B/43